Reliability of Optical Fibers and Optical Fiber Systems

Critical Reviews Series: Recent Titles

Reliability of Optical Fibers and Optical Fiber Systems (CR73, D. K. Paul)
Optical Metrology (CR72, G. A. Al-Jumaily)
Wavelength Division Multiplexing (CR71, R. T. Chen, L. S. Lome)
Heterogeneous Integration: Systems on a Chip (CR70, A. Husain, M. Fallahi)
Materials Characterization and Optical Probe Techniques (CR69, R. A. Lessard, H. Franke)
Sol-Gel and Polymer Photonic Devices (CR68, M. P. Andrews, S. I. Najafi)
Advanced Materials for Optics and Precision Structures (CR 67, M. A. Ealey, R. A. Paquin, T. B. Parsonage)
Advancement of Photonics for Space (CR66, E. W. Taylor)
Optoelectronic Devices and Systems for Processing (CR65, B. Javidi, K. M. Johnson)
Inorganic Optical Materials (CR64, P. Klocek)
Polymers in Optics: Physics, Chemistry, and Applications (CR63, R. A. Lessard, W. F. Frank)
Optoelectronic Interconnects and Packaging (CR62, R. T. Chen, P. S. Guilfoyle)
Standards for Electronic Imaging Technologies, Devices, and Systems (CR61, M. C. Nier)
Standards and Common Interfaces for Video Information Systems (CR60, K. R. Rao)
Technologies for Advanced Land Combat (CR59, G. R. Lane, T. A. DuBois, E. H. Vandiver)
Distributed Interactive Simulation Systems for Simulation and Training in the Aerospace Environment (CR58, T. L. Clarke)
Digital Signal Processing Technology (CR57, P. Papamichalis, R. Kerwin)
Defining the Global Information Infrastructure: Infrastructure, Systems, and Services (CR56, S. F. Lundstrom)
Adaptive Computing: Mathematics, Electronics, and Optics (CR55, S.-S. Chen, H. J. Caulfield)
Millimeter and Microwave Engineering for Communications and Radar (CR54, J. C. Wiltse)
Glass Integrated Optics and Optical Fiber Devices (CR53, S. I. Najafi)
Handbook of Critical Dimension Metrology and Process Control (CR52, K. Monahan)
64- to 256-Megabit Reticle Generation: Technology Requirements and Approaches (CR51, G. Hearn, R. Naber)
Fiber Optics Reliability and Testing (CR50, D. K. Paul)
Diffractive and Miniaturized Optics (CR49, S. H. Lee)
Photorefractive Materials, Effects, and Applications (CR48, P. Yeh, C. Gu)
Optical Technologies for Aerospace Sensing (CR47, J. E. Pearson)
Optical Inspection and Testing (CR46, J. D. Trolinger)
Integrated Optics and Optoelectronics (CR45, K.-K. Wong, M. Razeghi)
Fiber Optic Sensors (CR44, E. Udd)
Optomechanical Design (CR43, P. R. Yoder, Jr.)
The Aviation Security Problem and Related Technologies (CR42, W. H. Makky)
Lens Design (CR41, W. J. Smith)

Reliability of Optical Fibers and Optical Fiber Systems

Dilip K. Paul
Gould Fiber Optics
Editor

Proceedings of a conference held
20–21 September 1999
Boston, Massachusetts

Sponsored by
SPIE—The International Society for Optical Engineering

SPIE OPTICAL ENGINEERING PRESS

A Publication of SPIE—The International Society for Optical Engineering
Bellingham, Washington USA

Library of Congress Cataloging-in-Publication Data

Reliability of optical fibers and optical fiber systems: proceedings of a conference held 20-21 September 1999 Boston, Massachusetts / Dilip K. Paul, editor; sponsored by SPIE—The International Society for Optical Engineering.
p. cm.—(Critical reviews of optical science and technology; v. CR73)
Includes bibliographical references.
ISBN 0-8194-3324-1 (pbk.)
1.Optical Fibers–Reliability Congresses. 2. Optoelectronic devices–Reliability Congresses. I. Paul, Dilip K. II. Society of Photo-Optical Instrumentation Engineers III. Series.
TA1800.R47 1999
621.382'75—dc21

99-35695
CIP

Published by
SPIE—The International Society for Optical Engineering
P.O. Box 10
Bellingham, Washington 98227-0010 USA
Telephone 360/676-3290 (Pacific Time) • Fax 360/647-1445

Printed in the United States of America.

Contents

Preface

As you will note, the 1999 Critical Review is an update of an earlier Review, SPIE Volume CR50, *Fiber Optics Reliability and Testing*, which was published in September 1993. CR50 included fiber optic components, subsystems, systems, and networks—the state-of-the-art technical performance and reliability at component and systems levels and its impact on systems/network performance, specifications, redundancy requirements, and size/weight/cost issues. All systems such as terrestrial, undersea, and space photonics were considered in 1993.

Since our last Critical Review in 1993, many new developments have been reported in optical transmission network architectures with applications to robust DWDM and extensive network of undersea cables. The late 90s' trend to include technologies such as optical amplifier, DWDM, and gratings combined with a variety of new services such as the Internet, broadband network, and multimedia, can only make this inroad much deeper in the millennium. An ultimate goal, beyond the point-to-point/multipoint transmission and distribution capability achieved thus far, is to provide a fault-tolerant reliable connectivity in multiuser networks where the electrical bottlenecks will be avoided altogether by performing all functions—e.g., switching, Mux/Demux—transparently in the optical domain. Technology for all-optical networking is still evolving with much development and engineering work remaining.

To assure low-maintenance long-lifetime systems, reliable fiber optics are being developed at a considerable effort. It is true that extensive use of fiber optics in both terrestrial and submarine communications has made even the skeptics believe in the reliability of lightwave systems. However, concern for reliability still remains in yet unexplored areas, such as space and other harsh environments, and for new products and applications such as DWDMs. Furthermore, developing high-reliability components and fail-safe systems at a competitive cost remains a daunting challenge.

This year we succeeded in lining up 16 invited speakers to address the topics of interest, which included 5 papers in fiber and fiber device, 6 on active devices (light sources, detectors, integrated optic modulators, MEMs, etc.), 3 on systems, and 2 on passive devices such as splitters and connectors. For the 1999 Critical Review, the focus has been on WDM/DWDM type of components and systems, and the invited speakers, each distinguished in his field of expertise, are drawn from around the world.

Another important distinction we have been able to achieve in 1999 is having this publication ready for distribution at the time of the SPIE Photonics East meeting in Boston. I would like to thank SPIE staff for all their assistance in reminding/cajoling authors to submit their manuscripts and for their work in bringing this volume together.

I am confident that we shall have a wonderful day and a half of critical reviews of the fiber optic developments that have taken place since 1993.

Dilip K. Paul

Fiber Optic Reliability

Advancements in Mechanical Strength and Reliability of Optical Fibers

G. Scott Glaesemann*

Corning Incorporated, Corning, NY 14831

Abstract

This paper reviews recent advancements in the area of mechanical reliability of optical fiber, building upon previously published reviews in this area.[1-5] Of particular significance is progress in the area of establishing reliability programs. This review considers efforts made to reduce early failures and to establish reliability programs with an emphasis on functional reliability.

Keywords: Optical fiber, strength, reliability

1. Introduction

The mechanical reliability of optical fiber has been studied for several decades. In the early days concerns over delayed failure were mitigated by the advent of the loose tube cable design. In recent years those concerns persist due to the proliferation of constricted cable designs like tight buffer and ribbon. In these cases, fiber can experience a complex array of applied stresses. Even for loose tube cable designs fiber is being asked to carry more of the load. Recent changes to GR-20 state, "*With the cable subjected to the rated installation load for one hour, fibers shall exhibit a maximum tensile strain of 60 % of the fiber proof strain.*"[6] Recent strength testing results show that a proof stress level flaw will fail in minutes if loaded to 60% of the proof stress![7] This demonstrates the need for consensus on mechanical reliability issues. In addition to delayed failure concerns is the issue of fiber failures due to handling induced damage. The explosion of fiber-based components and devices has resulted in significant lengths of fiber being handled on the assembly floor and placed into ever-smaller packages. There is a need for mechanical reliability programs that address both delayed failure issues and fiber handling procedures.

The diagram in Figure 1 shows the classic "bathtub" reliability diagram. The goal is to reduce or eliminate early failures and to maximize the useful life of a product by accurately predicting the onset of wear-out. The primary wear-out phenomenon for optical fiber, from a mechanical point of view, continues to be subcritical crack growth or fatigue. Subcritical crack growth studies have traditionally focused on determining crack growth parameters for in-service lifetime predictions. Recently, high speed dynamic fatigue testing has allowed for a clearer understanding of fatigue during typical fiber processing events.[7] High-

* Correspondence: Email: glaesemags@corning.com; Telephone: 607 974 3736; Fax: 607 974 2364

speed processing events like proof testing also are important because they establish the strength distribution that will be subjected to in-service stresses.

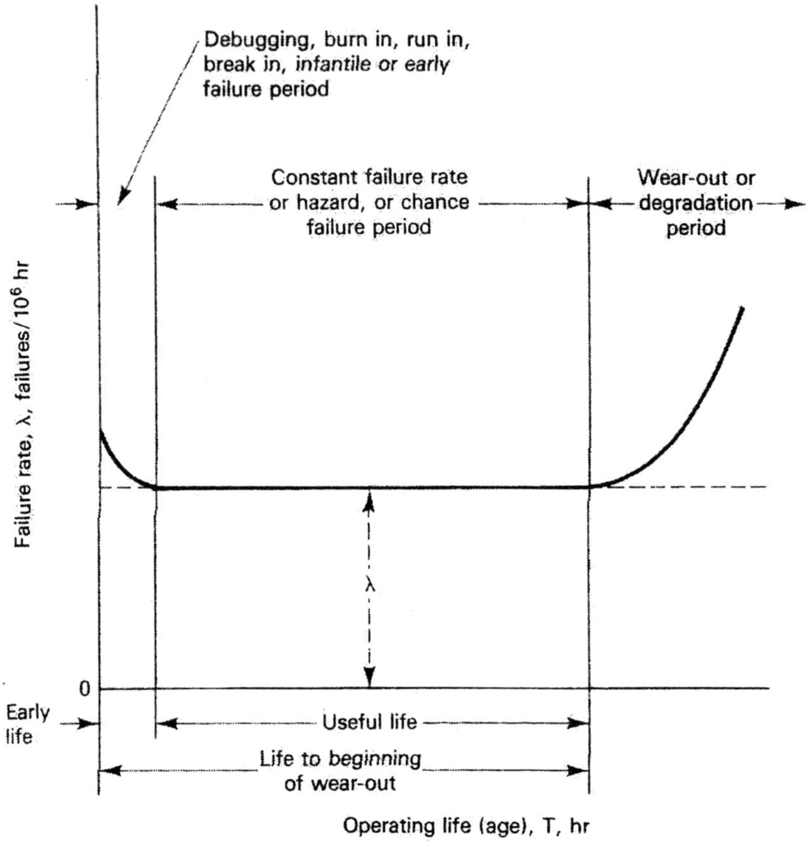

Figure 1. The "Reliability Bathtub" diagram from reference 8

There is continued research in characterizing manufacturing and handling induced flaws in optical fiber. Artificially induced flaws continue to be used as a means for understanding their mechanical behavior for two reasons: (1) manufacturing induced flaws are rare and (2) handling induced flaws, though more frequent, do not have sufficiently tight strength distributions.

Anyone intimately familiar with manufacturing and handling processes for optical fiber and optical fiber based components knows that most mechanical reliability issues are not related to the technical issues surrounding failure probability predictions from fatigue or aging. The primary issue is one of premature fiber failure caused by handling induced damage or excessive stress. The impact of premature fiber failure for these reasons is shown in Figure 1 as early failures. Early failures have become an increasingly important issue with the high volume of fiber being handled during cabling and cable installations and the growth of the active and passive fiber component industry.

Finally, with new fiber-based telecommunications products being introduced at a staggering pace it seems appropriate to discuss how mechanical reliability programs are established. The key ingredients for establishing a fiber reliability program are presented along with guidance for reducing early failures.

2. Advancements in Basic Understanding

2.1. Subcritical Crack Growth

At the core of the optical fiber lifetime prediction is the crack velocity model for subcritical crack growth or fatigue. The power law has been used extensively over the past 20 years, but exponential forms are attractive from a fundamental chemical kinetics point of view. As Matthewson[9] points out, fatigue data does not point to one model over another. Dynamic strength testing at ultra-slow stressing rates or long-term static fatigue tests on proof stress level flaws would be useful in resolving this dilemma. This testing necessitates the use of surrogate flaws due to the infrequency of naturally occurring flaws. The advantages and disadvantages of surrogate flaws will be discussed later. Semjonov et al.[10] have demonstrated the ability to create artificial proof stress level flaws using cube-corner indents suitable for static fatigue testing.

Over the last few years there has been increased interest in tensile testing optical fiber at high loading rates.[7,10-15] High speed strength testing of proof stress level flaws in fiber gives direct knowledge about strength degradation during high speed processing events like proof testing. Figure 2 shows high-speed strength test results from draw-abraded fiber measured under ambient conditions. Of particular importance is the non-linear behavior at the higher speeds. This behavior also has been observed on bare fibers with flaws produced by cube-corner indents (see Figure 3).[10] Thus, this curvature is not purely a coating related phenomenon. Hanson and Glaesemann[15] suggest that this behavior is indicative of region II type crack growth. However, from a flaw size point of view, region II behavior is unexpected[16] and has not been observed on direct crack velocity experiments on optical fiber using larger flaws.[17] Nevertheless, region II type behavior should be accounted for when modeling proof testing and other high speed processing events. Regardless of the mechanism at work modeling fatigue during proof testing and other high speed processing events now is a matter of interpolation rather than extrapolation.

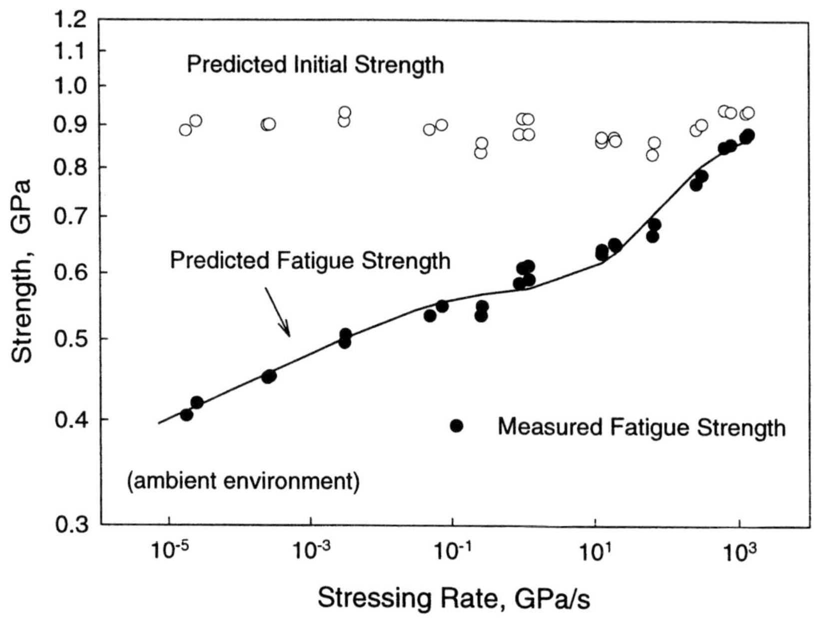

Figure 2. Measured strength and predicted initial and fatigue strength for abraded optical fiber tested at high speeds[7]

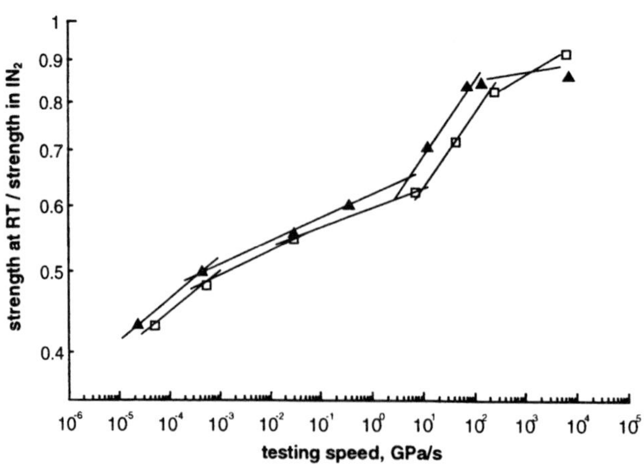

Figure 3. Median strength of indented fibers with high-speed tester.[10]
□ - 1 gram indent; ▲ - 2 gram indent

Tomozawa and colleagues[18-20] have determined that the strength and fatigue behavior of optical fiber is dependent on fictive temperature. They cite molecular dynamics calculations showing glass strength decreasing with decreasing fictive temperature and affirm this with the strength data shown in Figure 4. Water interacts with glass at flaw tips so as to cause structural relaxation, thereby, reducing the fictive temperature of the glass in that region. Lower fictive temperatures, in turn, allow more water vapor to enter the glass. So for long times under load in the presence of water, the fictive temperature will lower and flaws will be weaker. Volotinen et al.[21] have used this theory to explain the weakness of fusion splices.

This theory is also said to explain the well-known dependence of strength on stressing rate or static load. Figure 5 shows silica glass with the higher fictive temperature to have greater fatigue resistance than that of a lower fictive temperature. However, the fatigue parameter n of optical fiber, with a high fictive temperature, is typically half that of bulk glass having a lower fictive temperature.[1]

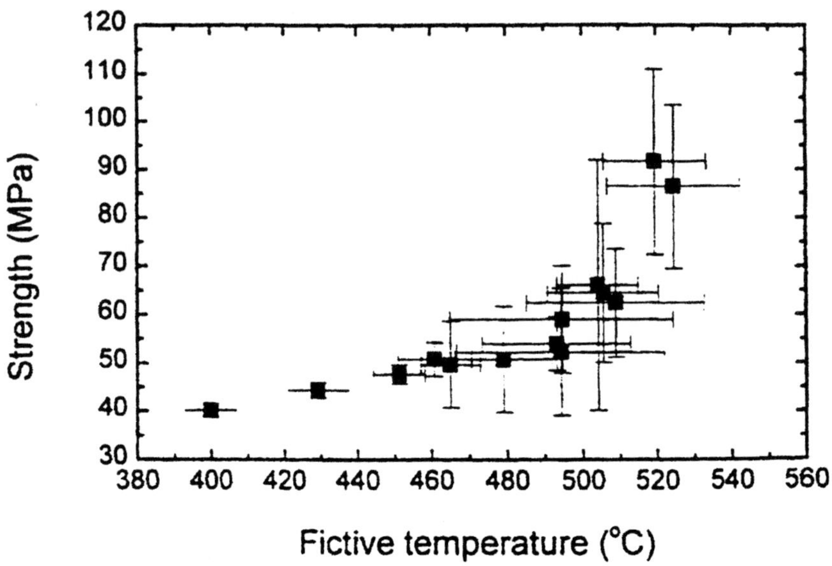

Figure 4. Mechanical strength versus surface fictive temperature of soda-lime glasses[20]

Perhaps there are factors affecting the fatigue difference between bulk and fiber forms of glass. Michalske[22] attributes this difference to the form of the assumed crack velocity model. High strength fiber tests the lower crack velocity range and lower strength bulk glass tests a higher crack velocity region. Glaesemann[23] showed that flaws with the same strength can have different fatigue behavior and by decreasing the strength slightly, large increases in fatigue resistance were observed. Furthermore, localized residual stresses are known to play a role in measured fatigue values.[24,25] The effects of material properties, flaws and stress on fatigue will continue to be an area of research and discussion.

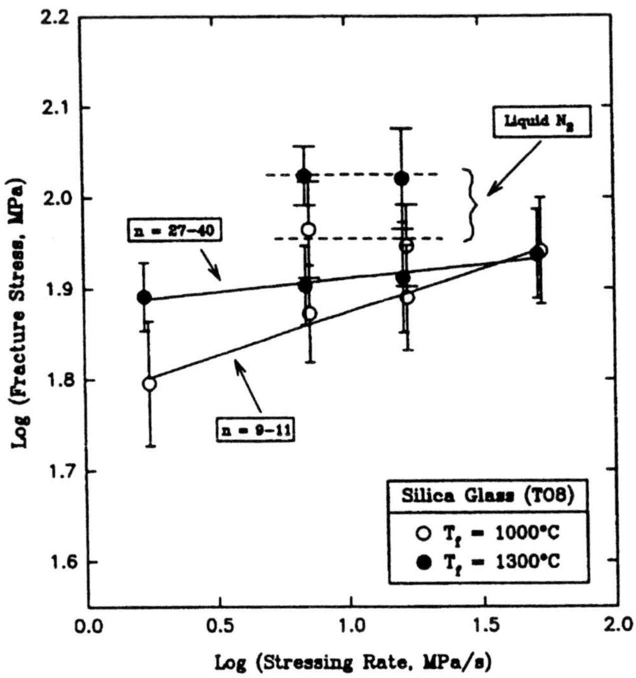

Figure 5. Dynamic fatigue and inert strength of T08 silica glass with different fictive temperatures[18]

2.2. Flaw Characterization

Flaws in optical fiber have been studied for many years.[26,27] Researchers continue to develop understanding about these flaws in an effort to better understand and predict their behavior. The fiber surface can be damaged before the coating is applied during the draw process and the resulting flaws can be sufficiently small to survive typical proof test stresses. There is little information in the literature detailing the morphology of these flaws, however, the basic manner in which fiber is drawn suggests contact events that can best be described as rubbing or scratching the fiber. The more common form of surface damage is through the fiber coating during post-proof test processing of fiber. Processing induced damage involves fracture of the polymer coating as a precursor to damaging the glass surface. The flaw formed in this manner has a morphology dependent on the handling environment and equipment. For example, abrasions caused by contaminants in a coloring die will be different than those generated by a damaged pulley. Wiederhorn and Lawn[28] state the key variables involved in generating an abrasion flaw to be the relative material properties of the abrasive and target material, the geometry of the abrasive, the localized residual stress generated by the impact event, the fracture toughness of the target material and the kinetic energy imparted during the damage event. Thus, the mechanical damage during processing or handling steps, like stripping, cleaning, cleaving and splicing, can have unique morphologies since the abrasion events vary widely. More recently, Ritter et al.[29] claim that for large abrasion flaws in bulk silica glass, the radial crack will always control failure. It is not known if the smaller abrasion type flaws in optical fiber can be generalized in this fashion. A deeper understanding of flaw morphologies related to specific handling events is needed.

The other basic flaw type for optical fiber is embedded particles.[30] Particulate can be wholly contained within the fiber preform or exist partially embedded on the fiber surface. The origin for such contamination can be the draw environment or surface contamination during splicing events.[31] The distinction between internal and external flaw is important in that internal flaws will not grow subcritically over time due to the absence of molecular water. The frequency of these flaws has decreased over the years as manufacturing processes improve, nevertheless, understanding their mechanical behavior is important. Recently, Wissuchek[32] modeled residual stresses and stress intensity factors for some common particulate compositions as a first step toward establishing a fracture mechanics model for these flaw types. He demonstrates the residual stress dependence on particle shape and elastic modulus. Furthermore, he concludes that the toughness and fatigue properties of the particle/silica interface rather than the host silica are key in determining the fiber lifetime for this flaw type. Continued research in this area is needed to develop the level of detail necessary for predicting the reliability of this flaw type.

2.3. Model Flaws

2.3.1 Glass

To study the weakest flaws in glass products like optical fiber, researchers have used artificially induced flaws. As-manufactured flaws occur too infrequently and have too much variability for strength and fatigue studies. Surrogate flaws were thought to provide reproducible flaws that replicated as-manufactured ones. In his review paper Matthewson[3] gives a detailed assessment of the success of surrogate flaws in modeling the behavior of actual flaws in fiber. Of particular significance are surrogate flaw behaviors that depart from the traditional fracture mechanics model[α] employed in most mechanical reliability predictions. "Crack

---

[α] Most optical fiber lifetime models assume the stress intensity factor of the life limiting flaw to be dependent on applied stress and flaw depth according to the well known relationship,

$$K_I = Y\sigma\sqrt{\pi a} \quad (1)$$

pop-in" is one such behavior. Here the strength of a flaw produced by indentation suddenly drops due to rapid formation of a radial crack from the indentation impression.[35,36] The driving force for this flaw formation is localized residual stress formed during the plastic deformation of the glass during indentation. Matthewson[3] contends that this phenomenon should be taken more seriously in the modeling of fiber reliability and uses the argument of similarity between indentation flaws and actual flaws in optical fiber. Recently, Ritter et al.[29] downplayed the significance of this event by citing silica's ability to accommodate the indenter or abrasion volume through densification.

From a field experience point of view, "pop-in" does not appear so threatening. For over a decade, the author has examined hundreds of post-proof fiber breaks and has never observed a single break that can be attributed to the "pop-in" of flaws as observed with indentation flaws. The vast majority of fiber breaks occur due to excessive stress or damage through the coating.[37]

The strength and fatigue effects of surface contamination have been studied by deliberately introducing contamination into the draw process. Early work by DiMarcello et al.[38] demonstrated dramatic strength decreases by introducing zirconia particulate into the draw furnace environment. Svensson and Breuls[39] have attempted to seed fiber preforms with zirconia particles in an attempt to measure the fatigue of fibers with this contaminant. They observed what could best be described as anomalous results with strength actually increasing at slower stressing rates. The difficulty with this type of testing is generating a strength distribution with sufficiently low variability.

Flaws produced by mechanical abrasion, usually by particles, have traditionally been used as a type of model flaw. They bring the flaw morphology closer to that of actual flaws than indentation flaws. Mechanical abrasions on optical fiber have been produced using a variety of abrasion methods. In several cases the flaws were created to replicate a particular surface abrasion created during actual manufacturing or handling. Table I summarizes some of these abrasion methods and the manufacturing and handling process they reflect.

Strengths near typical proof stress levels can be attained with abrasion flaws. Also, the variety of methods used to produce these flaws covers a wide range of actual fiber handling events. Therefore, the behavior of these flaws should translate well to that of actual flaws. The results from these studies have been shown to be useful in the study of fiber fatigue and process handling on strength. Higher Weibull m values are needed in order to make these flaw types more useful in distinguishing between crack velocity models.

To date, model flaws provide qualitative insight into the behavior of actual flaws.[3] It is believed that continued comparison between model flaws and actual flaws will narrow the gap between laboratory behavior and industry experience.

---

where Y contains the flaw shape and loading factors and a is the flaw depth. A semi-elliptical surface flaw will have a constant stress intensity factor over the entire crack front when $Y = 0.73$.[33,34]

Table I. Methods for Deliberate Abrasion of Optical Fiber Surfaces

Abrasion Method	Simulated Manufacturing or Handling Damage	σ_f (kpsi)	m	Ref.
Particulate in coating	Surface abrasion from coating contamination	178 131 <100	8 6.5	40 40 41
On-draw pulley	On-draw abrasion from normal contact during draw process	50	~16	42
On-draw abrasion by static object	Sliding contact abrasion during draw and post-draw processing	100 100 65	15 20 26	43 44 24
Particle abrasion	Contact damage during processing, handling and termination	70 50-160 100	11 9 8	45 46 24
Abrasive pulley	Post-draw fiber processing over pulleys	-	-	47 48

2.3.2 Coating

As mentioned previously, fracture of the polymer coating surrounding the glass fiber is a precursor to most glass surface abrasions. Whereas the fracture of polymers has been studied for many years, the fracture or abrasion resistance of optical fiber coatings as they exist over the glass fiber has not received much attention. Wissuchek et al.[49] have shown that the fracture properties of these polymer coatings can be studied with the use of controlled flaws. They developed coating fracture tests using both indentation and particle abrasion. Figure 6 illustrates the crack systems formed during indentation. An energy-based method was used to capture the complex response of coatings to fracture. Of particular importance is accounting for plasticity of the coating during the damage event.

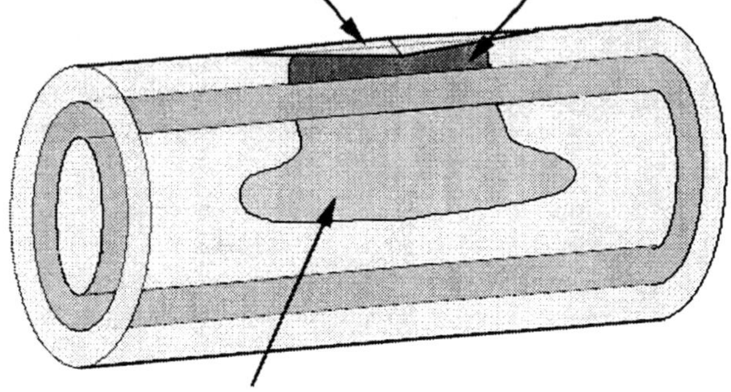

Figure 6. Schematic of an indented fiber showing the plastic indentation impression, secondary cracks, and coating/glass debonding[49]

3. Establishing a Reliability Program

Optical fiber has a proven track record of being reliable. One factor in establishing this fact was the use of the loose tube cable design in the early days of optical fiber. The primary applied stress was that generated by bending to create excess fiber length in the cable. This is not the case today. Constricted cable designs like tight buffer and ribbon cables have been used. These cable designs create both tensile and bending loads on the fiber. Furthermore, Bellcore recently has allowed the loading of fibers to 60% of the proof stress during cable installation.[6]

In addition to increased stress on fiber today, the amount of fiber being handled has increased dramatically. The shear volume of fiber being cabled and popularity of passive and active fiber based communication technologies has greatly increased the opportunities for handling induced damage. In order for fiber to maintain its reputation as a reliable technology engineers and scientists have to go beyond studies on the basic science of strength and flaw growth and implement *programs* that produce reliable products.

A common blueprint for a reliability program is simply conformance to GR-20[6] for fiber, GR-1209[50] and 1221[51] for fiber based components, or GR-458[52] for active devices. Fiber Optic Test Procedures (FOTPs) are employed as testing procedures in the mechanical testing program.[β] Whereas this approach provides the necessary information for trade and commerce purposes it doesn't

[β] An excellent review of mechanical testing methods can be found in reference 4.

necessarily link to *functional* reliability. A program focused on *functional* reliability attempts to find and eliminate sources of infant mortality and to generate specific knowledge needed to make a lifetime prediction for known wear out mechanisms. The diagram below describes the necessary ingredients for creating such a program. In this section these ingredients for establishing a reliability program based on product functionality are discussed.

The illustration in Figure 7 shows the basic ingredients for a mechanical reliability program for optical fiber and optical fiber based components and devices. Before an actual reliability prediction can be made these programs should be considered.

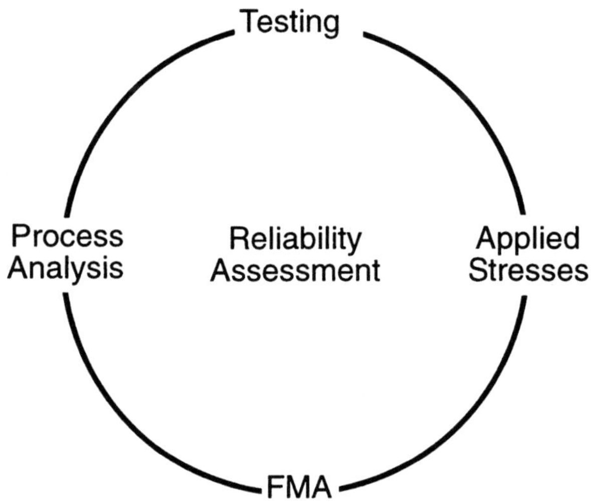

Figure 7. Ingredients for a mechanical reliability program

3.1 Testing

The testing part of the reliability program focuses on three needs. The need for passing a performance standard (e.g., GR-20, 1221, 1209, or 458), the need for finding failure modes, and the need for knowledge of material properties. Information about standardized tests for measuring against GRs is easily obtainable and outside the scope of this paper.

Testing in an effort to find failure modes in a fiber or fiber based products requires that the product be stressed until failure occurs. The strategy is to determine what it takes to cause failure and to compare the failure stress with what might be expected in service. Tensile testing, environmental exposure, shock and vibration, or extreme handling can impart stress. The test method itself depends on the product or process being investigated. For example, splices, fusion or mechanical, are simply loaded to failure by tensile testing.[31,53] Several years ago thin-coated optical fibers where proposed as a means for increased fiber density in high-count ribbon cables.[47,48] Several tests were created in an effort to find failure modes relevant to the processing of thin-coated fibers. Testing for failure modes is more than passing a specified level of performance. It gives a feel for the performance limits of the basic design and should be performed early in the development effort.

Processes for manufacturing and handling fiber should be examined for their ability to generate failure modes as well. In most cases fiber-handling equipment is qualified after engineers have carefully tuned the equipment so that it is examined under the best of conditions. However, to test processing equipment for possible failure modes, one would run the equipment at its extremes for speed, alignment, stress, etc. A simple "rule of thumb" is to set these extremes 2X away from the nominal operating condition. Break ends are collected for further examination and the fiber can be rescreened or strength tested in long lengths[54] to look for process induced damage. Again, the point here is to push the equipment until something "bad" happens in an effort to determine the limits of the fiber handling process.

The testing part of the reliability program also is where one obtains knowledge of material properties and fatigue parameters. Material properties like Young's modulus, cure shrinkage and glass transition temperature are needed to model environmentally induced in-service stresses. Thermal proof testing can be done provided these material properties are known and incorporated into the applied stress model. Measurements of fatigue behavior have been covered extensively in recent reviews and publications.[1,2,3,4,23] Of more recent significance is the need for appropriate crack growth parameters for high speed processing events.[7]

3.2 Applied Stresses

A predictive lifetime model for optical fiber is only as good as the applied stress data or model. Loose tube cable designers have an established set of independent variables for determining the fiber bend configuration and the strain window of the cable. When new cable designs are introduced these variables usually are addressed.[55] Loose tube cables usually are considered to be low risk for fiber failures. However, Nagata et al.[56] recently examined optical failures in a loose tube cable and concluded that unless the material properties of the loose tube are well characterized the tubes can shrink excessively, thereby placing fibers under excessive bending stresses. Fiber in cable experiences stress even in the most benign of cable designs and analytical models are not always sufficient to ensure tolerable stress levels. Finite element analysis (FEA) has recently been employed to model both cable[57] and cabled fiber.[58] This type of modeling, though difficult, holds promise for delivering applied stress distributions in complex cable structures.

Applied stress measurements for fiber in cable have been around for many years. Time-of-flight and phase-shift methods are used routinely to measure cabled fiber strain (see reference 59 for a recent example). These methods measure the average tensile stresses over a relatively long length of fiber. They should be used when the fiber subjected to the same tensile stress over the entire cable length being measured. Some have found these optical methods difficult to use and have developed a direct mechanical method for measure tensile strain.[60]

Fibers in a typical stranded loose-tube cable experience a considerable range of stresses, even over a few centimeters. Similarly, fibers in ribbon cables experience significant tension, compression and bending within a single strand length. In these cases caution must be exercised when using the common optical strain measurement techniques. Brillouin OTDRs can resolve an average stress over several meter fiber lengths,[61,62] but this is not sufficient for mechanical reliability assessments of these more complex cable designs. Direct applied stress measurements using in-fiber sensors hold promise for measuring more localized tensile stresses. For example, Bragg gratings written directly into fiber and deployed in a cable or component package would be capable of delivering accurate tensile stress measurements at a discrete location.[63] Several gratings strategically placed in a cable would allow for tensile stress mapping. Etched fibers[64] also could be used as an embedded strain gauge. What is needed is centimeter length resolution on applied bending and tensile stress measurements for cabled fiber in order to develop a distribution of applied stresses.

It is generally assumed that fiber based components and devices have low reliability risk from fiber failure due to the short lengths employed. However, it is common to remove the polymer coating during splicing, connecting and packaging and expose bare glass to handling induced damage.[65,66] Once this is done, the earlier proof testing performed during fiber manufacturing no longer has relevance. Furthermore, fibers can experience tight bends to accommodate the package design.[67-70] It is these situations that pose the greatest reliability risk for fiber components and devices. More research is not the answer here. Rather, knowledge and control over applied stresses coupled with strength testing or proof testing and a set of generally accepted best practices for handling fiber are needed.

For coated fiber, the long-term applied stress should be no greater than 1/5th the proof stress unless one performs an extensive investigation into the probability of encountering a proof stress level flaw.[15,71] Fiber stripped of its coating should be proof tested to five times the applied stress and protected from further damage.

3.3 Failure Mode Analysis (FMA)

Failure mode analysis is simply the task of determining the cause of failure and is a crucial step in analyzing test results as well as early failures during processing. The analysis of the fiber breaks traditionally is called Break Source Analysis (BSA). Skilled persons are able to read coating and glass fracture surface features to determine the cause of failure.[72,73] The primary tool is an optical microscope. Scanning Electron Microscopy (SEM) is of value in determining the composition of contaminants. Atomic Force Microscopy (AFM) is a proven research tool[74,75] and has found some use in FMA or BSA investigations.

Optical fiber has been deployed for over two decades and retrieval of fibers, components and devices from the field provides an important feedback loop into failure modes of fiber based technologies. Several FMA studies on early fibers showed evidence of strength degradation. Fibers exposed to steam explosions[76] or left unprotected in the cable yard[77] experienced strength degradation. Some early fibers designed to have low mechanical adhesion were found to be difficult to handle after prolonged exposure to wet conditions.[78] It also was found that one must take care to avoid coating damage while removing fibers from a cable, since such damage can make it difficult to interpret the post-aging strength distribution. Several studies on field aged fibers have been conducted on more modern fiber types[79-82] and in each case no strength degradation or handleability issues were observed. It is important to continue this form of examination as the fiber network ages. In 1993 Lisle[37] published an extensive review of failure modes for deployed cable. To no one's surprise, "dig-ups" were the largest source of premature failure. One can find an excellent review of field failure modes for components that make up DWDM transmission systems in reference 83.

One of most important requirements for performing FMA or BSA is having the failed component or fiber as close to the as-failed state as possible. For example, a fiber break in a buffer tube is best left in the tube until the fractologist begins the dissection. Similarly, a break in a component package should be left undisturbed. The position of the fiber and the surrounding materials are extremely important. Many times the fiber coating is removed before the fractologist begins the BSA, but more than half of the failure history is found in the coating. Mechanically stripping the coating during the dissection introduces more flaws and can mask the original cause of failure. Sometimes only a few centimeters of fiber are provided or the break ends are covered with tape. Great care is needed in preparing for FMA and BSA.

A common method for determining tensile failure stress from fiber fracture surface features is to obtain the mirror radius by measuring radially across the fiber as shown in Figure 8a. This method avoids the effects of surface features that can disturb the mirror region near the glass surface.[84] However, the more common method for measuring fracture mirrors for glass in general is the chord method shown in Figure 8b.[85] The chord method has less influence from bending induced stresses which elongate the mirror in the radial direction and result in an underestimation of the applied stress at failure. Furthermore, since the fiber surface is usually unblemished away from the actual damage site, the mirror region near the surface is less apt to be altered by surface effects. Baker and Glaesemann[86] reaffirmed this in their recent comparison of the two methods.

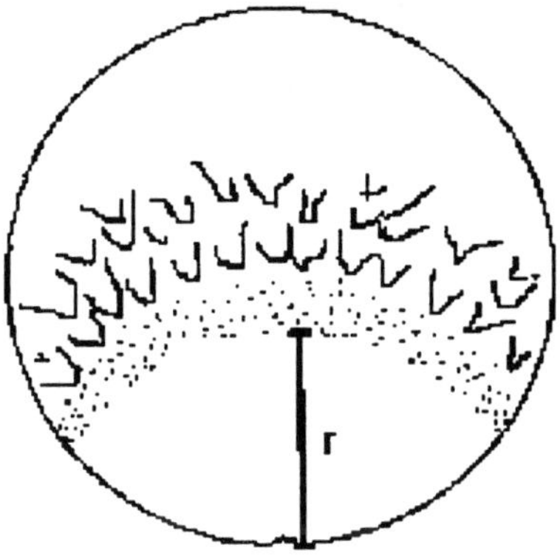

a. Traditional Radius Measurement

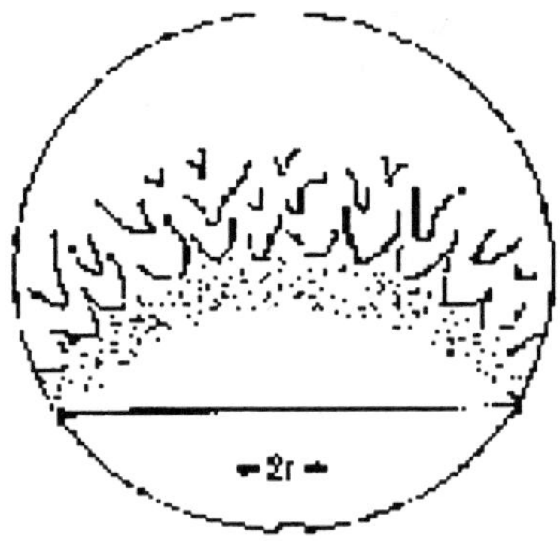

b. Chord Radius Measurement

Figure 8. Mirror measurement methods[84]

3.4. Process Analysis

The “process analysis” portion of a reliability program is where one obtains data on what the manufacturing process actually produces in the way of induced damage or applied stresses. Also, the strength distribution that is used in the reliability model is obtained directly from the manufacturing process. This is important because the flaw distribution usually is unique to the manufacturing process.

In “process analysis” one also seeks potential failure modes and implements corrective action. Common sources of failure modes for fiber processing originate at pulleys, coloring dies, guide wires, and mishandled fiber reels. Similarly, fiber failure modes from the manufacturing of components and devices originate from poor termination practices, contaminated work surfaces, and handling of pigtails or leads. These sources for failure modes seem rather obvious, but can be overlooked in today's fast paced manufacturing environments. For more admonitions and background in this area see references 31, 65, 74, and 87-90.

Proof testing is an effective means of truncating the fiber strength distribution during fiber manufacturing. It should always be considered during the development of component and device manufacturing processes. Proof testing also is recommended for the fast growing fiber sensor market.[91]

3.5. Reliability Assessment

A reliability assessment is an event where a cross-functional team, representing manufacturing, development and FMA, gathers to make an overall assessment of the reliability of the fiber based product. The assessment covers two basic areas related to mechanical reliability; namely, infant or early failures and wear out failures. Early failures are eliminated by careful examination of every fiber handling event and the results of the testing program. The probability of failure from wear out mechanisms like fatigue is assessed by inputting applied stresses, strength distributions and fatigue parameters into the reliability model. For long-length applications, the allowable stress should be based on a percentage of the proof stress. This is called a “minimum strength” design. Basically, one treats the fiber as being no stronger than the proof stress level. A search of reliability models reveals applied stress to proof stress ratios of 1/3rd to 1/6th. As stated earlier, the author recommends the 1/5th rule for 20 to 40 year terrestrial applications. This methodology also should be used in fiber termination as well. That is to say, anytime the coating is removed the fiber should be reproof-tested and the minimum strength design employed. The applied stress design rule for submarine applications usually is more conservative than that for terrestrial applications due to the greater consequences for an undersea fiber failure.

A failure probability model can be used provided the strength distribution is representative of the in-service application. For example, long-length applications require strength distributions consisting of many kilometers of data.[92] Some reliability models use the fiber break rate during proof testing as a means for incorporating the strength distribution into the failure probability prediction.[93] Paul and Glaesemann[94] questioned the underlying assumption for this model that the flaws failing proof testing come from the same distribution as those that survive. That is to say, the distribution of flaws that pass proof testing, and are placed in service, cannot be predicted from those that fail proof testing. They performed extensive testing and found little correlation between proof test break rate and the post proof strength distribution. Coated fiber pigtails are excellent candidates for a failure probability design since the probability of finding a proof stress level flaw is low. It still is important to have a long-length strength distribution even for these applications.

In the case where regions of the fiber are exposed to different stresses or the strength distribution varies, one divides the product into regions for reliability assessment. The reliability of each region then is determined and the total reliability for the entire product is obtained by simply multiplying the reliability's of all regions,[8]

$$R_T = \prod_{i=1}^{n} R_i \qquad (2)$$

This equation is for regions in series, which usually is the case for fiber applications. One can clearly see that the region with the lowest reliability governs the reliability of the total product. An example of such a reliability assessment for wear out failures can be found in reference 70.

4. Summary

This article reviewed some of the recent research on the mechanical behavior of flaws in optical fiber. High speed testing of proof test level flaws enables more accurate modeling of high speed processing events. Although there appears to be a correlation between fictive temperature and fatigue behavior the proposed theory does not explain differences in fatigue behavior between large flaws in bulk glass and small flaws in optical fiber. Surrogate flaws continue to be used for modeling the mechanical behavior of as-manufactured flaws in optical fiber. Indentation flaws are simpler than those produced by mechanical abrasion; however, abrasion induced flaws better represent actual as-manufactured flaws. New research into the fracture behavior of protective polymer coatings was reviewed. This is a first step toward a more comprehensive fracture mechanics model for abrasion resistance.

The key ingredients for a successful mechanical reliability program were presented. It is recommended that the testing part of the program go beyond simply testing to a standard. Rather one attempts to seek failure modes by testing the limits of the product or process. Applied stresses are just as important as strength and fatigue information. Finite Element Analysis is beginning to be used to model complex cable structures. Increased spatial resolution is needed for measuring applied stresses for cabled fiber. Failure Mode Analysis (FMA) and Break Source Analysis (BSA) are the tools one employs to discover the cause of fiber failure. A good optical microscope is needed as well as a user who is experienced in fractography. The process analysis step is where one learns what the process generates by way of induced damage and applied stresses. Recommendations for minimizing early failures were given for both fiber processing and fiber handling. Proof testing remains an effective tool in controlling the strength distribution. The actual reliability assessment consists of an in depth review of modes for early failures and predictions for wear out failures.

References

1. C.R. Kurkjian, J.T. Krause, and M.J. Matthewson, "Strength and Fatigue of Silica Optical Fibers," *J. Lightwave Tech.* **7** [9] 1360-1370 (1989).
2. C.R. Kurkjian, P.G. Simpkins, and D. Inniss, "Strength, Degradation, and Coating of Silica Lightguides," *J. Am. Ceram. Soc.* **76** [5] 1106-1112 (1993).
3. M.J. Matthewson, "Optical Fiber Reliability Models," *Fiber Optics Reliability and Testing*, Edited by D.K. Paul, **CR50**, 3-31, Proceedings of the SPIE, Bellingham, WA, 1993.
4. M.J. Matthewson, "Optical Fiber Mechanical Testing Techniques," *Fiber Optics Reliability and Testing*, Edited by D.K. Paul, **CR50**, 32-59, Proceedings of the SPIE, Bellingham, WA, 1993.
5. C.R. Kurkjian, "Critical Issues in the Mechanical Reliability of Lightguide Fibers," *Reliability of Photonics Materials and Structures*, Edited by E. Suhir,

M. Fukuda, and C.R. Kurkjian, **531**, 39-44, Proceedings of the Materials Research Society, San Francisco, 1998.
6. Bellcore GR-20-CORE, Issue 2, July 1998.
7. G.S. Glaesemann, D.A. Clark, T.A. Hanson, and D.J. Wissuchek, "High-Speed Strength Testing of Optical Fiber," *Reliability of Photonics Materials and Structures*, Edited by E. Suhir, M. Fukuda, and C.R. Kurkjian, **531**, 249-260, Proceedings of the Materials Research Society, San Francisco, 1998.
8. D. Kececioglu, *Reliability Engineering Handbook volume 1*, Prentice-Hall, Upper Saddle River, NJ, 1991.
9. M.J. Matthewson, "Chemical Kinetics Models for the Fatigue Behavior of Fused Silica Optical Fibers," *Reliability of Photonics Materials and Structures*, Edited by E. Suhir, M. Fukuda, and C.R. Kurkjian, **531**, 143-153, Proceedings of the Materials Research Society, San Francisco, 1998.
10. S.L. Semjonov, G.S. Glaesemann, C.R. Kurkjian, and M.M. Bubnov, "Modeling of Proof Test Level Flaws Using Cube Corner Indents," *Proceedings of the 47th International Wire and Cable Symposium*, 928-932, Philadelphia, PA, 1998.
11. P.T. Garvey, T.A. Hanson, M.G. Estep, G.S. Glaesemann, "Mechanical Reliability Predictions: An Attempt at Measuring the Initial Strength of Draw-Abraded Optical Fiber Using High Stressing Rates," *Proceedings of the 46th International Wire and Cable Symposium*, 883-887, Philadelphia, PA, 1997.
12. T. Svensson, "Evaluation of B-Values of Telecom Fibers, Objectives and Methods," *Reliability of Photonics Materials and Structures*, Edited by E. Suhir, M. Fukuda, and C.R. Kurkjian, **531**, 47-52, Proceedings of the Materials Research Society, San Francisco, 1998.
13. A. Gouronnec and N. Evanno, "High Speed Axial Strength Setup for the Measurement of the "B" Value," *Proceedings of the 45th International Wire and Cable Symposium*, 906-913, Reno, NV, 1996.
14. T. Volotinen, A. Breuls, N. Evanno, K. Kemeter, C. Kurkjian, P. Regio, S. Semjonov, T. Svensson, and G.S. Glaesemann, "Mechanical Behavior and B-Value of an Abraded Optical Fiber," *Proceedings of the 47th International Wire and Cable Symposium*, 881-890, Philadelphia, PA, 1998.
15. T.A. Hanson and G.S. Glaesemann, "Incorporating Multi-Region Crack Growth into Mechanical Reliability Predictions for Optical Fibers," J. Mat. Sci., **32**, 5305 – 5311 (1997).
16. H.C. Chandan, R.C. Bradt, and G.E. Rindone, "Dynamic Fatigue of Float Glass," *J. Am. Ceram. Soc.* **61** [5-6] 207-210 (1978).
17. M. Muraoka, K. Ebata, and H. Abe, "Effect of Humidity on Small-Crack Growth in silica Optical Fibers," *J. Am. Ceram. Soc.* **76** [6] 1545-1550 (1993).
18. H. Li, A. Agarwal, and M. Tomozawa, "Effect of Temperature on Dynamic Fatigue of Silica and Soda-Lime Glasses," *J. Am. Ceram. Soc.* **78** [5] 1393-1396 (1995).
19. Y.L. Peng, A. Agarwal, M. Tomozawa, and T.A. Blanchet, "Radial Distribution of Fictive Temperatures in Silica Optical Fibers," *J. Non-Cryst. Solids* **217**, 272-277 (1997)
20. B. Varughese, Y.-K. Lee, and M. Tomozawa, "Effect of Fictive Temperature on Mechanical Strength of Soda-Lime Glass," *J. Non-Cryst. Solids* **241**, 134-139 (1998).
21. T. Volotinen, M. Zimnol, M. Tomozawa, Y.-K. Lee, and K. Raine, "Effect of Mechanical Stripping and Arc-Fusion on the Strength and Aging of a Spliced

Recoated Optical Fiber," *Reliability of Photonics Materials and Structures*, Edited by E. Suhir, M. Fukuda, and C.R. Kurkjian, **531**, 163-168, Proceedings of the Materials Research Society, San Francisco, 1998.
22. T.A. Michalske, W.L. Smith, and B.C. Bunker, "Fatigue Mechanisms in High-Strength Silica-Glass Fibers," *J. Am. Ceram. Soc.* **74** [8] 1993-1996 (1991).
23. G.S. Glaesemann, "The Mechanical Behavior of Large Flaws in Optical Fiber and Their Role in Reliability Predictions," *Proceedings of the 41st International Wire and Cable Symposium,* 689-704, Reno, NV, 1992.
24. D.B. Marshall and B.R. Lawn, "Flaw Characteristics in Dynamic Fatigue: The Influence of Residual Contact Stresses," *J. Am. Ceram. Soc.* 63 [9-10] 532-536 (1980).
25. M. A. Isreal, J.E. Ritter, Jr., K. Jakus, "Strength and Dynamic Fatigue Behavior of Indented Borosilicate and Soda-Lime Glasses," *Physics Chem. Glasses* **28** [4] 133-138 (1987).
26. R.D. Maurer, "Behavior of Flaws in Fused Silica Fibers," *Strength of Inorganic Glass*, Edited by C.R. Kurkjian, 291-308, Plenum Press, New York, 1985.
27. D.G Holloway, "The Fracture Behavior of Glass," *Glass Tech.* **27** [4] 120-133 (1986).
28. S.M. Wiederhorn and B.R. Lawn, "Strength Degradation of Glass Impacted with Sharp Particles: I, Annealed Surfaces," J. Am. Ceram. Soc., **62** [1-2] 66-74 (1979).
29. J.E. Ritter, K. Jakus, and R.P. Panat, "Impact Damage and Strength Degradation of Fused Silica," *Reliability of Photonics Materials and Structures*, Edited by E. Suhir, M. Fukuda, and C.R. Kurkjian, **531**, 53-58, Proceedings of the Materials Research Society, San Francisco, 1998.
30. K. Yoshida, T. Satoh, N. Enomoto, T. Yagi, H. Hihara, and M. Oku, "Fracture Origins of Optical Fibers Fabricated by Hybridized Process," *J. Lightwave Tech.* **14** [11] 2506-12 (1996).
31. J.T. Krause, "Factors Affecting Arc Fusion Splice Strengths," *SPIE* **2611**, 98-109 (1994).
32. D.J. Wissuchek, "Effect of Refractory Particles on the Strength of Optical Fibers," *Reliability of Photonics Materials and Structures*, Edited by E. Suhir, M. Fukuda, and C.R. Kurkjian, **531**, 187-192, Proceedings of the Materials Research Society, San Francisco, 1998.
33. J.C. Newman and I.S. Raju, "An Empirical Stress Intensity Factor Equation for the Surface Crack," *Eng. Fract. Mech.* **15** [1-2] 185-192 (1981).
34. G.S. Glaesemann, K. Jakus, and J.E. Ritter, Jr., "Strength Variability of Indented Soda-Lime Glass," *J. Am. Ceram. Soc.* **70** [6] 441-444 (1987).
35. T.P. Dabbs and B.R. Lawn, "Strength and Fatigue Properties of Optical Glass Fibers Containing Micro-Indentation Flaws," *J. Am. Ceram. Soc.* **68** 563-569 (1985).
36. M.J. Matthewson, B. Lin, and A.P. Stanzeski, "Modeling of Weak Optical Fiber by Using Vickers Indentation," *OFC'94 Tech. Digest*, **5** 245-246 (1994).
37. S.V. Lisle, "The History, Prevention, and Impact of Fiber Cable Failures,", *Proceedings of the National Fiber Optic Engineers Conference*, 223-235, NFOEC, San Antonio, TX, 1993.
38. F.V. DiMarcello, A.C. Hart, Jr., J.C. Williams, and C.R. Kurkjian, "High Strength Furnace-Drawn Optical Fibers," *Fiber Opt.*, 125-135 (1979).

39. T. Svensson and A. Breuls, "Strength and Fatigue of Different Kind of Weak Spots from the Manufacture of Optical Glass Fibres," *SPIE* **2290**, 211-219 (1994).
40. F.A. Donaghy and D.R. Nicol, "Evaluation of the Fatigue Constant n in Optical Fibers with Surface Particle Damage," *J. Am. Ceram. Soc.* **66** [8] 601-604 (1983).
41. R.G. Huff and F.V. DiMarcello, "Critical Particle Size of Contaminants in High and Low Modulus Coatings for High-Strength Optical Lightguides," *J. Lightwave Tech.* **3** [5] 950-953 (1985).
42. S. Sakaguchi and Y. Hibino, "Fatigue in Low-Strength Silica Optical Fibres," *J. Mat. Sci.* **19**, 3416-3420 (1984).
43. S.T. Gulati, J.D. Helfinstine, B. Justice, J.S. McCartney, and M.A. Runyan, "Measurement of Stress Corrosion Constant *n* for Optical Fibers," *J. Am. Ceram. Soc. Bull.* **58** [11] 1115-1117 (1979).
44. K. Abe and R. Ernst, "Static and Dynamic Fatigue Tests of Abraded Optical Fibre," *Elec. Lett.* 21 [20] 926-928 (1985).
45. S.P. Craig, W.J. Duncan, P.W. France, and J.E. Snodgrass, "The Strength and Fatigue of Large Flaws in Silica Optical Fibres," *Proceedings of the 8th ECOC*, Cannes, France, 1982.
46. H.H. Yuce, P.L. Key, and D.R. Biswas, "Investigation of the Mechanical Behavior of Low Strength Fibers," *Fiber Optics Reliability: Benign and Adverse Environments III*, SPIE **1174**, 272-278 (1989).
47. K. Oishi, W. Katsurashima, T. Kakuta, Y. Matsuda, and S. Tanaka, "Coating Design of Thin Coated Fiber for Ultra-High-Count Optical Fiber Cable," *Proceedings of the 42nd International Wire and Cable Symposium*, 687-693, St. Louis, MO, 1993.
48. G.S. Glaesemann, "Process Handleability of Thin-Coated Optical Fibers," *OFC'94 Tech. Digest*, **5** 243-244 (1994).
49. D.J. Wissuchek, D.J. Walter, D.A. Clark, and G.S. Glaesemann, "Fracture and Abrasion Resistance Tests for Optical Fiber Coatings," *Reliability of Photonics Materials and Structures*, Edited by E. Suhir, M. Fukuda, and C.R. Kurkjian, **531**, 309-314, Proceedings of the Materials Research Society, San Francisco, 1998.
50. Bellcore GR-1209-CORE, Issue 2, February 1998
51. Bellcore GR-1221-CORE, Issue 2, January 1999.
52. Bellcore GR-468-CORE, Issue 1, December 1998.
53. A. Dwivedi and J.L. Smith, "Characterizing Field Handleability of Optical Fiber through the Measurement of Strength after Mechanical Stripping," *Proceedings of the 45th International Wire and Cable Symposium*, 381-390, Philadelphia, PA, 1996.
54. G.S. Glaesemann and D.J. Walter, "Method for Obtaining Long-Length Strength Distributions for Reliability Prediction," *Opt. Eng.* **30** [6] 746-748 (1991).
55. For example, see J. Horska, "Design of Loose Tube Fiber Optic Cable with Adjusted Contraction and Strain Windows," *Proceedings of the 43rd International Wire and Cable Symposium*, 50-58, Atlanta, GA, 1997.
56. H. Nagata, K. Ooto, M. Shiroishi, and J. Ogiwara, "Failure Case Study for Loose Tube Jacketed Optical Fibers," *Opt. Fiber Tech.* **4**, 480-489 (1998).

57. J. B. Stevens, P. Elission, B. Risch, and C. Bastide, "Finite Element Modeling of Optic Fiber Cable Crush Performance," *Proceedings of the 47th International Wire and Cable Symposium*, 25-33, Philadelphia, PA, 1998.
58. E. Opel, A. Stingl, M. Schuebbe, "Bending Simulation of a Slotted Core Cable," *Proceedings of the 46th International Wire and Cable Symposium*, 810-818, Philadelphia, PA, 1997.
59. A. Makiyama, J. Ohta, and A. Nishimura, "A Study on Strain Characteristics of Fiber Ribbons in a SZ-Grooved Spacer caused by Cable Bending," *Proceedings of the 47th International Wire and Cable Symposium*, 216-219, Philadelphia, PA, 1998.
60. C.S. Ma, L.L. Blyler, G.E. Johnson, R.T. Traut, C.E. Murphy, M.W. Jones, F.C. Gibson, W.M. Keith, "Direct Mechanical Fiber Strain Measurement and the Cabled Fiber Strain of an Undersea Cable," *Proceedings of the 47th International Wire and Cable Symposium*, 241-248, Atlanta, GA, 1994.
61. C.J. Sandwith, W.D. McCormick, J.A. Thornton, D.R. Wise, R.I. Odom, "Fiber-Strain Measurement using Brillouin Optical-Fiber Time-Domain Analysis," *Proceedings of the 45th International Wire and Cable Symposium*, 415-427, Philadelphia, PA, 1996.
62. T. Horiguchi, T. Kurashima, M. Tateda, K. Ishihara, and Y. Wakui, "Brillouin Characterization of Fiber Strain in Bent Slot-Type Optical-Fiber Cables," J. Lightwave Tech. **10** [9] 1196-1201 (1992). and N. Okada, K. Watanabe, M. Miyanoto, "Strain Analysis of U-Groove Type Cable with Multi Fiber Ribbons," *Proceedings of the 43rd International Wire and Cable Symposium*, 211-218, Atlanta, GA, 1994.
63. G. Meltz, J.R. Dunphy, W.H. Glenn, J.D. Farina, and F.J. Leonberger, "Fiber Optic Temperature and Strain Sensors," *Fiber Optic Sensors SPIE* **798** 104-111 (1987).
64. M. Vaziri and C.-L. Chen, "Etched Fibers as Strain Gauges," *J. Lightwave Tech.* **10** [6] 836-841 (1992).
65. W.R. Wagner, "Failure Analysis of Fiber Optic Connectors," *Advances in Ceramics, Volume 22, Fractography of Glasses and Ceramics*, 389-402, American Ceramic Society (1988).
66. T. Wei, H.H. Yuce, C.H. Hasz, and P.L. Key, "Degradation of Fiber Strength during Coating Stripping," *Proceedings of the 38th International Wire and Cable Symposium*, 199-203, Atlanta, GA, 1989.
67. E. Suhir, "Bending Stress in an Optical Fiber Interconnect Experiencing Significant Ends Off-Set," *Reliability of Photonics Materials and Structures*, Edited by E. Suhir, M. Fukuda, and C.R. Kurkjian, **531**, 209-214, Proceedings of the Materials Research Society, San Francisco, 1998.
68. E. Suhir, " Bending of an Optical Glass Fiber 'Pigtail' in a Splice Box Structure," *Reliability of Photonics Materials and Structures*, Edited by E. Suhir, M. Fukuda, and C.R. Kurkjian, **531**, 215-222, Proceedings of the Materials Research Society, San Francisco, 1998.
69. J.M. Anderson, J.F. Malluck, M.M. Tabaddor, C.K. Sidbury, and T.E. McNeil, "Optical-Fiber and Entry-Cone Contact Stress," *Reliability of Photonics Materials and Structures*, Edited by E. Suhir, M. Fukuda, and C.R. Kurkjian, **531**, 223-229, Proceedings of the Materials Research Society, San Francisco, 1998.
70. L.K. Baker and G.S. Glaesemann, "A Mechanical Reliability Study Of Bare Fibers Under Stress," to be presented at the International Institute for

Connector and Interconnection technology symposium in Anaheim, CA, September 1999.
71. G.S. Glaesemann, "Design Methodology for the Mechanical Reliability of Optical fiber," *Opt. Eng.* **30** [6] 709-715 (1991).
72. V.D. Frechette, *Advances in Ceramics: Failure Analysis of Brittle Materials*, **28**, American Ceramic Society, Westerville, OH, 1990.
73. J.J. Mecholsky, "Fracture Surface Analysis of Optical Fibers," Ceramics and Glasses, *Engineered Materials Handbook*, **4**, edited by S.J. Schneider, Jr., 663-668 ASM 1991.
74. J.D. Mann, O.S. Gebizlioglu, and C.R. Kurkjian, "The Effect of Preparation Conditions on the Strength of Fusion Splices I: Use of AFM Imaging," *Proceedings of the 47th International Wire and Cable Symposium*, 589-596, Philadelphia, PA, 1998.
75. Q. Zhong, D. Inniss, K. Kjoller, and V.B. Elings, "Fractured Polymer/Silica Fiber Surface Studied by Tapping Mode Atomic Force Microscopy," *Surface Sci. Letts.* **290** 1688-1692 (1993).
76. H.H. Yuce, T. Wei, R. Estes, C.J. Wieczorek, B.T. Devlin, J.P. Kilmer, and J.P. Varachi, Jr., "Fiber Reliability Study of Field Aged Optical Cables," *Proceedings of the 41st International Wire and Cable Symposium*, 705-712, Reno, NE, 1992.
77. H.H. Yuce, C.J. Weiczorek, A. DeVito, J.P. Varachi, Jr., and J.P. Kilmer, "Effects of the Environment on an Unprotected Reel of Optical Fiber Cable," *Proceedings of the 40th International Wire and Cable Symposium*, 700-705, St. Louis, MO, 1991.
78. W. Griffioen, T. Volotinen, P. Wilson, A. Gouronnec, T. Svensson, "Handleability of Aged Optical Fibers," *Proceedings of the 44th International Wire and Cable Symposium*, 857-864, Philadelphia, PA, 1995.
79. A. Dwivedi, G.S. Glaesemann, and C.K. Eoll, "Optical Fiber Strength, Fatigue and Handleability after Aging in a Cable," *Proceedings of the 43rd International Wire and Cable Symposium*, 728-735, Atlanta, GA, 1994.
80. J.L. Smith, A. Dwivedi, and P.T. Garvey, "Mechanical Behavior of Optical Fibers Removed from a Field-Aged Cable," *Proceedings of the 44th International Wire and Cable Symposium*, 848-856, Philadelphia, PA, 1995.
81. M.L. Lundergan, B.D. Zimmerman, B. Waterman, "Mechanical and Optical Functionality of Field-Aged Optical Ground Wire Cable," *Proceedings of the National Fiber Optic Engineers Conference*, 449-458, NFOEC, 1996.
82. K. Houser and S. Chahanovich, "Verification of Optical Fiber and Cable Reliability," *Proceedings of the 47th International Wire and Cable Symposium*, 468-474, Philadelphia, PA, 1998.
83. S. Etemad, "Dense-WDM Components and Systems: A Reliability Overview," *Reliability of Photonics Materials and Structures*, Edited by E. Suhir, M. Fukuda, and C.R. Kurkjian, **531**, 27-38, Proceedings of the Materials Research Society, San Francisco, 1998.
84. E.B. Shand, "Breaking Stress of Glass Determined from Dimensions of Fracture Mirrors," *J. Am. Ceram. Soc.* **42** [10] 474-477 (1959).
85. J.J. Mecholsky, R.W. Rice and S.W. Freiman, "Prediction of Fracture Energy and Flaw Size in Glasses from Measurements of Mirror Size," *J. Am. Ceram. Soc.* **57** [10] 440-443 (1974).

86. L.K. Baker and G.S. Glaesemann, "Break Source Analysis: Alternative Mirror Measurement Method," *Proceedings of the 46th International Wire and Cable Symposium*, 933-937, Philadelphia, PA, 1997.
87. G. Kiss and E. Vogel, "High Yield Fusion Splicing In The Outside Plant: Using Fiber Meltback To Monitor Electrode Condition" *Proceedings of the National Fiber Optic Engineers Conference*, 819-835, NFOEC, 1996.
88. J.T. Krause, "Ultrahigh-Strength Fibre Splices by Modified H2/O2 Flame Fusion," *Elec. Letters* **22** [20] 1075-1077 (1986).
89. G. Kiss, "Reliability of Optical Splices," *Fiber Optics Reliability and Testing*, Edited by D.K. Paul, **CR50**, 118-154, Proceedings of the SPIE, Bellingham, WA, 1993.
90. W.W. King, B.G. LeFevre, and D.L. Stephenson, "Structural Mechanics Aspects of Fractures of Optical Fibers in Connectors," *Proceedings of the 46th International Wire and Cable Symposium*, 228-235, Reno, NV, 1996.
91. M. Komachiya, R. Minamitani, T. Fumino, T. Sakaguchi, and S. Watanabe, "Proof-Testing and Probabilistic Lifetime Estimation of Glass Fibers for Sensor Applications," *Appl. Optics* **38** [13] 2767-2774 (1999).
92. G.S. Glaesemann, "Optical Fiber Failure Probability Predictions from Long-Length Strength Distributions," *Proceedings of the 40th International Wire and Cable Symposium*, 819-825, St. Louis, MO, 1991.
93. Y. Mitsunaga, Y. Katsuyama, H. Kobayashi, and Y Ishida, "Failure Prediction for Long-Length Optical Fiber Based on Proof Testing," *J. Appl. Phys.* **53** [7] 4847-4853 (1982).
94. A. Paul and G.S. Glaesemann, "An Appraisal of Mechanical Reliability Predictions for Optical Fibers Based on Break Rates," *Proceedings of the 46th International Wire and Cable Symposium*, 896-901, Philadelphia, PA, 1997.

Reliability of Fluoride Fiber Module for Optical Amplifier Use

Yoshiki Nishida, Kazuo Fujiura, Terutoshi Kanamori, Makoto Shimizu
NTT Photonics Laboratories
Tokai, Naka, Ibaraki, 319-1193 Japan

Makoto Yamada, Yasutake Ohishi
NTT Electronics Corporation
Tokai, Naka, Ibaraki, 319-1193 Japan

ABSTRACT

The reliability of a fluoride fiber module generally depends on three factors. They are the lifetime of the fluoride fiber under practical environmental conditions, the reliability of the splice between fluoride and silica fiber in terms of low-loss and low-reflection, and the effectiveness with which the packaging prevents moisture damage. The first factor depends strongly on the fluoride fiber strength. We describe a technique for fabricating fluoride fiber with improved strength and then estimate its lifetime using a static fatigue test. We evaluate the remaining two factors by applying the recommended reliability test program to the fiber modules. We describe the structure of our fluoride fiber module and present the reliability test results with reference to the Bellcore Technical Advisory TA-NWT-001221.

Keywords: Reliability, Fluoride fiber, Fiber module, Optical fiber amplifier

1. INTRODUCTION

The erbium-doped fiber amplifier [1] (EDFA) has advantages that have allowed us to make considerable progress on optical fiber communication systems. It has been used to replace electrically regenerative repeaters and thus allowed us to extend the repeater span in trunk-line and submarine systems. It has also made it possible to develop wavelength-division-multiplexing (WDM) systems which are currently being installed throughout the world to handle the rapid increase in demand for telecommunication capacity. Recent progress on optical fiber amplifiers, especially those for WDM application, has been related to the amplification bandwidth. The demand for increased capacity has led to an increase in channel number and this has made it necessary to broaden the amplification bandwidth or reduce the channel spacing. A conventional EDFA provides a 30 nm (1530-1560 nm : C-band) wavelength region for WDM signal amplification, therefore, to narrow the channel spacing requires that we improve the wavelength stability of laser diodes or the crosstalk in filtering components. In contrast, broadening the amplification bandwidth by means of installing a 1.58 μm band (L-band) EDFA [2,3] with a parallel configuration or using a tellurite-EDFA [4] which provides an amplification bandwidth of ~80 nm is an easy way of increasing the channel number and is promising for next generation WDM application. If we wish to increase the amplification bandwidth further,

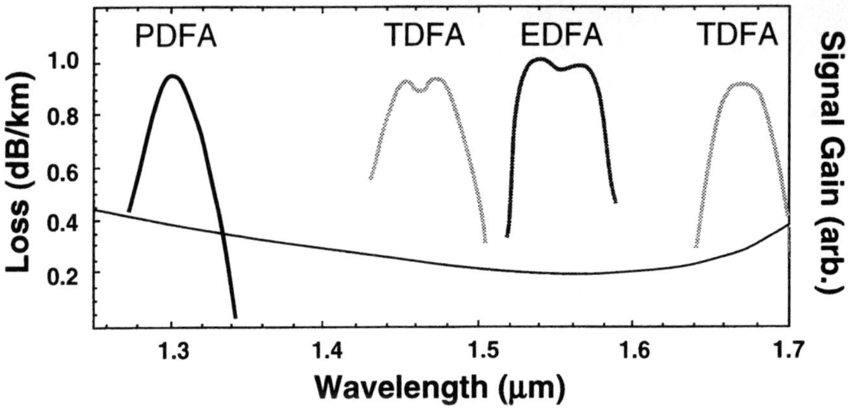

Fig. 1. Amplification band of rare-earth doped fiber amplifiers.

to cover the low-loss region of silica fiber, we need to consider using rare-earth ions other than erbium and combining them with host glasses other than silica glass.

Rare-earth-doped fluoride fiber has been recognized as an important alternative to silica-based fiber with regard to exploring amplification bands that are inaccessible when using silica-based fiber. This not only makes 1.3 μm amplification possible with praseodymium doping [5], it also makes 1.4 and 1.6 μm amplification possible with thulium doping [6,7] and these amplification bands are attractive with a view to extending the 1.5 μm transmission band beyond 80 nm. These bands are shown schematically in Fig. 1. The most desirable use of the fluoride fiber amplifier is as a praseodymium-doped fluoride fiber amplifier (PDFA). We have improved the amplification characteristics of the PDFA by various methods such as optimizing the fiber parameters, pumping configuration and amplifier structure. Moreover, we have developed a new host glass for efficient amplification and constructed an LD-pumped plug-in type PDFA module which can be used with rack-mounted shelves. [8] A successful demonstration of 10 Gbit/s digital and AM-VSB analog signal transmissions has confirmed the effectiveness of using PDFAs in advanced 1.3 μm optical telecommunication systems. [9]

One concern that must be met before the PDFA can be practically employed is to ensure its reliability. Fluoride glass is susceptible to humidity and the maximum theoretical strength of the fiber is estimated to be 3.3 GPa which is less than that of silica fiber (20 GPa). [10,11] This makes it critical to estimate its lifetime under practical environmental conditions. The intrinsic characteristics of fluoride fiber mean that it may require measures to be taken to protect it against moisture damage and ensure its long-term reliability under harsh environmental conditions. Moreover, the large difference between the melting temperatures of fluoride glass and silica glass gives rise to a problem in terms of splicing. The difficulty of fusion splicing these fibers means that a mechanical splicing technique must be developed that offers low loss and low reflection values. This makes splice reliability critical to the overall PDFA reliability. This paper reviews the current state of fiber module technology and presents our reliability study results.

2. RELIABILITY REQUIREMENTS

The reliability of an optical fiber amplifier (OFA) depends on various factors including the reliability of the active fiber, the pump laser and other optical components, and the reliability of the module, itself. With regard to the active fiber, practical applications require that a fiber never breaks down even when coiled around a compact spool in an amplifier module under any given service condition. The lifetime estimation theory established for silica fibers provides such mechanical reliability information. Another requirement for active fiber is that there should be no increase in transmission loss that leads to a degradation in amplification performance. The potential for optical property degradation depends on such parameters as temperature, humidity, H_2 loading and γ-ray irradiation. [12-14]

To demonstrate the long-term capability of OFAs to perform their functions reliably, they must pass a comprehensive reliability test program that attempts to duplicate the stresses that fibers are expected to experience during 20 years of operation. The qualification test requirements for OFAs are established in the Bellcore Generic Requirement GR-1312-CORE Issue 2, 1996 "Generic Requirements for Optical Fiber Amplifiers". It is specifically stated that this GR applies not only to conventional EDFAs, but also to PDFAs, however, this GR considers the OFA to be a "black box" and hence does not provide reliability criteria for its constituent components. In contrast, an international reliability standard for OFAs is currently being established by IEC Sub-Technical Committee No. 86C, Working Group 3. [15] However, since this standard is still being discussed we think that it is appropriate to ensure fluoride fiber module reliability by ascertaining that it conforms to the Bellcore Technical Advisory TA-NWT-001221 Issue 1, 1991 "Generic Reliability Assurance Requirements for Passive Optical Components".

3. FLUORIDE FIBER FABRICATION TECHNIQUE

Here, we describe the fluoride fiber fabrication technique which is different from that used for silica fiber. [16] This section also provides necessary background to fluoride fiber and module reliability. ZrF_4-based fluoride glass (ZrF_4-BaF_2-LaF_3-AlF_3-NaF) has potential for use in fiber fabrication. It has a glass transition temperature around 260 °C and a melting temperature around 700 °C. All its ingredients are bonded ionically, therefore this glass has a relatively low phonon-energy which gives it unique amplification characteristics as an optical fiber amplifier. There have been some attempts to fabricate a fluoride fiber preform by means of the chemical vapor deposition technique as used for silica fiber fabrication. [17] However, such difficulties as obtaining the necessary volatile raw materials, controlling the reaction, and the contamination caused by metal-organic raw materials make this technique impractical for fiber fabrication. The best approach for fabricating fluoride fibers is based on the crucible (glass melt) and jacket-drawing technique.

Figure 2 shows the suction-casting method used for preparing the primary fiber preform.[18] A core/cladding structure is formed by pouring cladding glass melt and then core glass melt into a mold. During solidification, there is a significant contraction in volume at the center of the cladding glass which has a suction effect and draws the core glass melt into the center of the cladding glass cylinder. Figure 3 shows the single-mode fiber fabrication

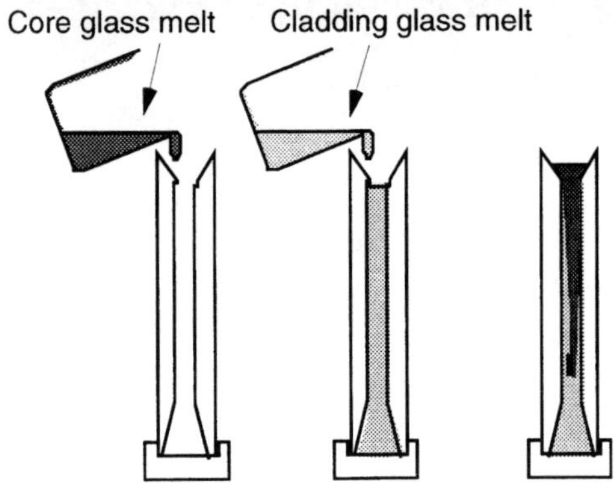

Fig. 2. Schematic diagram of suction-casting method.

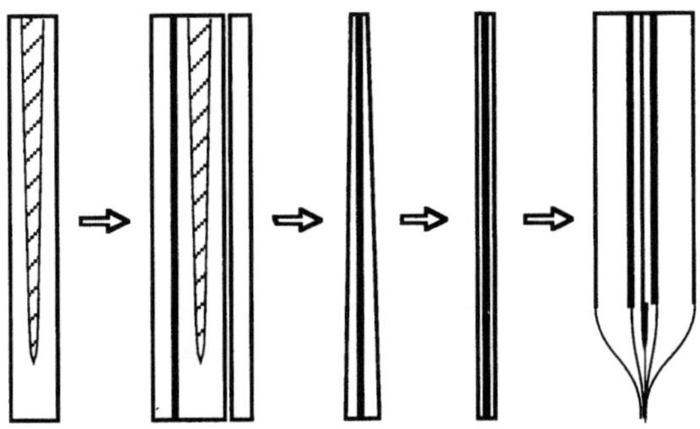

Fig. 3. Schematic diagram of jacket-drawing method.

technique which uses the tapered-elongation and jacket-drawing method. Fiber preforms and jacketing tubes are prepared by the suction-casting and rotational-casting methods, respectively. The obtained straight preform with a tapered core is elongated together with the jacketing tube into a secondary preform with a smaller straight core. This is achieved by programming the elongation speed as a function of time. The secondary preform is reshaped into a cylinder and drawn together with another jacketing tube into a fiber coated with UV-curable acrylate. Although the basic techniques needed for jacket-drawing, such as diameter control, capstan driving, and plastic coating, are based on standard methods developed for oxide glass fibers, some important differences exist. Elongation and drawing are undertaken at a much lower temperature of 300°C to 400°C using an electric furnace. The temperature of these processes must be precisely controlled because the viscosity of fluoride glasses has a strong temperature dependence and the working range between the deformation and crystallization temperatures is very narrow. The key to fluoride fiber fabrication lies in eliminating the moisture which induces devitrification through hydration and dehydration reactions at the glass surface at all stages of the process. If these conditions are not realized, bulk and surface crystallization will occur during elongation and drawing and result in high fiber loss and poor mechanical strength.

4. FLUORIDE FIBER STRENGTH

4.1 Mechanism of fluoride fiber strength degradation

The mechanism which causes fluoride fiber strength degradation is now well understood and has been found to depend strongly on the surface state of the fiber preform and the atmosphere surrounding it during fiber drawing. [19,20] Fiber strength has been improved by mechanically polishing the surface of the fiber preform. [21] However, the polishing compound and scratches which remained on the preform surface have caused flaws on the fiber surface. Water on the glass surface induces the formation of a hydroxide layer by a hydrolysis reaction such as

$$ZrF_4 + 2H_2O \rightarrow ZrF_2(OH)_2 + 2HF.^{22} \quad (1)$$

When heated in the fiber drawing furnace, this hydroxide layer can lead to the formation of oxide crystals,

$$ZrF_2(OH)_2 \rightarrow ZrO_2 + 2HF \quad (2)$$

which also cause flaws on the fiber surface. Therefore, the fabrication of high strength fluoride fiber requires a preform with a flaw free surface and the suppression of the surface hydrolysis reaction caused by atmospheric humidity during fiber drawing. [23,24]

There have been many approaches to high strength fluoride fiber fabrication. [20-26] Schneider *et al.* reported a chemical etching technique which is an important method for obtaining a preform with a flaw free surface. [20] This technique has been used to fabricate fluoride fiber with a mean tensile strength of 310 MPa. Techniques for treating preform surfaces with active fluorine and HF gas have also been reported, and both are capable of improving the tensile strength by 20-30 %. [23,26] Moreover, the addition of a reactive gas such as $SOCl_2$ or NF_3 to the drawing atmosphere has enabled the fabrication of a fiber with a mean tensile strength of 400 MPa. [24] Infrared absorption spectra observations have indicated that there are hydroxides on the fluoride glass surface after chemical etching. [24] To improve the fiber strength further, it is important to explore the possibility of efficiently eliminating these hydroxides. We have developed a surface treatment technique using HF gas as a reactive agent. To determine the effect of this technique, we performed a surface state analysis using secondary ion mass spectroscopy (SIMS).

4.2 HF gas treatment effect

The fluoride glass samples we prepared were pellets which had a composition of $41ZrF_4$-$23BaF_2$-$4LaF_3$-$2.5AlF_3$-$3YF_3$-$20NaF$ (mol%). The surface of each sample was mechanically polished and chemically etched using a solution of 0.4 mol $ZrOCl_2 \cdot 8H_2O$ dissolved in 1 mol HCl + 1 L H_2O as the etchant. [20] After the chemical etching, the samples were heated in an HF gas atmosphere with HF and Ar gas flowing at 200 and 100 cc/min, respectively, at 150, 200, and 250 °C for 2 hours. Since it is difficult to analyze the hydroxide directly, we analyzed the hydrogen and oxygen concentrations individually.

Figure 4 shows the oxygen concentration profile for samples which were heated at 150, 200 and 250 °C for 2 hours in an HF gas atmosphere. The concentration profile after chemical etching is also plotted for comparison. A high oxygen concentration of $1.6x10^{21}$ at/cm^3 was detected from the surface after chemical etching to a depth of 0.01 μm. It decreased to $9x10^{18}$ at/cm^3, which is about 100 times lower than that at the surface, at a depth of 0.1 μm below the surface. The oxygen concentration at the surface decreased with increasing temperature. The surface concentrations after HF gas treatment at 150, 200 and 250 °C were respectively 8, 13 and 30 times lower than those before treatment. In contrast, the hydrogen exhibited different concentration profiles, as shown in Fig. 5. The surface hydrogen concentration was $3.6x10^{20}$ at/cm^3 after chemical etching. It had a peak at a depth of about 0.015 μm and decreased sharply to $1.2x10^{19}$ at/cm^3 by a depth of 0.1 μm. The hydrogen concentration after HF gas treatment at 150, 200 and 250 °C decreased steadily with depth and reached $1x10^{19}$ at/cm^3 at depths of 2.7 to 10.7 μm. This is considered to be caused by the HF diffusion during the HF gas treatment. The hydrogen diffused more deeply as the treatment temperature was increased.

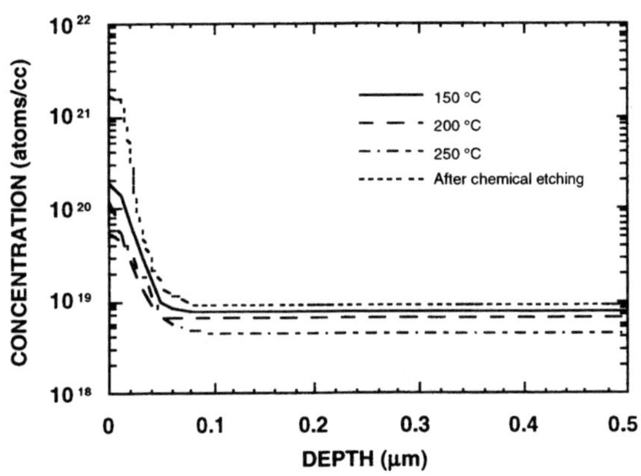

Fig. 4. SIMS depth profiles of oxygen after HF gas treatment at 150, 200 and 250 °C for 2 hours.

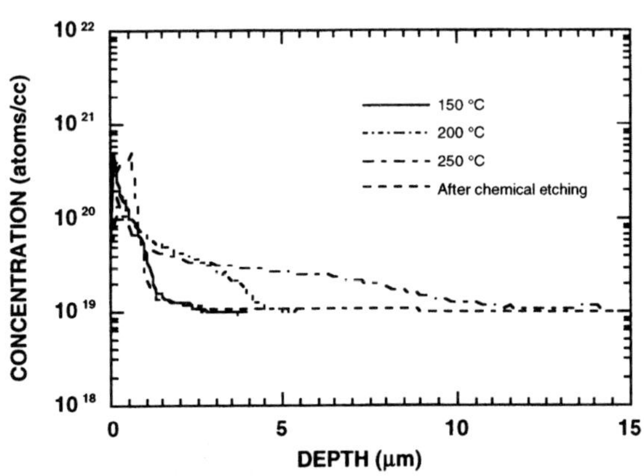

Fig. 5. SIMS depth profiles of hydrogen after HF gas treatment at 150, 200 and 250 °C for 2 hours.

Our SIMS analyses clearly revealed the effect of HF gas treatment on the removal of the surface hydroxide. Although oxygen diffusion was still observed, the amount of oxygen impurity at the surface was much less than before the treatment. However, the decrease in the hydrogen concentration at the surface is not as clear, and the analysis showed that hydrogen diffused into the glass matrix during the HF gas treatment. The hydrogen diffused more deeply as the HF gas treatment temperature increased and after treatment at 250 °C for 2 hours, it had reached a depth of 10 μm.

One possible explanation for the decrease in the oxygen concentration at the glass surface and hydrogen diffusion into the glass matrix during HF gas treatment is shown in Fig. 6. The reaction is summarized as

$$ZrF_2(OH)_2 + 4HF \rightarrow ZrF_4 + 2H_2O + F_2 + H_2 . \qquad (3)$$

During the first step of this reaction, the OH impurities on the glass surface react with HF gas to form H_2O (glass) and F_2 (gas). Since the HF gas surrounding the glass is provided continuously, HF reacts with H_2O on the surface and dissolves in it with ionized state H^+ and F^-. The F^- ions bond with the $(ZrF_2)^{2+}$ species to form ZrF_4 and the H atoms enter the glass matrix.

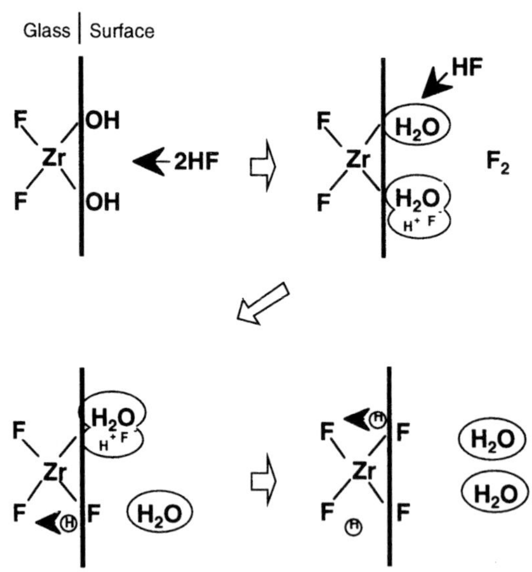

Fig. 6. HF gas treatment reaction mechanism.

There have been many reports on hydrogen diffusion into silica fiber associated with an increase in infrared transmission losses.[27,28] The diffusion of hydrogen in vitreous silica has been found to occur at a molecular level.[27] According to a structural study, it is predicted that 1.8×10^{27} sites/m^3 are available for hydrogen molecules in vitreous silica.[29] On the other hand, the packing density of ZrF_4-based fluoride glass is higher than that of silica and $5x10^{24}$ sites/m^3 are predicted for hydrogen molecules. [30] The results in Fig. 5 reveal hydrogen diffusion in the fluoride glass. This diffusion has not been well understood in terms of whether it occurs at molecular or atomic level. We estimated the diffusion coefficient for atomic hydrogen by fitting the concentration profile in Fig. 5 to the expression,

$$C(x,t)=C(0)[1 - \mathrm{erf}(x/K)], \tag{4}$$

where $K=(4Dt)^{1/2}$; x is the depth below the surface, D is the diffusion coefficient at a specific temperature, and t is the treatment time. [31] The diffusion coefficients for hydrogen are plotted in Fig. 7 as a function of 1000/T. By fitting the equation $D(T)=D_0 \exp(-E_a/kT)$, we obtained $D_0=1.45x10^{-1}$ cm^2/s and $E_a=0.959$ eV. Therefore, the hydrogen coefficient, D, in ZrF_4-based fluoride glass at room temperature is obtained as 1.1 $x10^{-17}$ cm^2/s. This value is 10^6 times lower than that in vitreous silica which is $D=1.5x10^{-11}$ cm^2/s. [27,32] We think that the diffusion of hydrogen in ZrF_4-based fluoride glass is a much less serious problem than that in vitreous silica.

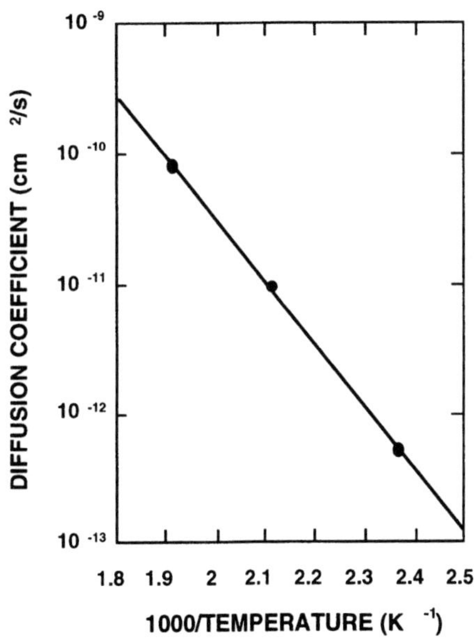

Fig. 7. Hydrogen diffusion coefficient as a function of 1000/T.

4.3 Fabrication of stronger fluoride fiber

We employed an HF gas treatment technique for our fluoride fiber fabrication. The preform was mechanically polished, chemically etched and then heated in an HF gas atmosphere at 200 °C for 2 hours. The preform was placed in a drawing furnace and drawn into 120 μm-diameter fibers at a speed of about 20 m/min. A UV-curable acrylate was in-line coated on the fiber surface to form a fiber 260 μm in diameter. These processes were carried out in a dry inert gas atmosphere. Figure 8 shows a Weibull plot of the tensile strength of this fiber. The gauge length was 350 mm and the strain rate was 5 mm/min. The fiber strength ranged from a low of 420 MPa to a maximum value of 600 MPa. We successfully achieved a mean tensile strength of 550 MPa and an m-value of

25. This fiber is about 150 MPa stronger than fibers reported in previous studies, and about 5 times higher in m-value.[23,24,26] These results show that the HF gas treatment is useful for fabricating a high strength fiber.

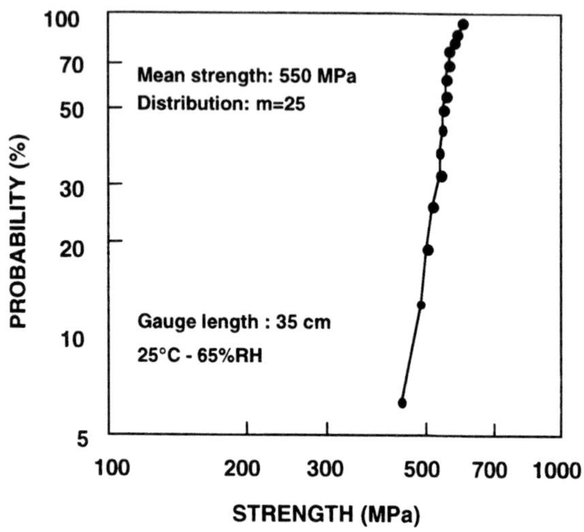

Fig. 8. Weibull plot of UV-coated fluoride fibers drawn from an HF gas treated preform.

5. COATING

With silica fiber, since the resistivity to moisture is relatively high compared with multi-component silica fibers and fluoride fibers, it seems that the coating materials may have little effect on reliability when the fiber is used for amplifiers. With regard to fluoride fibers, several kinds of material, such as polymer, glass overcladding, SiO_2, TiN and BN have been coated on both fibers and bulk glasses.[33-37] With regard to polymer coating, UV-curable epoxy acrylate yields a higher strength than Teflon FEP coating. This result correlates with the degree of adhesion to the glass surface and the moisture permeability. Glass overcladding provides useful moisture protection, but high-strength fiber has not yet been fabricated because of the crystallization of the glass interface. Hermetic coating offers the best moisture protection, but it appears to be difficult to coat such materials on the fiber because of the low drawing temperature of fluoride fiber. Consequently, the use of UV-curable epoxy acrylate is currently considered to be the most realistic solution.

6. LIFETIME ESTIMATION

A rare-earth-doped fluoride fiber for amplifier use must be coiled around a bobbin in an amplifier module. Hence, the fiber surface experiences a tensile stress corresponding to the bobbin diameter. In addition, since fluoride glasses are susceptible to moisture damage as described in the previous section, both stress corrosion and zero-stress aging will occur simultaneously in the presence of environmental moisture and this will reduce the fiber lifetime in a practical environment.

The strength degradation with time caused by zero-stress aging can be empirically fitted to a simple polynomial as follows.[38]

$$\sigma_f = \sigma_{f0} / (1 + t/\tau)^p \tag{5}$$

Here, σ_{f0} is the measured strength at time $t = 0$. τ and p are constants which vary with aging condition and fiber material.
With the static fatigue test, the lifetime can be estimated as follows. A differential equation has been formulated which combines zero-stress aging and stress corrosion.[39]

$$d(\sigma/\sigma_i)/dt = x/\tau(\sigma/\sigma_i)^{(x-1)/x} - 1/[B\sigma_i^{n-2}(n-2)]\sigma_a^n(\sigma/\sigma_i)^{3-n} \tag{6}$$

Here, σ_i is initial strength, B is a constant containing material and process parameters, n is a stress corrosion exponent, and $x = -(n+1)/(n-2)p$. The differential equation was solved numerically, and then the relationship between the lifetime and applied stress was derived from (6).

Strength is distributed along the length of practical fibers. This can be expressed by a Weibull survival distribution:

$$P(S, L) = \exp[-(S/S_0)^m (L/L_0)] \tag{7}$$

Here, $P(S,L)$ is the cumulative survival probability of the fiber length L up to the inert strength S, S_0 is the strength corresponding to a survival probability of e^{-1}, measured at a gauge length L_0. This Weibull distribution can be transformed into the following equations for static fatigue[40]:

$$P(t_f, \sigma_a, L) = \exp[-(t_f/t_0)^b], \qquad (t_0 = t_f(\ln 2)^{-1/b}) \tag{8}$$

where b is the Weibull parameter, t_f is the median lifetime. This Weibull distribution makes it possible to estimate the relationship between the lifetime and the allowable failure probability.

We estimated the fluoride fiber lifetime taking both stress corrosion and the zero-stress aging effect into account.[41,42] In a zero-stress aging experiment, the fiber strength was measured at 25 °C and 60 %RH (relative humidity) as a function of storage time at 80 °C and 50 %RH. We performed a static fatigue test in which tensile stress was applied to the fibers and the time to failure was measured at 80 °C and 50 %RH. Figure 9 shows the strength degradation resulting from zero-stress aging at 80 °C and 50 %RH. The mechanism of zero-stress aging for fluoride fibers is different from that for silica fibers. With fluoride glass fibers, this is mainly caused by H_2O which forms a hydrate or an oxide on the fiber surface and these crystals may be the origin of fractures. By fitting the results to (5), τ and p were found to be 2.2×10^{-3} years and 3.6×10^{-2}, respectively. To obtain static fatigue parameters such as B and n, we measured the relationship between lifetime and applied stress in the high stress region where the effect of zero-stress aging on the lifetime might be negligible. Figure 10 shows the dependence of lifetime on the applied

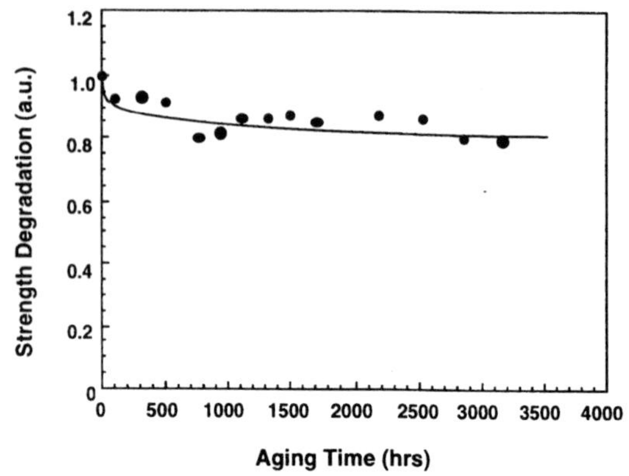

Fig. 9. Measured strength degradation with aging time at 80 °C and 50 %RH.

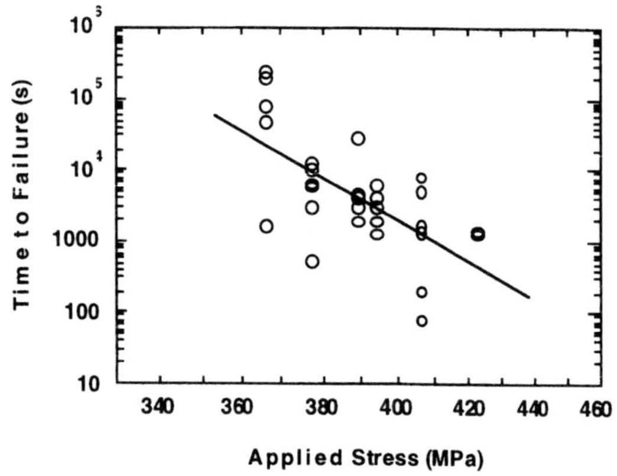

Fig. 10. Time to failure measured by static fatigue tests.

stress. From the slope and the intercept, we obtained n of 26.6 and $B\sigma_i^{n-2}$ of 2.87×10^{67}. The Weibull distributions of the lifetime at each applied stress indicated that the worst Weibull parameter b was 0.033. This worst parameter was used to estimate the lifetime with an allowable failure probability. Figure 11 shows the relationship between lifetime and applied stress estimated from (6) taking both zero-stress aging and stress corrosion into account. In this figure, the dashed line indicates the lifetime estimated only from stress corrosion. The difference between the estimated results in the figure suggests that the lifetime is mainly determined by zero-stress aging with practical stresses of lower than 100 MPa. Since a lifetime of over 25 years can be estimated in the practical stress region at 80 °C and 50 %RH, fluoride fiber can be expected to withstand a practical environment. Figure 12 shows the estimated lifetime with several allowable failure probabilities at 80 °C and 50 %RH. If we adopt an allowable failure probability of 10^{-6}, a lifetime of 25 years can be guaranteed for a fiber under an applied stress of 47 MPa, which corresponds to a bobbin diameter of 75 mm at 80 °C and 50 %RH. Therefore, we have confirmed that our fluoride fiber has a long-term reliability of over 25 years when using a fiber bobbin with a diameter of over 75 mm in a practical environment.

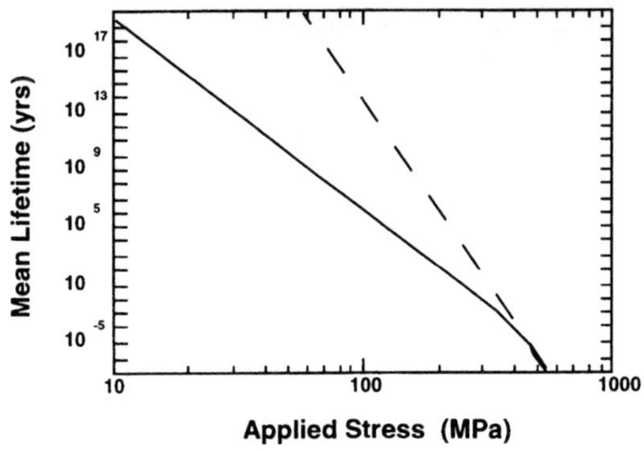

Fig. 11. Estimated lifetime at 80 °C and 50 %RH.

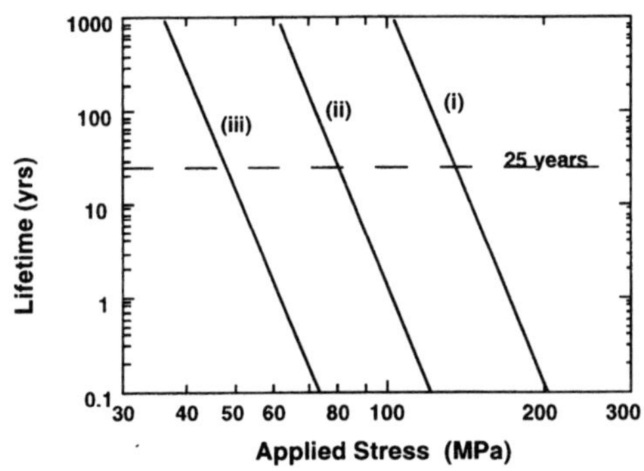

Fig. 12. Estimated lifetime with several allowable failure probabilities at 80 °C and 50 %RH. (i) allowable failure probability F=10^-2. (ii) F=10^-4. (iii) F=10^-6.

7. FLUORIDE FIBER MODULE RELIABILITY

7.1 Module structure

For practical amplifier use, it is desirable to accommodate the rare-earth-doped fluoride fiber in a sealed package with ports for silica fiber leads. We developed a fluoride fiber module consisting of fluoride fiber, a fiber bobbin, V-groove connections [43], silica fiber and an aluminum case. The key to its preparation lies in the packaging that protects the fluoride fiber from moisture damage. A 15 m long fluoride single-mode fiber with a Δn of 2.5%, a core diameter of 2.2 μm, a cut-off wavelength of 1.0 μm and a praseodymium concentration of 1000 ppm (weight) was coiled around a bobbin 75 mm in diameter and 5 mm high. Two V-groove connections were fixed inside the bobbin with epoxy adhesive. This fiber bobbin was inserted into a specially designed package with feed-through silica fibers and sealed in an inert gas atmosphere. We housed this package in an aluminum case which we filled with epoxy resin.

We used the mechanical splicing technique to splice these two fibers together. Figure 13 illustrates this mechanical splicing (V-groove connection) technique, schematically. [44]

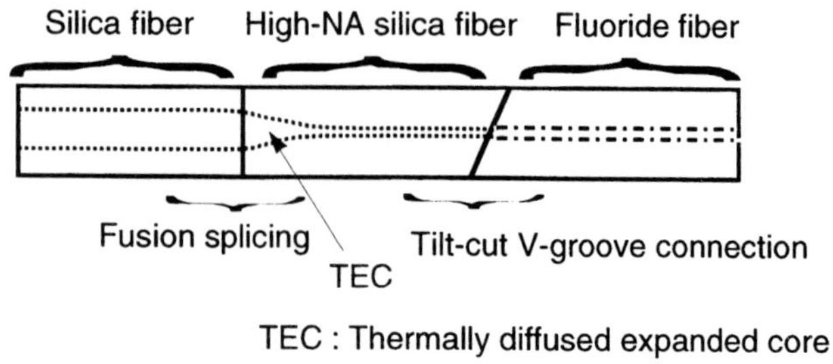

Fig. 13. Connection between fluoride and silica fiber.

To reduce the splicing and reflection loss, first, we used UV curable adhesive to connect the fluoride fiber to a high-NA silica fiber with a matching mode field diameter. Then the high-NA silica fiber was fusion spliced to a standard silica fiber. Tilt-cut V-grooves and the thermally diffused expanded core (TEC) technique were also used for the fluoride to high-NA silica fiber connection and the high-NA to standard silica fiber connection, respectively. These connection techniques reduced the splicing loss and reflection to less than 0.3 dB and less than -60 dB, respectively.

7.2 Reliability test

We tested the reliability of our fluoride fiber module with reference to the Bellcore Technical Advisory TA-NWT-001221.[45] This advisory is generally used for ensuring the reliability of passive optical components such as splitters, filters and WDM couplers. Since no appropriate reliability standard has been established for fluoride fiber modules, we followed this advisory based on the fact that the fluoride fiber module itself is a passive optical component without pump power.

We recorded the loss change of the samples with an on-line monitor during a temperature and temperature-humidity cycling test. We selected a wavelength of 1200 nm for the loss measurement to avoid the praseodymium absorption loss. With the damp heat storage, low-temperature storage, heat shock and vibration tests, we collected the loss values with the cut-back method in the interim downtime and after the tests had been completed. We consider that a sample had failed if there was a loss change of ±0.2 dB. Figure 14 shows the loss change in a temperature-humidity cycling test (-40°C to 75 °C, 90 %RH) against elapsed time. In this figure, the dashed line shows the temperature profile with reference to the above-mentioned advisory. There was a loss change of less than 0.2 dB during the test. Figure 15 shows the results of a damp heat storage test (75 °C, 90 %RH for 5000 hrs) against elapsed time. We had 5000 hours of trouble-free storage for 11 samples. We also had 5000 hours of trouble-free storage for 11 samples in a low temperature storage test (-40 °C). Table 1 compares the loss before the temperature cycling test and after 500 cycles. The loss change was less than 0.2 dB. The variations in the loss values of the 11 samples were caused by the loss differences of the fluoride fiber we used. Our reliability test results, including those for mechanical tests based on Bellcore requirements, are summarized in Table 2. The tests revealed that the loss changes were small and there were no fluoride fiber modules which did not meet the Bellcore criteria.

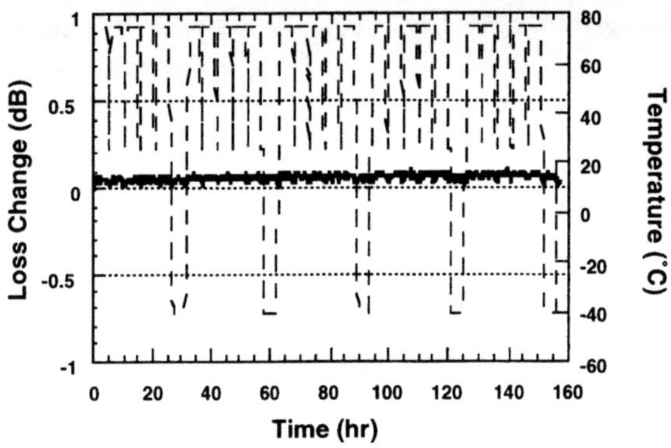

Fig. 14. Loss change in temperature-humidity-cycling test.
(Dashed line shows temperature profile.)

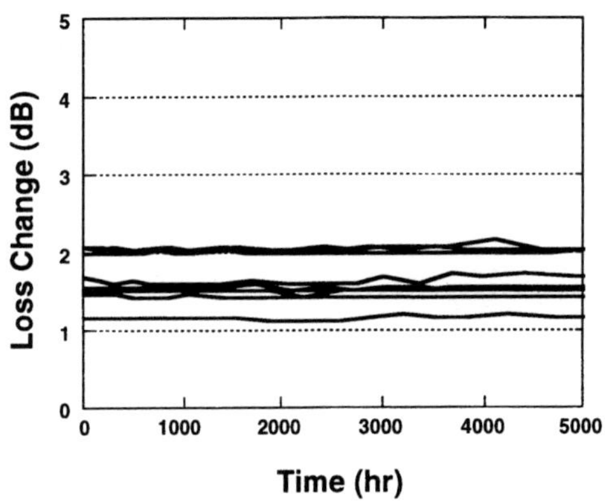

Fig. 15. Loss change at 75 °C, 90 %RH.

Table 1. Temperature cycling test results.

No.	Initial loss (dB)	After 500 cycles (dB)
1	1.7	1.7
2	2.4	2.4
3	2.4	2.5
4	1.9	2.0
5	1.5	1.5
6	2.2	2.3
7	2.3	2.4
8	1.2	1.2
9	1.5	1.7
10	1.1	1.3
11	1.2	1.2

If the fluoride fiber were to be exposed to damp heat conditions without appropriate protection, the fiber surface would be corroded by the water, and the fiber would eventually break during the test. Moreover, since we used UV curable adhesive at the interface between the fluoride and silica fibers, there is the potential for a loss increase if the adhesive reacts with moisture and deteriorates as the result of an improperly sealed package. However, our results for the damp heat storage and temperature-humidity cycling tests revealed that the sealed package we have developed can effectively protect fluoride fiber and V-groove connections from moisture damage.

The requirement for a large pump power caused by the low efficiency of PDFAs introduces new reliability issues related to optical power aging and transience. This is not only a problem with PDFAs. It is also critical for EDFAs and their passive constituents in terms of increasing the number of WDM channels and therefore increasing the pump power. A revised Generic Requirement, GR-1221-CORE, Issue 2, 1999, which was formerly TA-NWT-001221 will address this problem in Future Additions/Change, however no criteria regarding this issue are currently available. We have started a program to examine 11 fluoride fiber modules under 400 mW pumping and 0 dBm input signal power conditions with on-line gain monitoring. Figure 16 shows this experimental set-up schematically. Input signals and pump powers are individually launched into each sample through WDM couplers. After filtering the amplified spontaneous emission (ASE) components, the output signals are monitored using a 1 x N optical switch and a power meter. Pump

Table 2. Reliability test results.

Test	Requirements	Samples	Trouble
Temp. hum. cycling	-40 °C to 75 °C / 90 %RH, 5 cycles	40	0
Temp. cycling	-40 °C to 75 °C, 500 cycles	11	0
Damp	75 °C / 90 %RH, 5000 hrs	11	0
Low Temp.	-40 °C, 5000 hrs	11	0
Vibration	10 to 2000 Hz, 3 directions, 12 cycles	11	0
Heat shock	-40 °C to 70 °C, 20 cycles	11	0

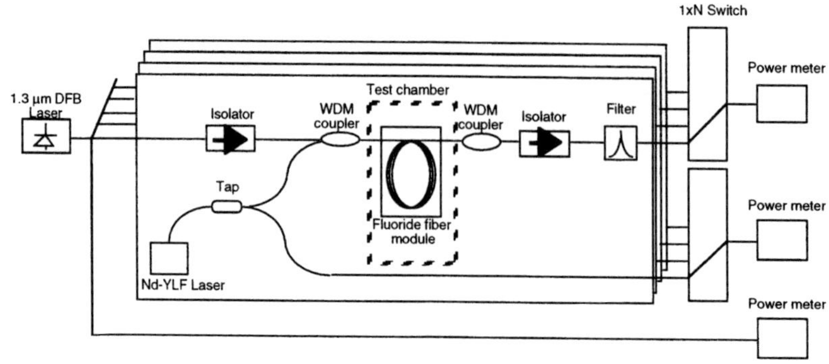

Fig. 16. Experimental set-up of high-power operation test.

power stability is monitored simultaneously by placing a tap in front of the WDM couplers. In this experiment the fluoride fiber module was tested in a 0 - 60 °C temperature cycle. The measurements showed that there is a temperature dependent gain variation caused by the temperature dependent change in the gain cross-section of the praseodymium. However, preliminary results suggest that our fluoride fiber modules are sufficiently reliable for high power operation. Further studies concerning this issue are underway.

8. CONCLUSION

We summarized our study on the reliability of a fluoride fiber module for practical amplifier use. We believe that the overall reliability of this fiber module is based on three factors; fiber strength, splicing and packaging. Fluoride fiber is inherently six times weaker than silica fiber and its strength must be increased to extend its lifetime. We described our approach to improving the fiber strength which involves developing a surface treatment technique. We carried out static fatigue experiments on our fluoride fiber and found it to have a long-term practical reliability of over 25 years provided that it is coiled around a 75 mm diameter bobbin. We established the V-groove connection technique for mechanically splicing fluoride and silica fibers. These V-grooves and fluoride fibers were housed inside a sealed package with feed-through silica fiber which can protect the constituents from moisture damage even under harsh conditions. We performed a reliability test on this module with reference to the Bellcore Technical Advisory TA-NWT-001221 and confirmed its long-term reliability in a practical environment.

ACKNOWLEDGMENTS

The authors thank K. Kobayashi, K. Shikano, K. Hoshino, T. Shimada, K. Oikawa, K. Kato, A. Mori, T. Sakamoto, H. Ono, K. Nakagawa, Y. Terunuma and Y. Hiraki for their help and useful discussions. They also thank T. Miya and S. Sudo for their encouragement.

REFERENCES

1. R.J. Mears, L. Reekie, I.M. Jauncey, D.N. Payne, "High-gain rare-earth-doped fiber amplifier at 1.54 μm," *OFC/IOOC'87, Reno, Nevada, Feb.* 1987, W12.
2. M. Yamada, A. Mori, K. Kobayashi, H. Ono, T. Kanamori, K. Oikawa, Y. Nishida, Y. Ohishi, "Gain-flattened tellurite-based EDFA with a flat amplification bandwidth of 76 nm," *OFC'98, San Jose, CA*, 1998, PD 7.
3. T. Sakamoto, M. Yamada, M. Shimizu, T. Kanamori, Y. Terunuma, Y. Ohishi, S. Sudo, "Thulium-doped fluoride fiber amplifiers for 1.4 μm and 1.6 μm operation," *OAA'96, Monterey, CA*, pp. 40-43, 1996.
4. A. Mori, Y. Ohishi, M. Yamada, H. Ono, Y. Nishida, K. Oikawa, S. Sudo, "1.5 μm broadband amplification by tellurite-based EDFAs," *OFC'97, Dallas, TX*, 1997, PD 1.
5. Y. Ohishi, T. Kanamori, T. Kitagawa, S. Takahashi, E. Snitzer, G.H. Sigel, Jr, "Pr^{3+}-doped fluoride fiber amplifier operating at 1.31 μm," *Opt. Lett.*, vol. **16**, no. 22, pp. 1747-1749, 1991.
6 T. Sakamoto M. Shimizu, T. Kanamori, Y. Terunuma, Y. Ohishi, M. Yamada, S. Sudo, "1.4 μm-band gain characteristics of a Tm-Ho-doped ZBLYAN fiber amplifier pumped in the 0.8 μm band" *Photon. Tech. Lett.*, vol. **7**, no. 9, pp. 983-985, 1995.
7. T. Sakamoto, M. Shimizu, M. Yamada, T. Kanamori, Y. Ohishi, Y. Terunuma, S. Sudo, "35-dB gain Tm-doped ZBLYAN fiber amplifier operating at 1.65 μm," *Photon. Tech. Lett.*, vol. **8**, no. 3, pp. 349-351, 1996.
8. Y. Nishida, M. Yamada, J. Temmyo, T. Kanamori, Y. Ohishi, "Plug-in type 1.3 μm fiber amplifier module for rack-mounted shelves," *OFC'97, Dallas, Texas*, Feb. 1997, WA6. and *IEEE Photon. Technol. Lett.*, vol. 9, no. 8, pp. 1096-1098, 1997.
9. M. Yamada, M. Shimizu, T. Kanamori, Y. Ohishi, Y. Terunuma, K. Oikawa, H. Yoshinaga, K. Kikushima, Y. Miyamoto, S. Sudo, "Low-noise and high-power Pr^{3+}-doped fluoride fiber amplifier," *IEEE Photon. Technol. Lett.* vol. 7, no. 8 pp. 869-871, 1995.
10. S.F. Cater, "Mechanical properties," *Fluoride Glass Optical Fibers*, ed. P.W. France, London, U.K.: Blackie, 1990, pp. 219-237.
11. M.A. Saifi, "Long-term fiber reliability," *Fiber Optics Reliability: Benign and Adverse Environments IV*, SPIE vol. 1366, pp. 58-69, 1990.
12. A. Tomita, P.J. Lemaire, "Hydrogen-induced loss increases in germanium-doped single-mode optical fibers: Long-term predictions," *Electron. Lett.*, vol. **21**, pp. 71-72, 1985.
13. P.J. Lemaire, "Hydrogen induced losses and their effects on optical fiber reliability," *Fiber Optics Reliability*, ed. D.K. Paul, Washington, USA: SPIE vol. CR50, pp. 80-104, 1993.
14. C. Fukuda, Y. Chigusa, T. Kashiwada, M. Onishi, H. Kanamori, "Effect of γ-ray irradiation on erbium-doped fibers," *Tech. Dig. of OEC'94*, vol. **4**, July, pp. 44-45, 1994.
15. IEC Publication 61290-8-1, "Reliability of optical fiber amplifiers," Version 4, Feb. 1999.
16. K. Fujiura, T. Kanamori, S. Sudo, "Fiber materials and fabrication," *Optical Fiber Amplifiers*, ed. S. Sudo, Boston, USA: Arteck House, 1997, pp. 193-403.
17. K. Fujiura, Y. Ohishi, S. Takahashi, "New process for fluoride glass preform

fabrication using chemical vapor deposition," *Japan J. Appl. Phys.*, vol. **28**, no. 12., pp. L2236-L2238, 1989.
18. Y. Ohishi, S. Sakaguchi, S. Takahashi, "Transmission loss characteristics of fluoride glass single-mode fiber," *Electron. Lett.* vol. **22**, no. 20, 1986, pp. 1034-1035.
19. J. Lau, A.M. Nakata, J.D. Mackenzie, "Influence of drawing temperature on the strength of fluoride glass fibers," *J. Non-Cryst. Solids.* vol. **70**, pp. 233-242, 1985.
20. H.W. Schneider, A. Schoberth, A. Staudt, Ch. Gerndt, "Fluoride glass etching method for preparation of infrared fibers with improved tensile strength," *Electron. Lett,* vol. **22**, no. 18, pp. 949-950, 1986.
21. T. Shibata, H. Takahashi, M. Kimura, T. Ijichi, K. Takahashi, Y. Sasaki, S. Yoshida, "Fabrication of high-strength, low-loss fluorozirconate glass optical fibers," *Mater. Sci. For.*, vol. **5**, pp. 379-386, 1985.
22. C.J. Simmons, J.H. Simmons, "Chemical durability of fluoride glasses," *Fluoride Glass Fiber Optics*, ed. Aggarwal and G. Lu., San Diego, USA,: Academic Press, Inc., 1991, pp. 283-287.
23. P.H. Klein, P.C. Pureza, W.I. Roberts, I.D. Aggarwal, "Strengthening of ZBLAN glass fibers by preform treatment with active fluorine," *Mater. Sci. For.* vol. **32-33**, pp. 571-576, 1988.
24. H.W. Schneider, A. Schoberth, A. Staudt, Ch. Gerndt, "Preparation of fluorozirconate glass fibers with improved tensile strength," *Mater. Sci. For.*, vol. **19-20**, pp. 321-326, 1987.
25. S. Sakaguchi, Y. Terunuma, Y. Ohishi, T. Kanamori, "Fluoride fiber drawing with improved tensile strength," *J. Mater. Sci. Lett.*, vol. **6**, pp. 1063-1064, 1987.
26. S. Sakaguchi, Y. Hibino, Y. Ohishi, S. Takahashi, "Surface flaws in ZrF_4-based fluoride optical fibers with improved tensile strength," *J. Mater. Sci. Lett.*, vol. **6**, pp. 1440-1442, 1987.
27. M. Fox, S.J. Stannard-Powell, "Attenuation changes in optical fibers due to hydrogen," *Electron. Lett.*, vol. **19**, no. 22, pp. 916-917, 1983.
28. K.J. Beales, D.M. Cooper, J.D. Rush, "Increased attenuation in optical fibers caused by diffusion of molecular hydrogen at room temperature," *Electron. Lett.*, vol. **19**, no. 22, pp. 917-919, 1983.
29. J.F. Shackelford, J.S. Masaryk, "The interstitial structure of vitreous silica," *J. Non-Cryst. Solids.* vol. **30**, pp. 127-139, 1987.
30. L.T. Hamill, J.M. Parker, "Hydrogen solubility in heavy metal fluoride glasses," *Phys. & Chem. Glasses*, vol. **26**, no. 2, pp. 52-54, 1985.
31. Paul G. Shermon: *Diffusion in Solids,* New York, USA: McGraw-Hill,1973.
32. J. Stone, A.R. Chraplyvy, C.A. Burrus, "Gas-in-glass a new Raman-gain medium: molecular hydrogen in solid-silica optical fibers," *Opt. Lett.* vol. **7**, no. 6, pp. 297-299, 1982.
33. L.J.B. Vacha, P.C. Schultz, C.T. Moynihan, G.N. Crichton, *Mater. Sci. Forum*, vol. **20**, p. 419, 1987.
34. M. Miura, I. Matsuda, K. Itoh, T. Yamashita, *Proc. 6th Int. Symp. on Halide Glasses, Germany*, p. 407, 1989.
35. G. Orcel, D. Biswas, *J. Am. Ceram. Soc.*, vol. **74**, p. 1373, 1991.
36. A. Mori, Y. Ohishi, S. Sudo, "Durability of phosphate glasses as overcladding materials for fluoride glass fibers," *J. of Non-Crystalline Solids*, vol. **194**, pp. 173-179, 1996.

37. P.C. Schultz, L.J.B. Vacha, C.T. Moynihan, B.B. Harbison, K. Cadien, R. Mossadegh, *Mater. Sci. Forum*, vol. **20**, p. 343, 1987.
38. P.W. France , W.J. Duncan, D.J. Smith, K.J. Beals, "Strength and fatigue of multicomponent optical glass fibers," *J. Mater. Sci.*, vol. **18**, pp. 785-792, 1983.
39. P. Hasløv, K.B. Jensen, N.H. Skovgaad, "Degradation of stressed optical fibers in water: New worst-case lifetime estimation model," *J. Ame. Ceram. Soc.*, vol. **77**, pp. 1531-1536, 1994.
40. F.P. Kapron, H.H. Yuce, "Theory and measurement for predicting stressed fiber lifetime," *Opt. Eng.*, vol. **30**, pp. 700-708, 1991.
41. K. Fujiura, K. Hoshino, T. Kanamori, Y. Nishida, Y. Ohishi, S. Sudo, "Reliability of fluoride fibers for use in fiber amplifiers," *OAA'95, Davos, Switzerland*, June, 1995. pp. 104-107.
42. K. Fujiura, Y. Nishida, T. Kanamori, Y. Terunuma, K. Hoshino, K. Nakagawa, Y. Ohishi, S. Sudo, "Reliability of rare-earth-doped fluoride fibers for optical fiber amplifier application," *IEEE. Photon. Tech. Lett.*, vol. **10**, no. 7, pp. 946-948, 1998.
43. Y. Yamada, F. Hanawa, T. Kito, T. Maruno, "Low-loss and stable fiber-to-wave guide connection utilizing UV curable adhesive," *IEEE Photon. Technol. Lett.*, vol. **4**, pp. 906-908, 1991.
44. M. Yamada, M. Shimizu, Y. Ohishi, J. Temmyo, T. Kanamori, S. Sudo, "One-LD-pumped Pr^{3+}-doped fluoride fiber amplifier module with signal gain of 23 dB," *Electron. Lett.*, vol. **29**, no. 22, pp. 1950-1951, 1993.
45. Y. Nishida, K. Fujiura, K. Hoshino, M. Shimizu, M. Yamada, K. Nakagawa, Y. Ohishi, "Reliability of fluoride fiber module for optical amplifier use," *OFC'99, San Diego, USA*, Feb. 1999, WM8.

The mastercurve method: analysis of diverse reliability issues in optical fibers and fiber gratings through a unified approach

SriRaman Kannan*, Michael LuValle and Paul Lemaire

Specialty Fiber Devices
Lucent Technologies, Bell Labs Innovations
25 Schoolhouse Road
Somerset, NJ 08873

ABSTRACT

This paper reviews the applicability of the *mastercurve* method for analyzing diverse reliability issues in fiber optics. Approaches to assess and predict small changes in optical properties of fiber gratings caused by thermally induced decrease in refractive index modulation, are presented and compared. The process of thermal *stabilization* to obtain gratings with excellent stability is explained. It is shown that the mastercurve approach when suitably modified could be used to analyze other reliability problems such as hydrogen induced loss and radiation induced effects in fibers.

Keywords: Optical fibers, reliability, accelerated testing, fiber gratings, thermal stability, hydrogen, radiation, mastercurve, induced loss.

1. INTRODUCTION

Optical fibers and fiber components serve as the backbone of telecommunications today. The evolution of the internet has placed enormous demand for bandwidth, accelerating developments in optical transmission and networking. The optical fiber is the core physical medium for transmission of information around the world. The explosive need for bandwidth has lead to innovative ways for transmitting information leading to extensive development and deployment of fiber based components such as fiber gratings. The applications of optical fiber and fiber based components are in systems that are both terrestrial and undersea, in terms of deployment. A high degree of reliability is warranted in these systems.

* Correspondence*:* *E-mail*: sri@lucent.com; *Telephone*: 732-748-7451; *Fax*: 732-748-7596 .

Approaches to assess the reliability of optical fibers and fiber-based components are critical. To this end, this paper describes an approach that has emerged in the past five years to address diverse reliability issues. The method has been variously called as *demarcation diagrams, aging curve* or *mastercurve method.* This method is applied to three different reliability issues:

1. Thermal stability of fiber gratings
2. Hydrogen induced loss in fibers
3. Radiation induced loss in fibers.

While it may appear that the above mentioned subjects are diverse, the underlying physical factor in each of the above is the same viz., stoichiometric defects in the glass. The measured parameters such as spectral width in fiber gratings and radiation or hydrogen induced loss in optical fibers are related to creation or modification of defects.

The *mastercurve* method was originally developed to deal with thermal stability of fiber gratings. Fiber gratings are refractive index patterns inscribed in the core of an optical fiber. The modulation of refractive index is due to creation of defects. In telecommunications, fiber based Bragg gratings are used in wavelength division multiplexing (WDM) as add/drop filter elements where wavelengths are added or dropped using the property of the grating to reflect selective wavelengths. Gratings are also used as gain equalizers in erbium doped fiber amplifiers (EDFA) wherein long period gratings are used to flatten the gain spectra of the amplifier, thereby facilitating greater bandwidth. Other applications of fiber gratings include laser stabilizers, pump reflectors and sensors, to name a few. In all the above stated applications, it is essential to use gratings that are very stable. Formation of gratings in fibers involves UV-induced defects. Section 2 of this paper deals with gratings and their thermal stability. Approaches to assess and predict thermal stability of gratings are crucial to many applications. It is shown that the mastercurve technique is a very general and powerful technique for analyzing and predicting grating thermal decay. The use of this technique in arriving at conditions at which gratings are thermally *stabilized* to ensure superior lifetime performance is illustrated.

The topic of hydrogen induced loss has been investigated from the time optical fibers were deployed in telecommunications. It is shown that the mastercurve approach when suitably modified can be used to address other reliability issues pertaining to optical fibers in analyzing hydrogen induced loss in optical fibers. The fundamental hydrogen absorption is at around 1420 nm. The tail of this peak stretches well in to the wavelength of interest in fiber optic applications (1550 nm). The applicability of the mastercurve method in analyzing and predicting hydrogen-induced loss in fibers is presented in section 3. Methods to minimize hydrogen-induced loss in optical fibers are also presented.

Radiation induced effects in glasses and optical fibers have been widely studied. A review of models applied applied to radiation induced loss in fibers is presented. It is shown that the mastercurve method can be used to address the radiation paradigm as well. Section 4 of this paper deals with the radiation paradigm.

2. FIBER GRATINGS

2.1. Fiber grating fundamentals

Gratings are UV-induced refractive index modulations inscribed in the core of photosensitive optical fibers. Photosensitivity in optical fibers was first reported[1] in 1978. Side writing of gratings[2] has made grating fabrication much easier. An elegant method of fabricating gratings in optical fibers is through the use of a mask. Figure 1 illustrates the fabrication of a Bragg grating through a phase mask.

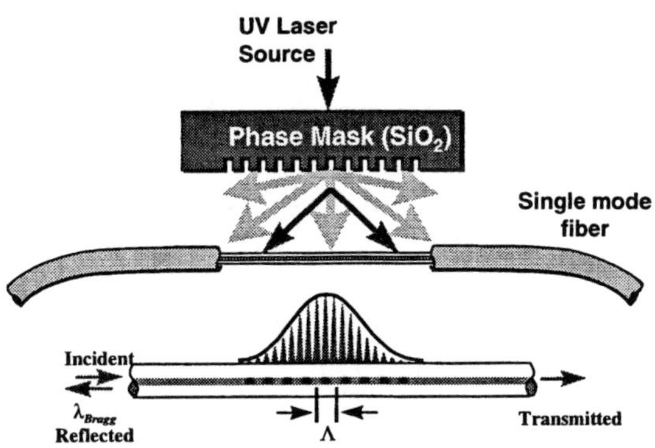

Figure 1: Illustration of fabrication of a Bragg grating using a phase mask.

The interference pattern of the UV laser (e.g.: excimer pumped dye laser at 242 nm) is imposed on the fiber through a phase mask. The induced refractive index pattern has a periodicity of Λ. Bragg gratings (also known as short period gratings) such as that shown in Figure 1 are governed by the equation:

$$\lambda_{Bragg} = 2\, n_{eff} \Lambda \qquad (1)$$

where λ_{Bragg} is the wavelength that is reflected and n_{eff} is the effective refractive index of the fiber grating. From the above equation it is easy to see how a Bragg grating may be used as a wavelength selective reflector. For a set of wavelengths incident on the grating, only those that satisfy the Bragg condition given by equation (1) will be reflected, while the remaining wavelengths will pass through the grating unaffected.

A second type of grating is the long period grating[3]. Unlike Bragg gratings, which typically have a periodicity (pitch) of less than 1 μm, these have much larger periodicity (100 - 500 μm). Long period gratings (LPG's) are fabricated using amplitude masks. LPG's couple selected wavelengths into forward propagating cladding modes and are hence used as wavelength selective loss elements (for instance, as gain equalizers in erbium doped fiber amplifiers in telecommunication transmission systems). The wavelength that is coupled into the cladding is given by :

$$\lambda = \Lambda(n_{core} - n_{clad}) \qquad (2)$$

The fibers in which gratings are written possess photosensitive cores usually doped with germanium. The photosensitivity has been ascribed to bleaching of native oxygen deficient sites in germanium doped silica[1]. Typically in germanium-doped silica, a refractive index modulation $\Delta n_{UV} < 5 \times 10^{-4}$ can be induced for 9% GeO_2. However increased photosensitivity has been achieved by several methods. These include the use of reducing conditions during preform fabrication[4], high temperature treatment of hydrogen preforms/planar waveguides[5,6] and flame brushing of fibers[7]. However, by far the most common method of increasing photosensitivity in optical fibers is through high pressure loading of fibers[8]. This process involves loading the glass with hydrogen or deuterium molecules by inert diffusion to the fiber core. On exposure to UV light during the grating writing, hydrogen related reactions are initiated contributing to permanent refractive index modulation. After the grating fabrication, the excess molecular H_2 or D_2 is removed. Greatly enhanced photosensitivity has been achieved in this manner in germanium and phosphorous doped silica fibers. The enhanced photosensitivity offered by deuterium/hydrogen loading makes it possible to achieve profound refractive index modulations (10^{-2} - 10^{-3}) as opposed to Δn_{UV} of 10^{-4} in unloaded fibers.

2.2. Thermal stability of gratings

Formation of gratings involves defects. This is true regardless of the type of the grating (Bragg or long period), fiber composition or factors such as hydrogen/deuterium loading. In the grating inscription process, there is "excitation" of the glass to a metastable state. An example of this is the UV-induced displacement of an electron to a trapping site or a structural change, which is quenched into the glass. When thermally induced decay is observed at elevated temperatures in accelerated aging experiments, this suggests that the UV-induced defects are not thermodynamically stable but rather are sites that can be thermally reversed resulting in a decay of the index modulation when adequate thermal energy is provided. Furthermore, the decay kinetics of gratings suggests a mixture of time constants.

It may be visualized that associated with the sites that provide the index modulation, there is an activation energy barrier E_{a1} as shown in Figure 2(a). Due to the inherent disorder in a glass, the activation energies assume a range of values in the grating formation stage (Figure 2(b)). It may be imagined that the defects "attempt" to decay with a characteristic vibrational frequency, ν. When a defect overcomes its activation energy barrier with thermal assistance, the contribution of the site to the refractive index modulation (Δn_{uv}) is lost resulting in grating decay. The range in activation energies in an amorphous material will lead to a subsequent decay phenomenon as shown in Figure 2[c]. The sites with low activation energy such as E_{a1} decay faster (smaller time constants) as compared to those with high activation energies such as E_{a3}, which correspond to larger time constants. The net decay is then a summation of all the individual components (Figure 2[c]), showing an initial rapid decrease followed by a progressively diminishing but a non-zero rate.

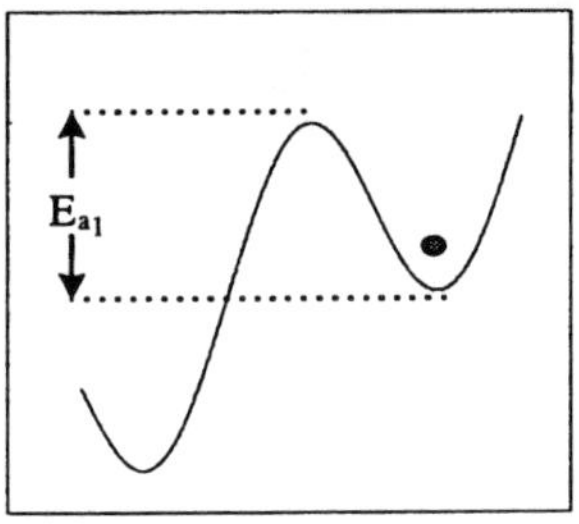

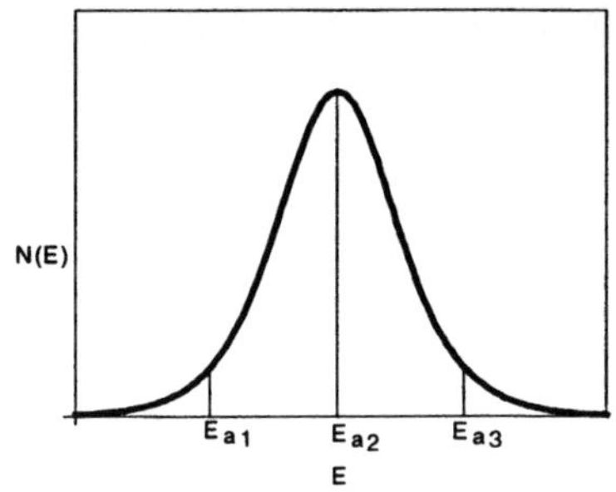

2(a) 2(b)

Figure 2: [a] Refractive index modulation in fibers is caused by UV-induced defect sites. [b] the amorphous nature of the glass leads to a distribution in the activation energies and [c] subsequently, the net decay of the grating is a composite of individual components with various time constants.

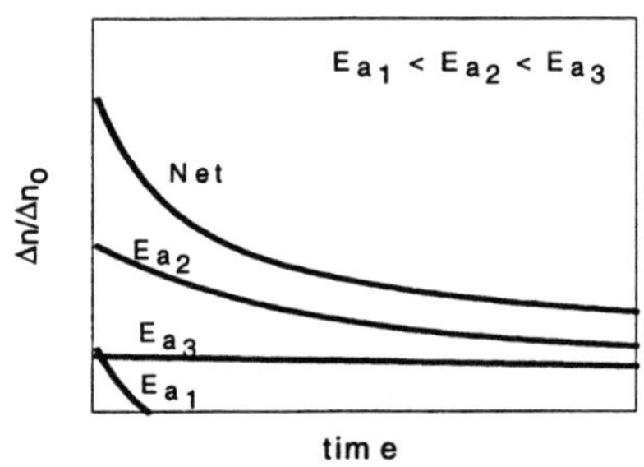

2[c]

2.3. Analysis and prediction of decay in Bragg gratings

Bragg gratings are deployed in applications such as in WDM add/drop reflectors in telecommunications. These applications typically involve long mission times. An accurate estimate of grating decay is crucial to many of these applications. Two approaches to achieve this end have been developed. Both methods are derived from the same physical basis and involve accelerated anneal experiments. However, the interpretation of the anneal data is different as will be shown in the following sections.

2.3.1. Accelerated aging experiments on Bragg gratings

In order to analyze decay behavior of any particular type of grating, accelerated anneal experiments are carried out at elevated temperatures (typically 125°C to 250°C). The tests are carried out on identical gratings of the particular type of interest. During the experiments, a quantity proportional to refractive index modulation of the grating is monitored as a function of time. In the case of simple strong Bragg gratings (reflectivity > 99%) this is the spectral width as shown in Figure 3. The figure illustrates the observed "narrowing" of the grating spectra as the anneal progresses at 200°C. In simple (unapodized) gratings the short wavelength edge of the grating spectrum is unchanged since it corresponds to zero index modulation. The long wavelength edge is defined by the UV-induced index modulation and hence when there is thermally assisted diminishing of Δn_{uv}, the edge moves towards the short wavelength side causing "narrowing" of the grating spectrum.

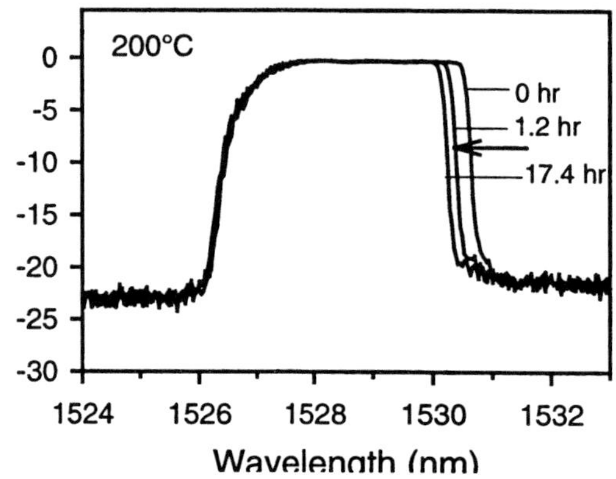

Figure 3: Aging behavior exhibited by a strong Bragg grating during an accelerated anneal test at 200°C

In the case of weak Bragg gratings (reflectivity < 96%), the integrated coupling coefficient (ICC) given by equation 4 is found to be proportional to the UV-induced refractive index change[9-11]. In accelerated anneal experiments, ICC is monitored as a function of time in the case of gratings.

$$ICC = \tanh^{-1}(R^{0.5}) \tag{4}$$

where R is the reflectivity of the grating.

2.3.2. The Mastercurve approach

The first approach through which thermal decay of Bragg gratings can be analyzed is the *mastercurve* approach[12]. This technique is also referred to as *demarcation energy diagram* or as the *aging curve* method[13].

In this approach, data from accelerated aging experiments at elevated temperatures are combined together to obtain a composite master-curve. In an accelerated aging experiment, the extent of aging on a grating at any temperature (T) and time (t) is described by an aging parameter, X (alternately referred to as the demarcation energy, E_d).

$$X = k\,T\,\ln(\nu t) \tag{5}$$

where k is Boltzman's constant and ν is the (attempt) frequency term. The change in refractive index modulation (Δn_{uv}) is plotted as a function of the aging parameter (X) to yield the mastercurve for that particular type of grating. When the data obtained at different temperatures are combined together in one plot, the individual data sets will align to fall on one smooth curve for a unique value of ν. This curve then defines the aging/decay behavior of the grating and can be used to make predictions pertaining to stability of that type of grating.

Figure 4(a) shows the mastercurve obtained on a set of strong gratings through accelerated aging experiments at three different temperatures in the 100-250°C range. The curve fit to this data yields an optimal ν value of 2.2 x 10^{10} /second. With this mastercurve, the decay for any combination of temperature (T) and time (t) can be predicted. For instance, knowing the operating and system tolerances of an application, the suitability of a particular type of grating for that application can be quickly ascertained by referring to its mastercurve. Figure 4(b) shows the predicted stability of the grating for which the mastercurve is given by Figure 4(a), at two different assumed temperatures (40°C and 70°C). It may be noted that the grating is predicted to exhibit excellent stability (a change in Δn_{uv} of less than 0.5 % after 25 years at 40°C). It was possible to obtain gratings with such superior stability through a process called *stabilization*, which will be explained later.

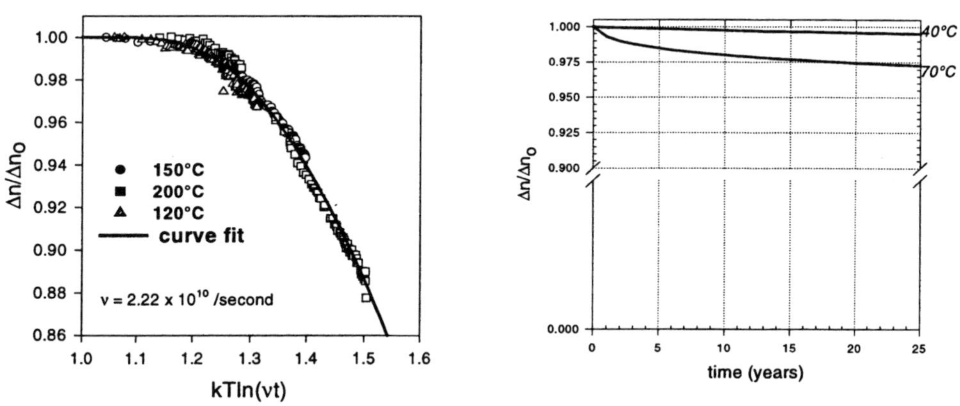

(a) (b)

Figure 4: (a) Mastercurve obtained through data from accelerated aging experiments on strong Bragg gratings and (b) predicted decay behavior of the grating at two different assumed operating temperatures.

The results presented above pertain to simple Bragg gratings where the short wavelength edge is determined by the unmodulated fringes in the refractive index pattern of the grating. Bragg gratings with a more complex index pattern have also been fabricated to remove Fabry-Perot modes and to increase bandwidth utilization. Such gratings have a refractive index pattern in which the short wavelength edge is itself determined by a raised pedestal (DC) and the long wavelength edge by a modulated pattern (AC) on this pedestal. These are known as *apodized* Bragg gratings. In this case, both the edges shift when there is thermal decay. Grating decay of this nature has also been analyzed successfully[14] using the mastercurve approach. Mastercurves have been obtained for both edges independently, from which predictions can be made for lifetime changes in both the AC and DC components of the refractive index pattern.

2.3.3. The Power-Law approach

The second approach to analyzing accelerated aging data of grating is what has come to be known as the *power law* method[12]. Here, the decay in the UV-induced refractive index change is expressed as:

$$\eta = 1/(1+A(t/t_1)^{\alpha}) \qquad (6)$$

where t is the time, t_1 is a time unit dimension factor. Ideally, according to this model, the factor A and exponent α should be related to temperature in the following manner :

$$A \propto \exp(T) \qquad (7)$$

$$\alpha \propto T \qquad (8)$$

The power law technique also involves carrying out accelerated aging experiments at several elevated temperatures similar to the aging curve approach. The data at each temperature is then fit to equation 6 to obtain A and α for that temperature. By repeating this at several temperatures, the exact A vs. T and α vs. T relationships are established for that grating. It is then possible to predict the stability of that grating type at any given operating temperature.

Earlier experiments on certain gratings yielded α and A, which appeared to satisfy the relationships given by equations (7) and (8) at high temperatures[12]. Power law fits have been attempted in a few other subsequent works[15-17]. However, it has been shown that the fit may not hold true for all types of gratings[18]. It has been further shown that forcing fits on experimental data yields values of α which result in predictions that are deviant from experimental results[19]. Figure 5 shows the results of the experiment in which predictions from alternate techniques on stabilized hydrogen loaded gratings are compared. The continuous and the broken lines represent decay predicted by the aging curve and the power law methods respectively at 200°C. The symbols represent data obtained from an actual experiment on the grating carried out at 200°C. It can be noted that the aging curve predictions show excellent agreement to the experimental data, while the power law prediction overestimates the expected decay.

The reason for the discrepancy in power law predictions arises from the fact that this approach assumes specific initial distributions of activation energies for the defect population. This distribution assumes a form in which the rising portion assumes an exponential nature[12]. The mastercurve approach on the other hand does not assume any specific initial distribution and hence can be applied more generally without any regard to the actual distribution. The above discussion does not imply that the power law is invalid for all types of gratings. It could be used for gratings with a certain type of activation energy distribution (for which, $\alpha \propto T$), whereas the mastercurve can be applied more generally. In this sense, the power law model can be viewed as a special case of the more general model. All things considered however, the mastercurve analysis is preferable since it is certain to work and involves just as much or less experimentation as compared to its alternative.

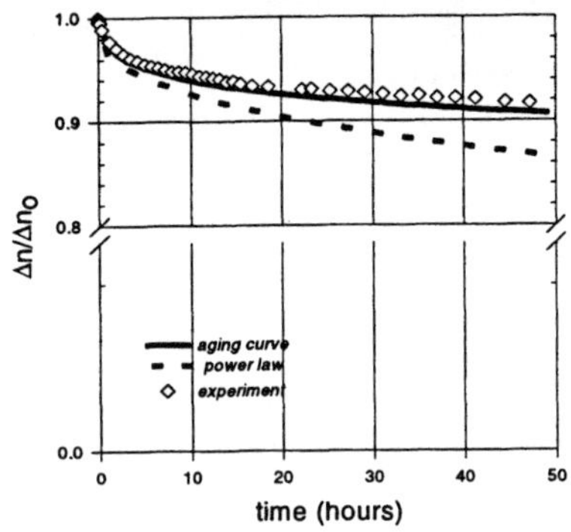

Figure 5: Comparison of reliability predictions by different methods of analysis to experimental data.

2.4. Stabilization of gratings

In the examples of gratings mentioned in the previous sections, it was stated that the gratings were *stabilized.* Stabilization is process in which the most unstable sites are removed by a pre-designed anneal. The anneal conditions can be obtained from a mastercurve derived from experiments on as-written gratings. The extent of this anneal will be determined by the decay behavior of the grating and system tolerance (operating temperature). Figure 6 illustrates the process by which the stabilization criterion is arrived at for a given type of grating. Figure 6(a) portrays the decay behavior of a particular grating. Then for a given system tolerance of index decay at end of life, the stabilization anneal (temperature T_{anneal} and duration t_{anneal} the grating will be designed such that :

$$k\, T_{anneal} \ln(\nu\, t_{anneal}) = E_{critical} \tag{9}$$

This anneal will result in the removal of the unstable sites shown in the shaded region in figure 6(a). When viewed in terms of a hypothetical plot showing the activation energy distribution of sites, this treatment will correspond to removing sites shown in gray in Figure 6(b).

In performing the stabilization anneal, it is recognized that a certain portion of the UV-induced refractive index modulation will be lost. However, this margin can be anticipated and built in during the grating writing process. It is hence advantageous to use fibers with enhanced photosensitivity so that excess Δn_{uv} can be provided for the purpose.

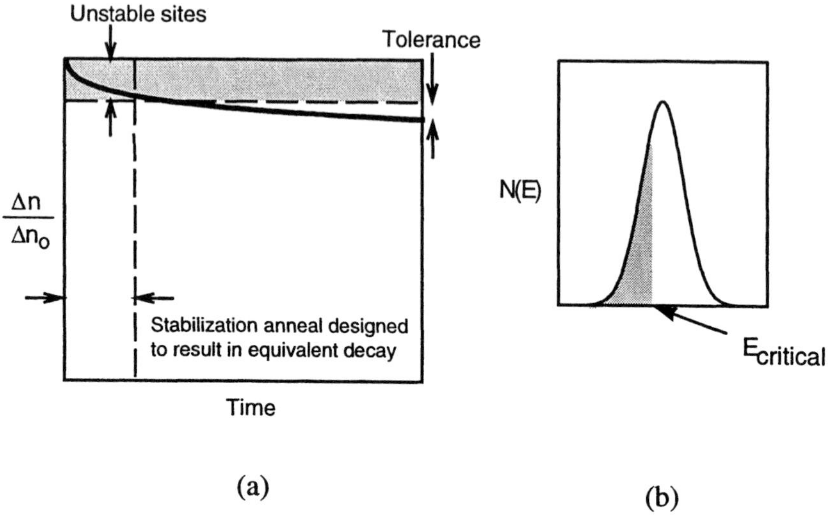

Figure 6: (a) Index decay behavior and design of appropriate stabilization anneal (b) the removal of the most unstable defect sites may correspond to removing a certain portion (the shaded region) in the activation energy distribution

Bragg gratings of simple and complex refractive index profiles have been obtained through properly designed stabilization treatments. Such stabilization treatments are even more important for long period gratings as will be discussed in the following section.

2.5. Long period gratings

The mastercurve approach to decay analysis can be well applied to long period gratings (LPG). As stated before in long period gratings, light is selectively coupled into forward propagating cladding modes and are used as wavelength dependent loss elements in equalizers for erbium doped gain spectra . Furthermore, the Δn_{uv} of a LPG is directly related to the central wavelength and minimum (known as isolation) in its transmitted spectrum. Thermal decay of refractive index modulation in long period gratings has the effect of (1) shift in the central wavelength of transmitted LPG spectrum and (2) a decrease in the isolation. Hence, the thermal aging of LPG's can be defined by monitoring either of the above two parameters. Figure 7 schematically shows the effect of thermal aging on a LPG in an accelerated aging experiment.

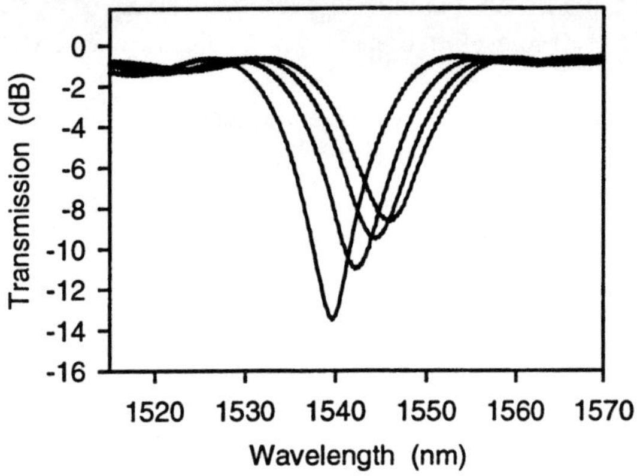

Figure 7: Schematic illustrating the effect of thermal aging on a long period grating in an accelerated aging test.

The mastercurve method has been successfully applied in characterizing $\Delta\lambda$ and isolation changes of long period gratings[20]. Using the mastercurve, predictions on anticipated changes in long period grating properties in system operating conditions have also been made[20]. The effect of such changes in the gain spectra of an erbium doped fiber amplifier where the LPG is used as an equalizer has been modeled using OASIX[21] simulation program[20,22]. Such effects on amplifiers can be minimized through appropriate stabilization of LPG's[20].

The importance of thermal stabilization techniques for successful use of Bragg gratings was explained in an earlier section. This is even more important in LPG's, because the wavelength at which the coupling from the core to the cladding modes takes place is directly dependent on the UV-induced refractive index change (Δn_{UV}) and on n_{core}-n_{clad} (Δn). A decrease in these can greatly shift the spectra in a LPG[20,23]. For example, thermal conditions that cause a 2% decrease in Δn_{UV} would have a negligible effect in an add/drop WDM Bragg grating ($\Delta\lambda < 10$ picometers). A similar decay in the case of a LPG would shift its center wavelength by about 0.5 nm, which is unacceptable for most applications (such as gain equalizers). However by appropriate stabilization treatments at conditions that are typically more aggressive as compared to that for Bragg gratings, it has been shown that long period gratings with superior stability can be obtained[20].

3. HYDROGEN INDUCED EFFECTS IN OPTICAL FIBERS

3.1. Hydrogen in fibers

It is well known that hydrogen induced losses can occur in silica based optical fibers. Losses can be induced in silicate fibers by molecular hydrogen. Hydrogen can also react at defect sites in optical fibers, giving rise to OH formation and other losses. Hydrogen related losses have been seen in field installed telecommunication cables, where use of dissimilar metal components has been observed to generate very large amounts of hydrogen[24]. While most standard single mode fibers have been shown to be relatively unreactive with hydrogen under normal conditions[25,26], erbium doped amplifier fibers were initially found to be highly reactive to even trace levels of hydrogen[27,28]. The topic of hydrogen related effects in erbium doped fibers is discussed elsewhere[29].

Several approaches can be considered when attempting to eliminate hydrogen-aging problems[29]. They are :

1. *Suitable host glass compositions*: In some cases, the hydrogen-induced losses can be minimized in this manner but this approach may not be possible always.
2. *Hermetic packaging*: The fiber can be packaged such that even traces of hydrogen do not reach the fiber. In practice however this may cause a problem since any hydrogen that is generated inside the package will be trapped (for instance, if galvanic cells exist inside the package).
3. *Hermetic fiber coatings*: When properly applied, carbon coatings can essentially eliminate hydrogen aging[30]. These coatings prevent hydrogen from reaching the fiber core and thus prevent loss increases. There are some extra processing steps involved and may warrant additional quality control procedures but this approach does offer a viable method to eliminate hydrogen aging in fibers.
4. *Deuterium Passivation*: In this method, a fiber is treated in deuterium gas at low temperatures in order to eliminate the reactive sites that would otherwise cause hydrogen related loss increases later in the fiber's lifetime. This is a highly attractive method and will be discussed later.

Regardless of whether a fiber is passivated/hermetically coated or not it is essential to have methods that can be employed to assess the effects of hydrogen that may be expected in system conditions. To this end, we show below that the demarcation energy diagram/master curve approach with suitable modification, can be used to assess hydrogen aging in fibers and also predict loss increases.

3.2. Hydrogen aging mastercurve

Typically the OH peak caused by hydrogen is centered at about 1420 nm (the exact peak wavelength depends on the composition). The tail of this peak could stretch into the 1550 nm wavelength region, thus increasing the loss at the system wavelength(s) in telecommunications. Our studies have shown that the ratio of loss induced at 1550 nm to that at the OH peak is about 0.25.

The long-term behavior of any fiber to hydrogen can be mapped through a hydrogen aging parameter :

$$X_{H2} = kT\ln(\nu C_{H2}t) \qquad (10)$$

where k is the Boltzmann's constant, T is the temperature, C_{H2} is the concentration of the hydrogen (related to hydrogen partial pressure P_{H2}) and t is the time. ν is a frequency term that is derived through experiments.

The experimental method involves carrying out experiments at elevated temperatures and controlled partial pressures of hydrogen. Fiber transmission spectra are recorded as a function of time in each test. The data thus obtained at different temperatures are combined together in a X_{H2} vs. loss increase($\Delta\alpha$) plot, where the individual data sets will align to fall on one smooth curve for a unique value of ν. Once this curve is established, predictions can be made on hydrogen induced loss for any assumed system P_{H2} and temperature. The hydrogen-aging curve could be in the form of either X_{H2} vs. $\Delta\alpha_{OH\ peak}$ or it could be X_{H2} vs. $\Delta\alpha_{1550nm}$. $\Delta\alpha_{1550nm}$ are $\Delta\alpha_{OH\ peak}$ are related by a fixed ratio (typically $(\Delta\alpha_{1550nm})/(\Delta\alpha_{OH\ peak}) = 0.25$).

Figure 8 shows a hydrogen aging mastercurve derived from three tests carried out at 215°C, 180°C and 155°C, respectively on a germanosilicate fiber. The partial pressure of hydrogen was maintained at 0.01 atm. in three tests. The transmission spectra were recorded as a function of time, from which the growth of the OH absorption was calculated in each case. Mastercurve (thus the value of ν) was derived from these tests. A value of 5×10^8 /second was obtained. A fourth test was carried out at 215°C and at a different P_{H2} (0.03 atm.). The data collected from this test were plotted on the same graph for the previously determined value of ν. This was done in order to validate our model. It may be observed that the latter data aligns very well with the mastercurve (figure 8).

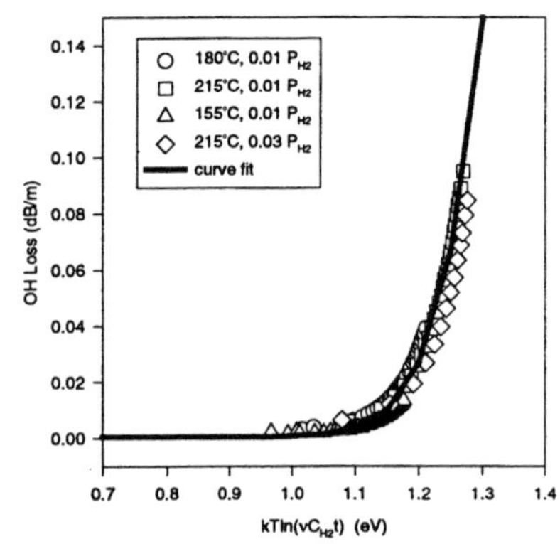

Figure 8: Hydrogen aging master curve of a germanosilicate fiber.

Using the hydrogen aging curve, predictions can be made for hydrogen induced loss at any assumed system conditions. For instance, using the above hydrogen aging master curve it is predicted that for a fiber exposed to 0.01 P_{H2} at 40°C for 25 years, the expected increase in loss is 3.2 dB/km at the OH peak. Since loss increase at 1550 nm is one-fourth of the above, the expected $\Delta\alpha$ at 1550 nm is about 0.8 dB/km.

3.3. Deuterium passivation of fibers

The basis for deuterium passivation lies in the fact that the rapid initial loss increases in fibers due to hydrogen are caused by a relatively small population of highly reactive sites. If the highly reactive sites can be eliminated, the fiber will not show significant hydrogen aging. Conceptually, passivation is very similar to the stabilization of gratings against thermal decay, in that, both involve pre-reaction of active species.

The hydrogen aging curve such as that shown in Figure 8 can itself be used to specify conditions needed to pre-react most sites with low activation energies (or in other words, the most reactive sites). Since same aging occurs for given values of X_{H2}, it is feasible to accelerate the aging in a known manner by using high temperatures and/or high P_{H2}'s. Therefore, by heating a fiber in hydrogen, it is possible to achieve in a short time, the same aging that would tend to occur over periods of years at normal operating conditions. Pre-reaction of the sites with H_2 does deplete reactive sites but this has the effect of improving reliability at the expense of raising OH losses and compromising initial fiber performance. By substituting deuterium for hydrogen in a passivation anneal, it is possible to deplete the most reactive sites, while avoiding the vibrational losses associated with OH formation. When deuterium is used for passivation, OD is formed instead of OH and the characteristic 1420 nm OH loss peak is shifted to 1950 nm. The latter wavelength is outside the window of application in telecommunications and hence does not affect system performance. The deuterium atoms occupy the reactive sites and prevent later reactions with hydrogen in the fiber environment. Figure 9(a) illustrates the effect of deuterium passivation in fibers. Figure 9(b) shows an experimental result comparing the loss increase in passivated fiber to another, which was not passivated, at an accelerated test temperature of 150°C and 0.012 P_{H2}. The substantial reduction in hydrogen related loss due to passivation is clear from this example.

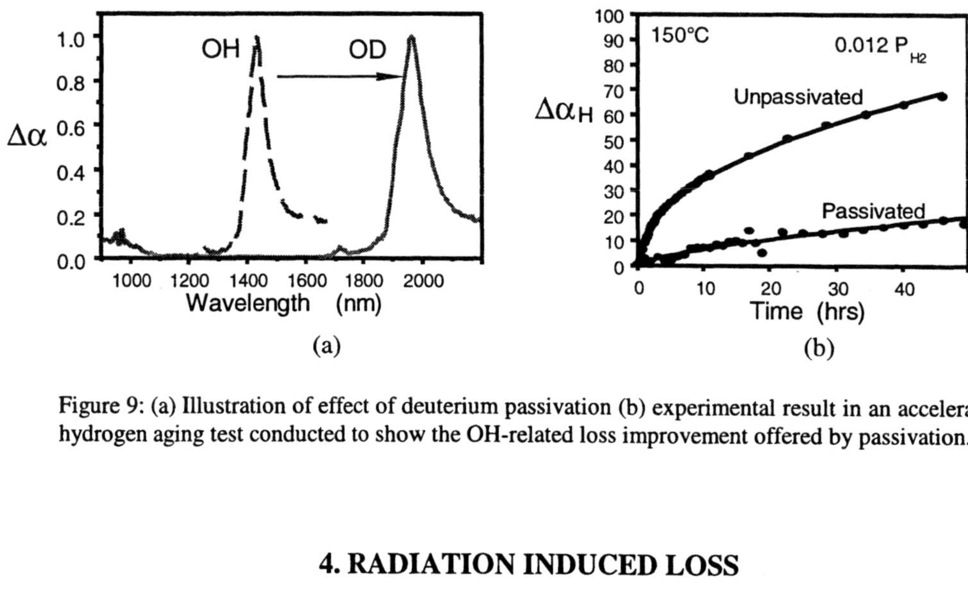

Figure 9: (a) Illustration of effect of deuterium passivation (b) experimental result in an accelerated hydrogen aging test conducted to show the OH-related loss improvement offered by passivation.

4. RADIATION INDUCED LOSS

4.1. Radiation effects on optical fibers

The study of radiation effects in silica glass almost always translates into a study of defects. Defects in silica are of two types: (1) those characteristic of flaws in the basic Si-O network called *intrinsic* defects and (2) those resulting from the presence of substantial impurities known as *extrinsic* defects. Sigel[31] has outlined the latter class of defects and their effects. The concentration of extrinsic defects can be minimized in an optical fiber by the use pure raw materials in the preform fabrication. The dominant factor *vis-à-vis* radiation is then the intrinsic defects in silica. The defects in this class include (1) *E'* center[32], (2) non-bridging oxygen hole centers[33-35], (3) peroxy radicals[36,37], (4) self-trapped holes[38] and oxygen deficient centers[39-43]. Several structural models, often competing with each other have been proposed for each of the defects.

There are two principle interactions of radiation with glass – ionization of electrons and direct displacement of atoms by elastic scattering[36]. When glass is subjected to ionizing radiation (X-ray or γ-ray photons, UV light or charged particles such as electrons or protons), electrons are ionized from the valence band if the energy of the radiation is greater than the band gap and the excess energy is converted to kinetic energy. The electrons could then move through the glass matrix and will either be trapped by pre-existing flaws to form defect centers in the glass structure or recombine with the positively charged holes. In glass optical fibers, such interactions manifest most prominently as induced loss at different wavelengths.

There have been many important works in the area of defects in silica based fibers and effects of radiation on the same. The degradation of several fibers on exposure to nuclear radiation was well characterized in the 0.8-0.9 μm wavelength region for the first generation optical communication systems[44]. The radiation response of pure fused silica fibers, Ge-doped silica fibers and Ge-alkali silicate fibers at 1.3 μm was studied and compared to that 0.82 μm[45]. The radiation response was in general found to be less at 1.3 μm as compared to 0.82 μm although the magnitudes depended on the core composition[45]. The effect of radiation on optical properties of pure silica glass cores as well those with dopants such as Ge, B, P and C was extensively reviewed by Dianov et al[46]. This work[46] concluded that the primary reason for radiation induced absorption in different fibers for doses $\leq 10^8$ rads is due to formation of color centers due to the capture of charge carriers (excited by ionizing radiation) by defects present in the glass before radiation. The work also concluded that for moderately low radiation dose ($< 10^7$) the effect of different type of radiation (protons, electrons or γ-ray) is similar[46]. A complete study[47] of P-doped silica/fibers identified the radiation induced defect centers and their corresponding optical transitions in that glass.

4.2. Power law fits to radiation effects in optical fibers

In studying radiation induced effects in glasses, power-law type growth has been observed for induced attenuation in 0.01-10 krad dose range in silica core fibers and over the 0.01 –100 krad range in Ge-doped silica core fibers[48]. Defect growth kinetics have been reported to take the power law form in several works[49-53].

$$C(D) = CD^f \qquad (11)$$

Where D is the radiation dose and C and f(<1) are empirical constants. Griscom et al[53,54] have discussed the origin of the power-law dependence of radiation induced defects in glasses. For the case of Ge-doped silica core optical fiber the works[53,54] propose and verified a general explanation for the power-law dependencies on dose for the case of a Ge-doped silica core optical fiber. The work also suggests that the 1.3 μm γ-ray induced absorption in Ge-doped-silica fibers arise from radiolytically produced interstitial O_2 molecules.

Williams et al[55] found the radiation induced losses in two commercially available erbium doped to also increase following a power law of dose with an exponent of about 0.8. The conclusion was independent of the type of radiation (γ-rays or proton), wavelength of measurement, dose rate or erbium concentration. From these observations, the work concluded that the radiation induced damage is primarily determined by the host glass and dopants such as aluminum. Further, the work[55] used power law models for dose rate dependencies[53,54] as well as a non-power law model[56] to describe the growth and recovery behaviors of loss. However it was inferred that all the above models overestimated the effect of varying dose rate on the incremental loss.

Liu and Johnston[56] proposed a model that is very different from the power law, to explain the growth and decay of radiation induced loss in an optical fiber. The model treats the radiation-induced defect generation and decay process as a series of superposable infinitesimal growth and decay events. The approach is appropriately termed as the *superposition model.* The work[55] tested the theory on a multimode optical fiber exposed to ^{60}Co at a dose rate of 15.5 rad/s and notes a reasonably good fit of the experimental data to theory. The compositional details of the fiber are not mentioned. However, when this model was applied to erbium doped fibers in another work[57], the fit was found to be poor except in the limit of extremely large kinetic orders, whereas most models proposed for describing radiation induced loss suggest low order kinetics (n=2).

4.3. Mastercurve approach to analyze radiation effects

A different method of analyzing radiation induced effects in glass optical fibers is through the mastercurve method. A specific formulation of this method was applied on erbium doped fibers[58]. The radiation (γ-ray) induced effects in rare doped fibers has two aspects viz., (1) the γ-ray induced defects causing increase in the loss of the fiber and (2) decrease in the induced loss due to thermal bleaching of the defects. This phenomenon is analogous to that seen in fiber gratings where, the two aspects are (1) formation of UV induced defects (and increase in loss) during the inscription of gratings in a photosensitive fiber and (2) thermal decay of such defects. It was hence expected that the technique used in the analysis of thermal decay of fiber Bragg gratings will apply to the thermal recovery of radiation induced loss in fibers[59]. The results of the work[59] validate this expectation.

Silica fibers (EY01) co-doped with erbium and ytterbium were used in this study. The fiber also contained phosphorus. When this fiber is used as a laser, the ytterbium is pumped (at around 915 nm). Energy transfer occurs between ytterbium and erbium. Lasing is achieved at around 1550 nm due to erbium. Phosphorus is thought to aid in the energy transfer between ytterbium and erbium.

Loss induced in the (EY01) Er-Yb fiber was monitored *in situ* while the fiber was radiated. The irradiation was done using a cobalt-60 source (gammacell). Figure 10 shows the increase in loss of 1 meter length of the fiber, as a function of dosage. It may be noted from the figure that the radiation induced loss in the fiber is about 1.6 dB/m at 1550 nm for 100 krad of exposure. The fiber was allowed to recover at room temperature following irradiation. It was observed that the recovery was less than 4 % even after 1 hour after the exposure to γ radiation. This behavior is indicative of defect formation in glass fiber that is relatively stable (thermally).

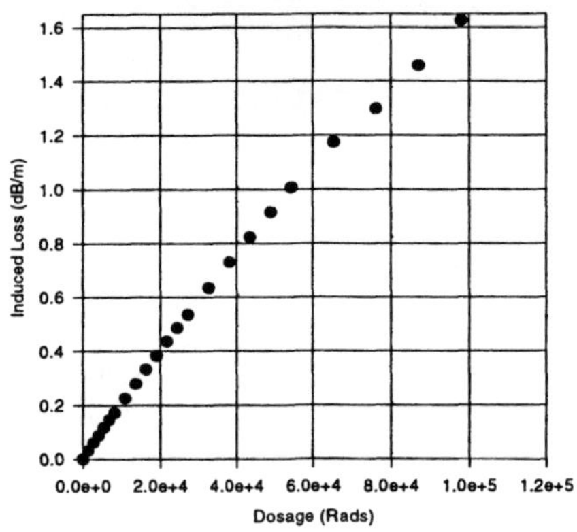

Figure 10: *In situ* measurement of radiation induced loss in Er-Yb doped silica fiber

The stability of defects was analyzed through post-irradiation anneal experiments. Three segments of the (EY01) Er-Yb fibers were irradiated to a dose of 100 krad each. The spectra of the fibers were measured immediately after the irradiation. The fibers were then annealed at three different temperatures, 200°C, 225°C and 250°C, respectively. Spectra of the fibers were recorded periodically over several hours to days, using a setup that included a white light source and an optical spectrum analyzer. It was noticed that the spectral changes in this fiber were broadly similar in the 1100 nm to 1600 nm wavelength region, that is, there was no creation or destruction of absorption bands and there was a recovery in transmission at all wavelengths during the thermal annealing of the fibers.

Figure 11 shows the recovery in loss of the EY01 fiber at the three temperatures at which annealing was carried out after exposure to radiation. It may be observed that at each temperature, the recovery in induced loss is rapid initially, slowing down progressively with time.

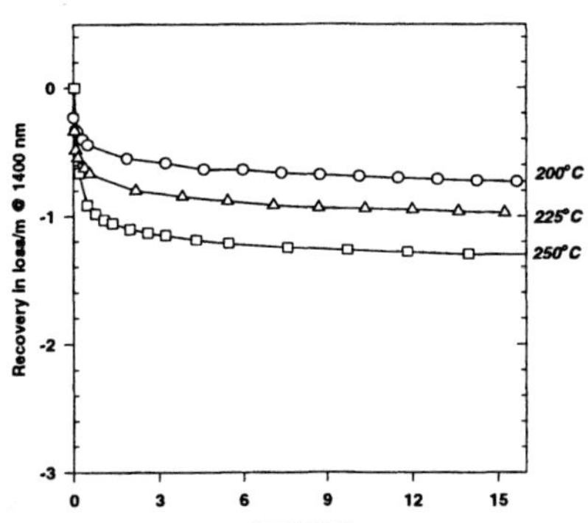

Figure 11: Recovery of Er-Yb fiber transmission at various annealing temperatures after exposure to 100 Krads of γ radiation

This behavior is very similar to that observed in the thermal decay of UV-induced fiber gratings and is suggestive of the presence of defect sites with a range of activation energies[12,19].

The loss recovery ($\Delta\alpha$) of the fibers during their annealing was plotted as a function of a parameter, X. The parameter X is given by equation 5. The term T is the temperature of the post-radiation anneal and t is the time the fiber had been subjected to the anneal. ν is a frequency term that is obtained by plotting the data from the three experiments on one graph and obtaining a fit. For the master curve analysis approach to be valid as in the case of thermal decay of gratings and hydrogen induced losses in fibers, the plot of $\Delta\alpha$ vs. KTln(νt) for the various sets of data should merge smoothly along one curve for a unique value of ν. Figure 12 shows that, this is indeed the case for the recovery data of radiation induced loss in EY01 fiber. It may be observed that the data from the three experiments fall along one curve for a ν value of 10^{23} /hour.

The $\Delta\alpha$ vs. KTln(νt) curve may be physically interpreted in the following manner. It may be visualized that the γ irradiation releases carriers from certain sites in the fiber which are trapped at other sites leading to a loss in transmission. Associated with each such site is an activation energy barrier. When the trapped carrier overcomes this barrier with thermal assistance, the loss in transmission caused by its entrapment is recovered. Due to the amorphous nature of glass, the activation energy assumes a range of values. The difference in the values of activation energies at different sites leads to defects with a range of stability. This leads to a recovery that is rapid initially but decreases gradually. The rate and the extent of the recovery depend on the temperature as shown in Figure 11. The term ν may be interpreted as related to the attempt frequency of the carrier when it is trapped in the activation energy barrier.

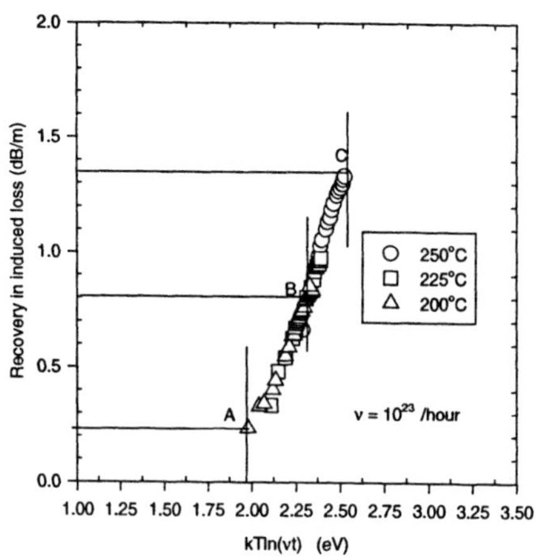

Figure 12: Master Curve derived by combining annealing experiments after 100 Krad of γ irradiation (see text for explanation on points A, B and C)

Points A, B and C in Figure 12 correspond to the points on the loss recovery curve, the fiber will be at if it was held at 80°C, 150°C and 200°C respectively at the end of 10 years, following 100 krad irradiation. It may be observed that at 80°C a recovery of 0.23 dB/m is expected. Similarly at 150°C and 200°C a recovery of 0.8 dB/m and 1.35 dB/m are predicted which correspond to 50 % and 85 % recovery of the 1.6 dB/m induced loss. An alternate way of looking at the stability of the defects is by plotting the decay time constants of the radiation induced defects at a chosen temperature. The time constant is given by the term:

$$\tau = (1/\nu)\exp(E_a/kT) \qquad (12)$$

Figure 13: Illustration of stability of radiation induced defects through a plot of time constants at a chosen temperature

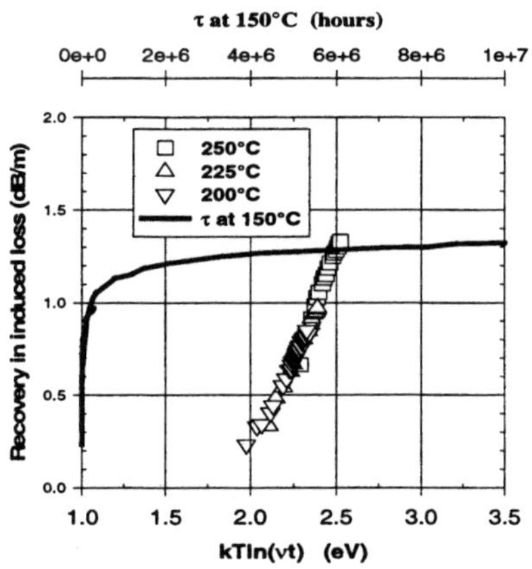

Demarcation energy diagrams obtained through post-irradiation anneal experiments such as that shown in Figure 12 can yield valuable information about the stability of defects formed due to γ radiation. This is especially useful when fibers with different compositions are compared for the effects of radiation on their performance. In choosing a fiber for performance in a radiation environment, the stability of the induced defects is as relevant as the magnitude of the loss (defects) induced by the radiation. Figure 14 illustrates the mastercurves of four fibers. Evidently, fiber 4 has least stable radiation induced defects as compared to the other three. Information such as this when used in combination with the magnitude of radiation induced loss can lead to appropriate choice for a given application.

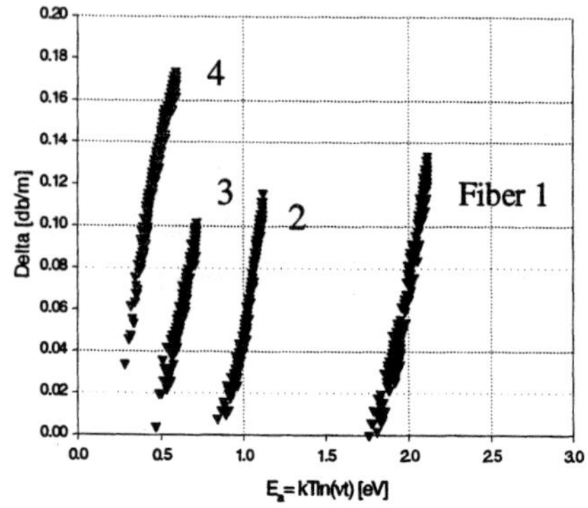

Figure 14: Comparison of mastercurves of different fibers

5. CONCLUSIONS

Fiber gratings are used in wide of variety of applications. Thermal stability of the gratings is vital in these. Alternate techniques to analyze and predict thermal stability of gratings are presented and compared. It is shown that the mastercurve technique is a powerful and general technique for the purpose. The approach can also be used to arrive at pre-treatment conditions that would yield gratings with superior stability. The approach when suitably modified can be used in the analysis of hydrogen induced loss in fibers – to predict induced loss as well as in arriving at passivation conditions resulting in insignificant changes at system conditions. The mastercurve method is also shown to be applicable for analyzing stability of radiation induced defects in optical fibers.

ACKNOWLEDGEMENTS

The authors thank a number of individuals at Lucent Technologies who contributed to this paper. Specifically, thanks are due to Joe Abate, Robert Ahrens, Matt Andrejco, Jim Auborn, John Borick, Darryl Brownlow, Leon Copeland, David DiGiovanni, Anthony Fields, Justin Judkins, Herman Presby, Carl Soccolich, Ken Walker, Paul Wysocki and Robert Windeler. Thanks are also due to the U.S. government for support of a portion of the work performed at Lucent that is presented in the paper

REFERENCES

1. K. O. Hill, Y. Fujii, D. C. Johnson and B. S. Kawasaki, *Appl. Phys. Lett.* **32**, pp. 647-649, 1978.
2. G. Meltz, W. W. Morey and W. H. Glenn, Opt. Lett. **14**(15), pp. 823-825, 1989.
3. M. Vengsarkar, P. J. Lemaire, J. B. Judkins, V. Bhatia, T. Erdogan and J. E. Sipe, *J. Light. Tech.* **14**(1), pp. 58-65, 1996.

4. D. L. Williams, B. J. Ainslie, J. R. Armitage, R. Kashyap and R. Campbell, *Electron. Lett.* **29**, p.45, 1993.
5. G. Meltz and W. W. Morey, *Proc. SPIE 1516,* p. 185, 1991.
6. G. D. Maxwell, B. J. Ainslie, D. L. Williams and R. Kashyap, *Electron. Lett.* **29**, p. 45, 1993.
7. F. Bilodeau, B. Malo, J. Albert, D. C. Johnson. K. O. Hill, Y. Hibino, M. Abe and M. Kawachi, *Opt. Lett.* **18**, p. 953, 1993.
8. P. J. Lemaire, R. M. Atkins, V. Mizrahi and W. A. Reed, *Electron. Lett.* **29**, p. 1191, 1993.
9. H. Kogelnik, *Bell. Syst. Tech. J.* **55**, 109, 1976.
10. P. Yeh, *Optical Waves in Layered Media*, Wiley Interscience, p. 190, 1988,
11. H. Patrick, S. L. Gilbert, A. Lidgard and M. D. Gallagher, *J. Appl. Phys.* **78**, pp. 2940-2945, 5 (1995).
12. T. Erdogan, V. Mizrahi, P. J. Lemaire and D. Monroe, *J. Appl. Phys.* **76**, p. 73, 1994.
13. S. Kannan, J. Guo and P. J. Lemaire, *Tech. Dig. OFC'96*, TuO4, pp. 84-85, 1996.
14. J. Guo, S. Kannan and P. J. Lemaire, *Tech. Dig. OFC'97*, ThJ6, pp. 285-287, 1997.
15. D. L. Williams and R. P. Smith, *Electron Lett.* **31**(24), pp. 2120-2121, 1995.
16. I. Riant, S. Borne and P. Sansonetti, *Tech. Dig. OFC'96*, TuO5, pp. 86-87, 1996.
17. R. J. Egan, H. G. Inglis, P. Hill, P. A. Krug and F. Ouellete, *Tech. Dig. OFC'96*, TuO3, pp. 83-84, 1996.
18. S. Kannan, P. J. Lemaire, J. Guo and M. J. Luvalle, *Tech. Dig. Of Conference on Bragg Gratings and Photosensitivity* **17**, BsuD3, pp. 52-54, 1997.
19. S. Kannan, J. Guo and P. J. Lemaire, J. Lightwave. Tech. **15**, pp. 1478-1483, 1997.
20. S. Kannan, L. Copeland, J. Judkins, M. Luvalle and P. J. Lemaire, *Tech. Dig. OFC'98*, ThG6, pp. 282-283, 1996.
21. OASIX^{TM} is a commercially available EDFA simulation software program from Specialty Fiber Devices group at Lucent Technologies.
22. S. Kannan, J. Judkins, P. F. Wysocki and P. J. Lemaire, in preparation.
23. F. Bakhti, J. Larrey and P. Sansonetti, *Tech. Dig. Of Conference on Bragg Gratings and Photosensitivity* **17**, BsuD2, pp.49-51, 1997.
24. W. T. Anderson, A. J. Johnson, J. P. Kilmer and R. M. Kanen, *Proc. 37th Int. Wire and Cable Symp.*, pp. 188-199, 1988.
25. Tomita and P. J. Lemaire, *Electron. Lett.* **21**(2), pp. 71-72, 1985.
26. N. J. Pitt and A. Marshall, *Electron. Lett.* **20**(12), pp. 512-514, 1984.
27. J. F. Marcerou, J. Hervo, S. Artigaud, H. Fevrier, P. Guitton and S. Landais, *Proc. ECOC'92*, pp. 497-500, 1992.
28. P. J. Lemaire, H. A. Watson, D. J. DiGiovanni and K. L. Walker, *Photonics Tech. Lett.* **5**(2), pp. 214-217, 1993.

29. P. J. Lemaire in *Fiber Optics Reliability and Testing*, edited by D. K. Paul, (SPIE Proc. **CR50**, Boston. MA), pp. 80-99.
30. P. J. Lemaire, *Opt. Eng.*, **30**(6), pp. 780-789, 1991.
31. G. H. Sigel, Jr., *Proc. Nineth Intl. Cong. On Glass:Vol.1*, pp. 801-809, 1971.
32. R. A. Weeks, *J. Appl. Phys.*, **27**, 1376, 1956.
33. J. W. H. Schreurs, *J. Chem. Phys.*, **47**, p.818, 1967.
34. D. L. Griscom, *J. Non-Cryst. Solids*, **31**, p.241, 1978.
35. K. Nagasawa, Y. Hoshi and K. Yahagi, *Proc. Conf. Elect. Insulation and Dielectric Phen.*, pp. 114-119, 1983.
36. E. J. Friebele and D. L. Griscom, *Treatise on Materials Sc. And Tech.: Glass II.* **17**, p.257, 1979.
37. Edwards A. H. and Fowler W.B., *Phys. Rev. B.* **26**(12), p. 6649, 1982.
38. D. L. Griscom, *J. Non-Cryst. Sol.* **149**, p.137, 1992.
39. K. Nagasawa, H. Mizuno, Y. Yamasaka, R. Tohmon, Y. Ohki and Y. Hama, *The Physics and Technology of amorphous SiO2*, pp. 193-198, 1988.
40. R. Tohmon, Y. Yamasaka, K. Nagasawa, Y. Ohki and Y. Hama, *J. Non-Cryst. Solids*, **95&96**, p.671, 1987.
41. A. R. Silin, and L. N. Skuja, *J. Non-Cryst. Solids.* **71**, p.443, 1985.
42. L. N. Skuja, A. N. Streletsky and A. B. Pakovich, *Sol. St. Comm.* **50**(12), p.1069, 1984.
43. M. Kohketsu, K. Awazu, H. Kawazoe and M. Yamane, Jpn. *J. Appl. Phys.* **28**(4), p.615, 1989.
44. E. J. Friebele, *Opt. Eng.* **18**, p.552, 1979.
45. E. J. Friebele, M. E. Gingerich and K. J. Long, Appl. Opt. **21**(3), pp.547-553, 1982.
46. E. M. Dianov, L. S. Kornienko, E. P. Nikitin, A. O. Rybaltovskif, V. B. Sulimov and P. V. Chernov, *Sov. J. Quantum Electron.* **13**(3), pp. 274- 289, 1983.
47. D. L. Griscom, E. J. Friebele, K. J. Long and J. W. Fleming, *J. Appl. Phys.* **54**(7), pp. 3743-3762.
48. E. J. Friebele, M. E. Gingerich and D. L. Griscom, *Proc. SPIE* **1791**, pp. 177-188, 1992.
49. R. L. Pfeffer, *J. Appl. Phys.* **57**, pp.5176-5180, 1985.
50. R. A. B. Devine and J. Arndt, *Phys. Rev. B.* **39**, pp. 5132-5138, 1989.
51. D. L. Griscom, *Nucl. Inst. & Methods.* **B46**, pp. 12-17, 1990.
52. E. J. Friebele, C. G. Askins, C. M. Shaw, M. E. Gingerich C. C. Harrington, D. L. Griscom, T. E. Tsai, U. C. Paek and W. H. Schmidt, *Appl. Opt.* **30**, pp. 1944-1957, 1991.
53. D. L. Griscom, M. E. Gingerich and E. J. Friebele, *IEEE Trans. on Nucl. Sci.*, **41**(3), pp. 523-527, 1994.
54. D. L. Griscom, M. E. Gingerich and E. J. Friebele, *Phys. Rev. Lett.*, **71**(7), pp. 1019-1021, 1993.
55. G. M. Williams, M. A. Putnam and E. J. Friebele, *Proc. SPIE* **2811**, p. 30.
56. D. T. H. Liu and A. R. Johnston, *Opt. Lett.*, pp. 548-550, 1994.
57. W. C. Goltsos, *Proc. SPIE* **2699**, pp. 304-309.

58. G. M. Williams and E. J. Friebele, *IEEE. Transac. On Nucl. Sci.* **45**(3), pp. 1531-1536, 1998.

59. S. Kannan, M. J. LuValle, R. Ahrens and P. J. Lemaire, *OFC. Tech. Dig.* THG5, 1999.

Materials Issues in the Development and Use of Lightguide Fibers, Cables, and Components

O.S. Gebizlioglu[1], C.R. Kurkjian, L.A. Reith[2]

Telcordia Technologies, Inc.
Formerly Bellcore
445 South Street
[1]Rm # 1A-140B
Morristown, New Jersey 07960

[2]Lucent Technologies (Work done while at Telcordia)
500-700 Mountain Avenue
Rm # 7B-204
Murray Hill, New Jersey 07974

ABSTRACT

Lightguide technology is now considered to be mature. Indeed, improvements in all aspects of this technology over the years have been most impressive – multimode to singlemode fiber, electronic to optical amplification and the newest and perhaps the most important: the use of WDM and DWDM technologies. In spite of these striking advances, the very nature of the silica lightguide material poses reliability issues which have not yet been fully resolved or in some cases even confronted. These issues by and large have to do with the intrinsic brittle nature of the glass material. In order to assure reliable performance of these lightguide fibers in telecommunications service environments, coating and cabling technologies developed over the past two decades have evolved to give robust fiber optic cables, devices and components. In this presentation, we review key materials issues in the development and use of lightguide technology in telecommunications. Furthermore, we analyze current trends and discuss major materials reliability issues that need to be resolved for further developments in future applications of optical fibers, fiber optic cables and fiber-based components.

Keywords: Materials Issues, Optical Fibers, Cables, Fiber-Based Components, Strength, Fatigue, Dimensional Stability

1. INTRODUCTION

Lightguide fibers, with their nearly infinite bandwidth, have penetrated 90 % of long-haul communications systems. In the distribution networks and local loop, their bandwidth remains unrivaled while the cost of deployment keeps fiber communications from reaching customer premises. However, rapidly rising popularity of internet access and major developments in enabling high-capacity high bandwidth communications have recently brought fiber optic communications to center stage. The key element of fiber optic telecommunications systems is the optical fiber. Telecommunications optical fibers are made of silica glass drawn into 125-μm diameter optically highly transparent strands. The recent demonstrations of very high bandwidth communications have been made possible by the development of erbium-doped optical fiber amplifiers, which enable longer-distance transmission of communications signals without expensive and intrinsically unreliable electronic signal regenerators, and wavelength division multiplexing technologies. In these high-bandwidth networks, fiber attenuation and dispersion are the key limiting factors that determine the regenerator spans and transmission rates. While fiber design[1] is focussed on minimizing these limiting factors, cable design[2] and device packaging are directed at both preserving the mechanical integrity/strength of silica glass transmission medium and isolating the fiber from external transmission-impairing (such as bending-induced attenuation) mechanical effects.

In the evolution of fiber optic telecommunications networks, many supporting technologies such as fiber stripping, splicing and cable installation have accompanied the developments in fiber and cable design and manufacturing to meet the deployment needs in the distribution and subscriber loops. Popular demand for internet access has generated a trend toward expanding deployment of high-fiber density (high-fiber count) cables and higher packing density in device packaging. These trends, in turn, have led to significantly reduced tolerances on the mechanical, thermal and transmission performance of optical fiber, cable and devices, and their components. Marketplace demands for highly reliable and survivable mass communications require robust high-fiber density cables and devices with very low transmission loss. These same demands have led to a variety of new network architectures that require cables with a broad range of designs, sizes and fiber counts. Cables with fiber counts as high as 864 fibers have already appeared with the transmission performance promise of only 0.2-0.3-dB/km attenuation in both 1310-nm and 1550-nm operations. Such impressive achievements, however, rely on the recognition of the fact that the physical layer of today's increasingly complex telecommunications networks are made of components and materials with a broad range of mechanical, thermal, and optical characteristics. In the following sections, we present an overview of key materials issues in three essential elements of modern fiber optic communications networks: These are lightguide fibers, fiber optic cables and fiber-based components such as connectors, splices, fiber Bragg gratings, couplers, splitters, and attenuators.

2. LIGHTGUIDE FIBERS

2.1 Strength, Aging and Fatigue

In its third decade of growth, fiber optic communications technology depends on the availability of optical fibers that are aging-resistant and fatigue-resistant. Despite all the advances in fiber design and manufacturing, basic physico-chemical aspects of fiber aging and mechanical strength degradation are not well understood. A key difficulty to understanding fiber strength degradation has been the unavailability of tools for direct assessments of the role and the condition of fiber coatings on aged fibers. Two such tools have been employed recently and will be described in the following sections.

Two time-dependent phenomena that strongly affect the lifetime of lightguide fibers are aging and fatigue. Aging refers to the degradation in mechanical strength that takes place when a fiber is maintained at zero stress at a given temperature and humidity. Fatigue, on the other hand, refers to delayed failure, that is, the reduction in the failure stress as the failure time is increased, or as the stressing rate is decreased. Both of these phenomena occur only in the presence of water. The basis of these processes has been shown to be the reaction of silica with water[3] with an activation energy of ~80 kJ/mol. The aging process of an optical fiber is usually described in a logarithmic plot of strength versus time, and it often involves an induction period with no apparent strength reduction. At the end of this period, however, strength degrades rapidly with the appearance of the well-known "knee."

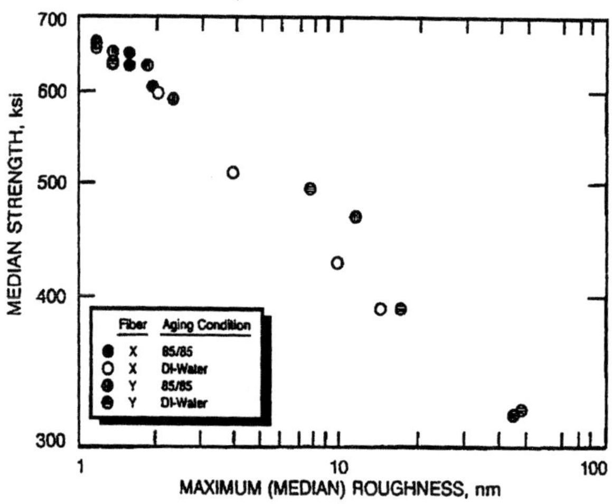

Fig. 2.1 Two-point bending strength as a function of surface roughness.

It has been suggested that the rapid strength degradation occurs because of the inhomogeneous solution of the silica fiber surface in liquid water. This pitting[4] is suggested by the increased surface roughness as median strength drops as shown in Fig.2.1. In fatigue, however, no induction period is required. Figure 2.2 shows that while fatigue occurs from essentially zero time, the knee in both processes appears at comparable times[5]. For both aging and fatigue to occur, the presence of 'water' at the silica surface is clearly necessary. Fatigue occurs through the formation of Si-OH bonds, and, thus, may also proceed in water vapor, while according to the above model, aging requires actual solution of the silica surface and, thus, it can only occur in liquid water. However, Matthewson and Yuce[6] have shown that similar aging can take place with a coated fiber in water vapor, although the appearance of the knee is delayed by an order of magnitude in time. Thus, either a liquid is formed by the reaction of the vapor with the coating, or possibly a different degradation mechanism is at work in this case. It should be noted that fatigue failure takes place in the presence of large stresses which may cause local mechanical delamination of the fiber coating from the fiber surface, thereby allowing the formation of Si -OH bonds, and thus fatigue.

Under typical conditions, the 10-second strength of a standard telecommunications optical fiber in air is about 5.5 GPa, while if a stress of half that value (2.25 GPa) is applied to the fiber, it will take about two weeks to fail. This time dependence of failure stress is described by the following expression:

$$n = -d\left(\frac{\ln t}{\ln \sigma}\right) \text{ where for silica fibers, n ~20}$$

It is generally agreed that this expression is only an approximation since the true physico-chemical aspects of fatigue weakening are not understood.[7] Moreover, a detailed understanding of the influence of coating chemistry variations on fatigue is also lacking. Recent work has suggested that changes in coating chemistry can have a major impact on both the short- and long-time fatigueand aging. In Figures 2.3a and 2.3b, we illustrate these effects.. In Fig. 2.3a, curves A (crosses) and B (triangles) are for two coated fibers which we have studied. Curve C (circles) is for bare silica fiber. The n value for the bare fiber over the entire experimental time is about 20. For fiber A at short times a similar value (n ~ 20) is observed, while at long times, flattening occurs and a value of n ~ 40 is calculated. For fiber B only the short time data are available, but here a value of n ~ 40 is calculated. It has been suggested that higher n values may be achieved by either (1) controlling the pH of water at the glass surface, or (2) reducing the permeability of the coating, and thus reducing the water activity at the glass surface. In an earlier investigation[8], we examined the effects of aging conditions on the strength of two high-n value fibers and studied the relationship of fiber aging to the aging of the fiber coating.

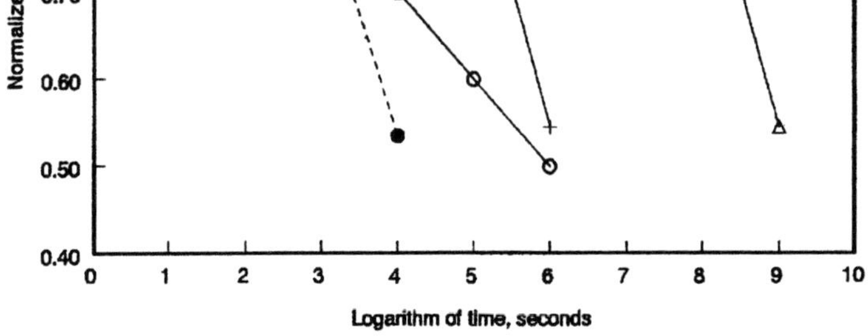

Fig. 2.2 Normalized strength (measured at room temperature) as a function of aging time in water at room temperature (triangles) and 85°C (crosses). Normalized static fatigue stress as a function of time in water at 85°C with (filled circles) and without (open circles) aging.

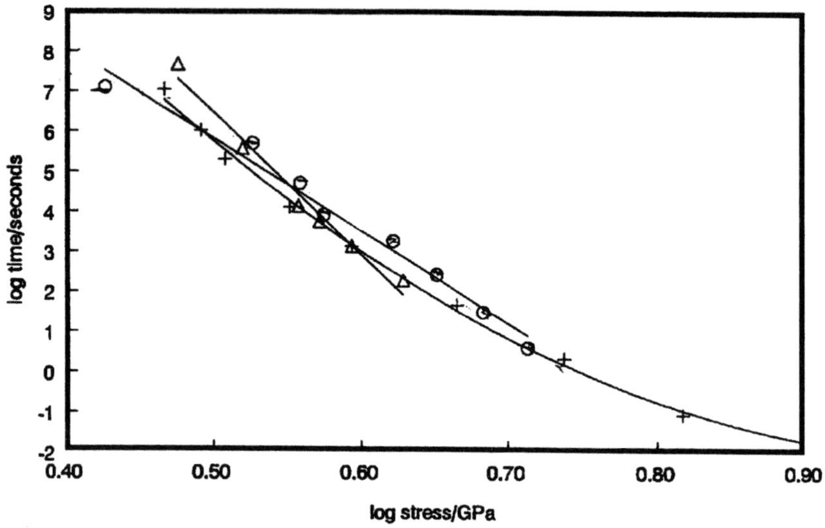

Fig. 2.3a Static fatigue in ambient humidity at room temperature for a bare silica fiber (circles) and fiber with coating A (crosses) and B (triangles).

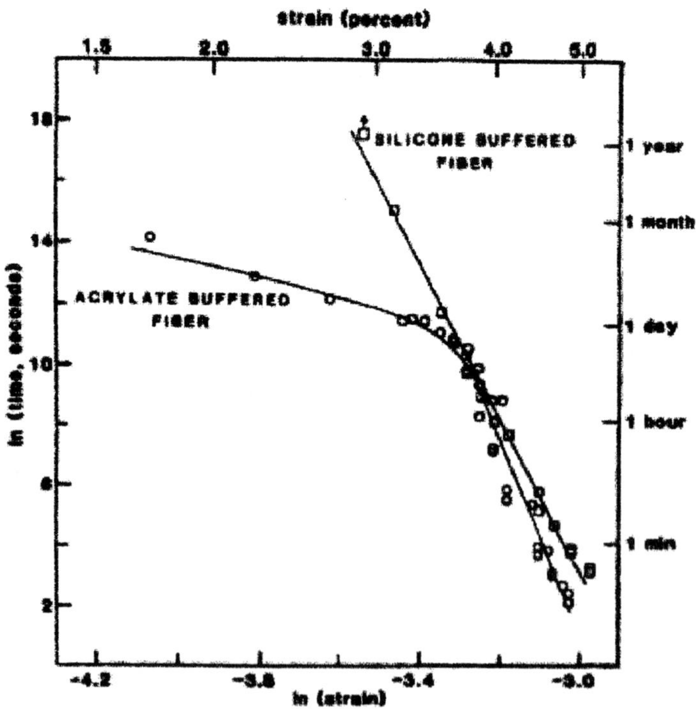

Fig. 2.3b Aging of fibers with two different coatings in 80°C water.

2.2 Aging of High-N-Value Fibers

In the analysis of high-n value fibers, the two-point bend strength, n value, and the dynamic mechanical properties of the fibers in the as-received condition and after a two-month aging treatment in DI (De-Ionized) water at 85°C were measured. The experimental equipment employed for strength testing has been described earlier.[9] Dynamic mechanical analysis (DMA) measurements on optical fibers were described in earlier publications.[10,11]

Table 2.1 presents the results of strength and fatigue measurements. Column 1 gives the fiber designation and column 2 indicates whether the sample was as-received (AR) or aged at the measurement time. The next two columns indicate the strength at the two stress-rates used for the evaluation of the fatigue constant, n, which is shown in the final column. All of the strength measurements were carried out at room temperature after equilibrating the fiber with DI water. Fiber strengths were measured as-received, and after a two-month aging in DI water at 85°C. For aged fibers, measurement results are presented for both coated fiber and bare fiber after the removal of its coating with hot sulfuric acid. For fiber A, three strength values are obtained after progressively longer acid stripping times of 10, 30 and 120 seconds. While the stripping time has no effect on the behavior of fiber B, fiber A exhibited strong dependence at the coating removal rate. This observed dependence suggests that the shorter stripping times are not sufficient to remove the effects of the coating. This, in turn, implies that the mechanism giving rise to the high n value in fiber A does not depend on the coating thickness, but, most likely, it resides at the glass-coating interface.

Table 2.1. Effects of Aging and Stripping on Fiber Strength and Fatigue

FIBER	CONDITION	STRENGTH 1000μ/s	STRENGTH 100μ/s	n
Standard	coated/AR	760	590	18
	coated/aged	280	248	38
	bare/aged	313	264	27
A	coated/AR	800	680	28
	coated/aged	736	613	25
	bare/aged			
	10 sec strip	743	630	28
	30 sec strip	730	610	26
	120 sec strip	734	587	20
B	coated/AR	780	600	18
	coated/aged	722	616	29
	bare/aged	730	590	21

2.3 DMA of Coatings by Direct Measurements on Fibers

In an attempt to investigate any correlation of the strength of these fibers to the mechanical condition of their coatings, we used DMTA (Dynamic Mechanical Thermal Analysis). Basic principles and details of the DMTA analyses were presented in earlier reports.[10,11] This analysis provides two profiles as shown in Fig. 2.4: Fiber composite modulus-temperature profile and coating loss tangent (or tan δ)-temperature profile. Of these two profiles, the loss tangent-temperature profile shows the condition of the fiber coating more directly; it indicates the softening temperature (commonly called the glass transition temperature, T_g) by the loss tangent peak and any change in the mechanical condition of the coating on the fiber. Figures 2.4 and 2.5 show that DI-water aging at 85°C for fiber A results in a broadening of the loss tangent peak. This may be due to a combination of the following changes in the coating: 1) coating may have taken up water and swollen, 2) coating may have a mild structural break-down while remaining on the fiber, and/or 3) coating may have delaminated from the glass fiber. For fiber B, Figs. 2.6 and 2.7 show the effects of aging in DI-water at 85°C. In this case, we observe that a lower-temperature tan δ peak develops while the area under the tan δ profile increases more dramatically than that for fiber A. Tables 2.2 through 2.4IV present a detailed comparison of trends shown in Figs. 2.4 through 2.7. They include characteristics of a standard telecommunications fiber to satisfactorily explain the role of coatings in affecting the aging performance of fatigue-resistant fibers. Table 2.2 illustrates one important difference between the aging performance of fatigue-resistant fibers A and B, and the standard fiber; DI-water aging affects the tan δ peak temperatures for both the low-temperature and high-temperature peaks in the standard fiber while it affects only the low-temperature peak significantly in fatigue-resistant fibers A and B. For the standard fiber, an almost two-fold drop in temperature occurred for the high-temperature peak while the low-temperature tan δ shifted upwards. A downward shift in the tan δ peak temperature indicates loosening of the coating structure while an increase in the peak temperature signals tightening and/or stiffening of the material volume undergoing the DMTA test. One plausible structural explanation is that DI-water aging may induce swelling in the secondary coating while it may simultaneously cause degradation by cross-linking of the primary coating. In Table 2.2, we observe that fiber B exhibits a more dramatic drop in the temperature of the low-temperature tan δ peak. This result suggests that the primary coating may swell significantly upon exposure to DI-water. Tables 2.3 and 2.4 show two other interrelated measures of structural change in the coatings of aged fibers: the tan δ peak height and the area under the tan δ - temperature curve. Typically, the area under the tan δ curve may change in parallel with the tan δ peak height. Therefore, we focus on the calculated area under the tan δ curve to further analyze the DMTA data. Table 2.4 shows that DI-water aging at 85°C results in nearly 30% reduction in the area, and, therefore, in the volume of the coating for the standard fiber. For the fatigue-resistant fibers, however, the area increases; this increase is a modest 2% for fiber A while it is a dramatic 82% for fiber B. As pointed out earlier, an increase in the tan δ peak area is proportional to an increase in the relaxation volume of the coating. The tan δ peak temperatures of Table 2.2 indicate that the secondary coating with its high T_g is not affected by the aging treatment while major changes occur in lower-temperature segment of the tan δ profile where the characteristics of the primary coating are represented. The downward shift in the temperatures for the low-temperature tan δ peaks with an increase in the coating volume as evidenced by an increase in the area under the tan δ curve suggests that the primary coating undergoes swelling. Whether this swelling is accompanied by delamination at the glass-coating interface is not established yet.

Table 2.2. Effect of Two-Month Aging in DI-Water at 85°C

FIBER	LOSS TANGENT PEAK TEMPERATURES (°C)
Standard Fiber	-3 / 70
Standard Fiber - Aged	10 / 39
Fiber A	9 / 76
Fiber A - Aged	1 / 77
Fiber B	30 / 70
Fiber B - Aged	5 / 65

Table 2.3. Effect of Two-Month Aging in DI-Water at 85°C

FIBER	LOSS TANGENT PEAK HEIGHTS
Standard Fiber	0.05 / 0.055
Standard Fiber - Aged	0.065 / 0.077
Fiber A	0.11 / 0.141
Fiber A - Aged	0.075 / 0.105
Fiber B	0.049 / 0.069
Fiber B - Aged	0.087 / 0.119

Table 2.4. Effect of Two-Month Aging in DI-Water at 85°C

FIBER	AREA UNDER THE LOSS TANGENT PROFILE (°C)
Standard Fiber	3.2
Standard Fiber - Aged	2.3
Fiber A	6.6
Fiber A - Aged	6.7
Fiber B	4.5
Fiber B - Aged	8.2

Thus, the results reviewed above indicate that the coatings of these high-n value fibers exhibit much greater resistance to two-month aging in DI-water than acrylate coatings used in standard telecommunications fibers. While a standard acrylate coating degrades and loses all its viscoelastic relaxations in DI-water aging, the two high-n value fibers retain or increase their viscoelastic relaxations through the DI-water aging process. In other words, these high-n value fibers possess coatings that are DI-water aging resistant.

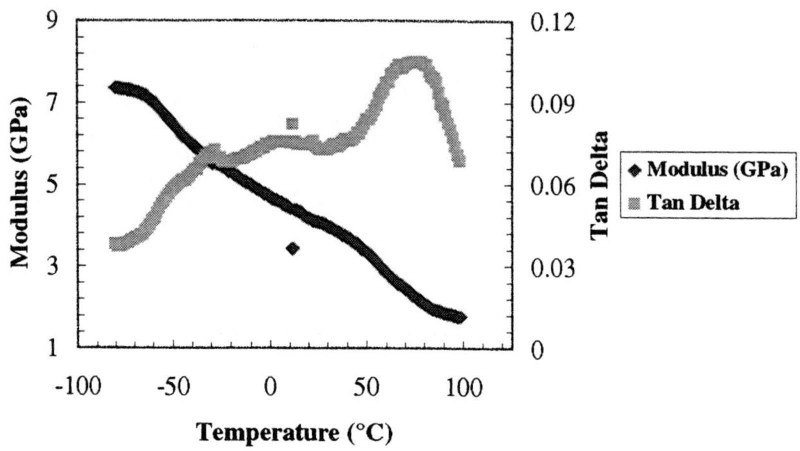

Figure 2.4 Unaged Fiber A

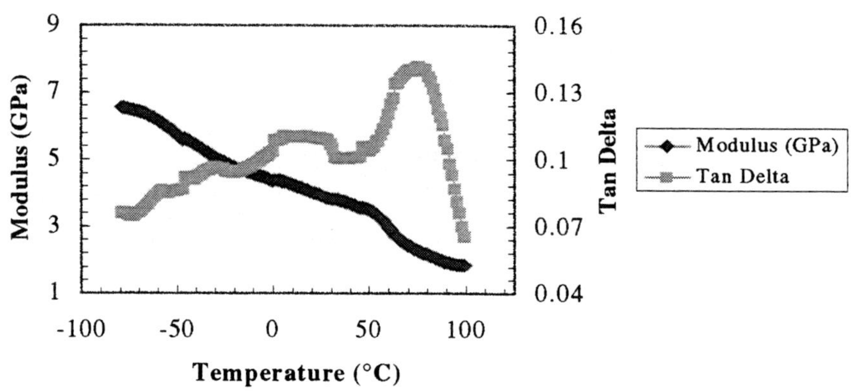

Figure 2.5 Fiber A – Aged 2 Months in DI-Water Water @ 85°C

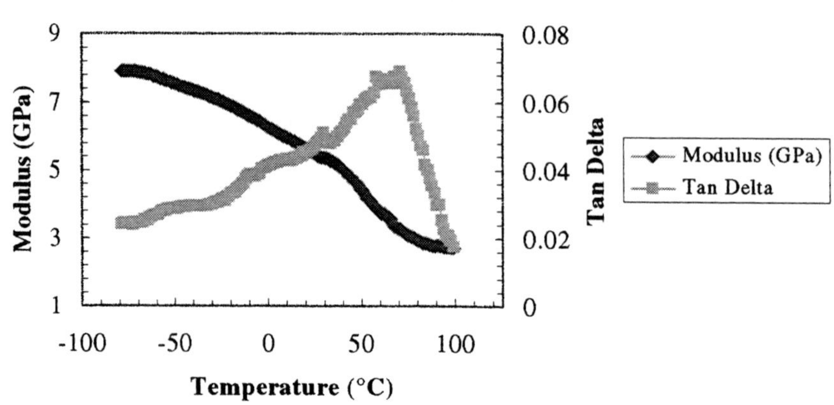

Figure 2.6 Fiber B -Unaged

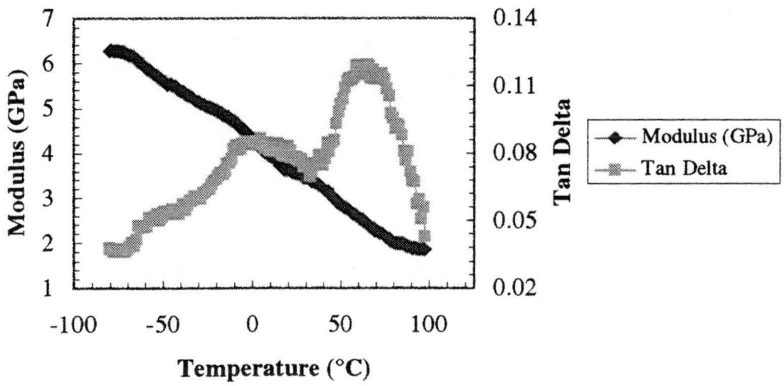

Figure 2.7 Fiber B – Aged 2 Months in DI-Water @ 85°C

2.4 Specialty Optical Fibers and Devices

While the development of low-loss singlemode silica fiber made the first generation lightwave communications systems possible, the recent development of erbium-doped fiber amplifiers (EDFA) has enabled economical delivery of high-bandwidth telecommunications signals over very long distances. From the beginning of the EDFA developments[12,13,14,15] to today, the greatest promise of EDFA technology combined with wavelength-division multiplexing components has been to provide communications service providers with the ability to upgrade system capacity without the cost of adding new fiber into their existing physical plant infrastructure. Despite all the advances on the system technology side[16], the brittle nature of silica glass fiber remains a major limiting factor in reducing the cost of fiber optic system deployment.

In addition to the brittleness of silica glass, its extremely low coefficient of thermal expansion (CTE) presents a major challenge in minimizing damage during fiber manufacturing, cabling, field installation and maintenance operations. In the packaging of fiber-based devices, thermally-induced stresses due to large differential thermal expansion of polymeric and other non-silica glass components need to be accounted for in the device design and manufacturing. The brittle nature of silica glass has received considerable attention due to very low damage tolerance of bare glass fibers,[17] as illustrated in Fig. 2.8.

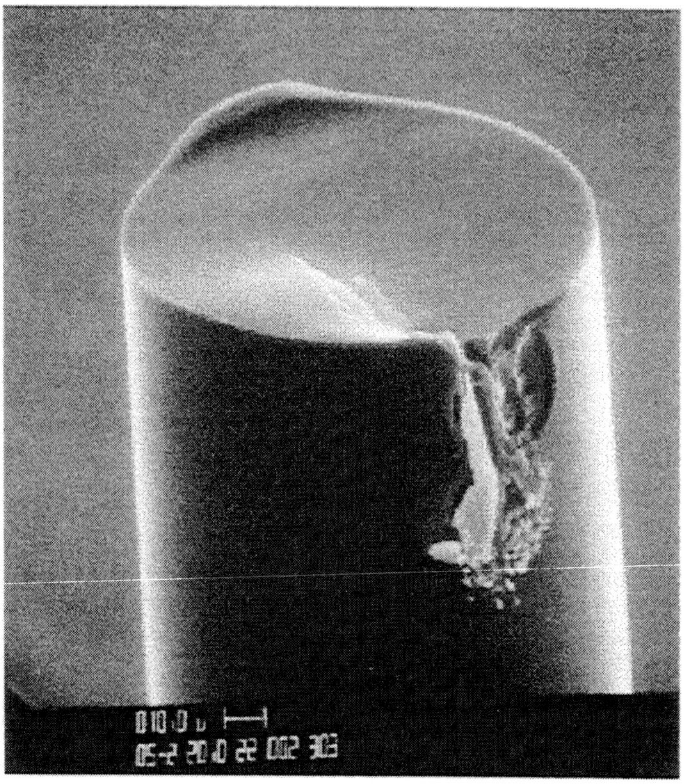

Figure 2.8. Fractograph of a broken fiber with surface damage.[17].

For a quantitative understanding of brittleness and damage sensitivity of silica glass, it will be useful to review the data presented in Fig.2.9. The plot of fiber strength against indentation load in Fig. 2.9[18] clearly shows the strong effect of indenter geometry (shape and size) on the strength of indented fiber. In this figure, cube-corner diamond indenter (included angle of 90°) is compared with Vickers diamond indenter, which has an included angle of 136.° An eightfold reduction in fiber strength from 5.5 GPa (for a pristine fiber) to 0.7 GPa (proof test strength) requires 200-gm load on Vickers indenter while such degradation occurs with only 0.2-gm load on the cube corner indenter. This indicates clearly the great sensitivity of the fiber surface to mechanical damage and consequent loss of strength. It is important to note that the Vickers plot consists of two straight lines. The first of these at the lower loads results only in the formation of a permanent impression, but no cracks. The second section and the entire loading range shown for the cube corner indenter represents cracking of the silica fiber. This two-section behavior with the Vickers indenter depends on the time scale and the environment. The indentation (or scratching) process induces stress during indentation and gives rise to residual stress after the indenter withdrawal. At high mechanical loads and in aggressive environments, surface cracks will 'pop-in' immediately or very soon after the indenter is withdrawn, while under less severe conditions, it takes time for the cracks to nucleate.[19] This is an important consideration in developing handling strategies to control delayed strength degradation that may follow handling-induced damage in manufacturing and installation of fiber optic cables and fiber-based devices. It suffices to conclude that the 'sharpness' of the indenter can reduce indentation load required to cause serious strength degradation in silica optical fibers.

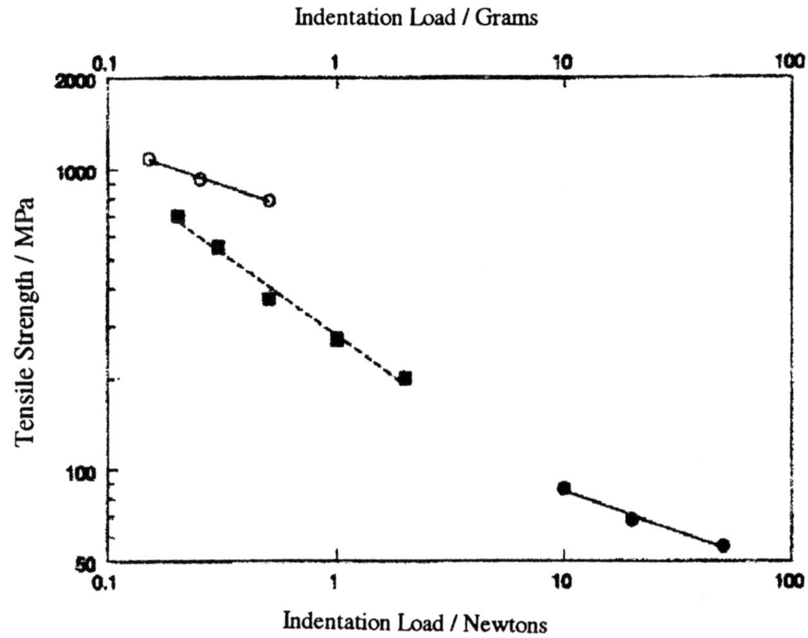

Figure 2.9. Fiber strength versus indentation load for two indenter geometries. Closed squares and circles represent Vickers indenter data while open circles refer to cube corner indenter results.

2.5 Splices

Field handling/installation operations such as splicing involve potentially strength-degrading steps such as mechanical stripping of coatings. The effectiveness of the stripping operation and the potential of damage to the fiber surface depend on stripping/tool parameters as well as coating material parameters.[20,21,22] Recent investigations[22] confirmed that the ease of stripping (magnitude of strip force) and the quality of stripping (cleanliness of the stripped fiber surface) are determined by the mechanical properties of the coating. Not only the strength of the stripped fiber, but also flaw distribution along the entire length gains importance in the success of splicing operations since all surface flaws may not heal uniformly during fusion splicing. Although "the strength of fusion splices is believed[23,24] to be totally dependent on the integrity of the protective sleeve and strength member", this is an extrinsic not an intrinsic issue. As shown earlier[25] by Krause and Kurkjian, it is possible to attain the "ideal, unattainable goal of fiber splicing to recreate the original fiber in all respects". The splice strength is determined by a) the extent of damage imparted to the fiber surface by the stripping tool blade and b) the amount of fiber coating residue and/or other foreign contaminant on the fiber surface. Surface contaminants result in major strength reductions in splices. Very careful cleaning is necessary to completely remove all of the residual coating. Detection of thin layers of remaining coating has been explored[26] by means of a new scratching technique employing an Atomic Force microscope (AFM). While earlier application of standard AFM imaging techniques did not conclusively determine the level of residual coating remaining, this new technique shows greater promise for this determination.

Although the strength of fusion splices usually degrades significantly during splice fabrication, the application of fusion protector sleeves recovers the strength of the splice[27,28,29]Only if the protector sleeve degrades during subsequent aging, are the fibers expected to be at risk. The types of protector sleeves most often used are heat-activated types, which consist of an inner tube of adhesive and an outer tube of heat-shrink material. When the protector sleeve is applied, the adhesive melts and adheres to the stripped section of fiber while the heat-shrink tube shrinks down around the fiber and central strength member. This design, as long as proper adhesives are chosen, has proven highly reliable and there have been very few reports of field failures. However, new types of clamp-on, or clam-shell types of protector sleeves have now come on the market, which offer convenience of assembly. The reliability of these new designs is still an open issue. It is likely that the strain relief provided is not as great as that of conventional protector sleeves, but it may be adequate for most applications. Additional testing is required to determine the resistance to environmental stresses.

Regardless of the quality of the protector sleeve, it is desirable to have splices with sufficient strength for easy handleability, as a back up in case the sleeves do fail, and also for applications where extremely high reliability is paramount, such as undersea, space, or military applications. As always, minimizing the damage to the fiber is the key to high strength and reliability. Chemical stripping, proper rinsing, and clean types of fusion arcs that do not deposit contaminants are all critically important[29].

Recent studies have shown that, for typical telecommunications applications in which electrodes are used to perform the fusion, electrode wear can adversely affect both the strength and the loss of the fusion splice. A technique of measuring the fiber meltback of fibers that are held stationary during the fusion process was found to be a good indicator of how well the fusion splicer was working.

2.6 Packaging of Specialty Fibers

Because of the very low thermal expansion coefficient of silica glass, packaging of fiber-based devices such as fused-tapered couplers and splitters requires substrates made from silica or other low-thermal expansion (CTE) materials. In the case of fiber Bragg gratings (FBG) used in dispersion compensators and wavelength division multiplexers / de-multiplexers, however, the temperature coefficient of refractive index, the term in dn/dT ($\sim 11\mathrm{x}10^{-6}$ / °C) controls the temperature sensitivity, with the thermal expansion term accounting for less than 10% ($\alpha \sim 0.52\mathrm{x}10^{-6}$ /°C). It has been suggested that the use of a glass-ceramic which has a compensating negative thermal expansion coefficient (-5 to –

$8x10^{-6}$ / °C) can be used to give an effective temperature coefficient of Bragg wavelength of less than 0.001 nm /°C compared to 0.012 nm /°C for uncompensated in $GeO_2 - SiO_2$ fibers.[30]

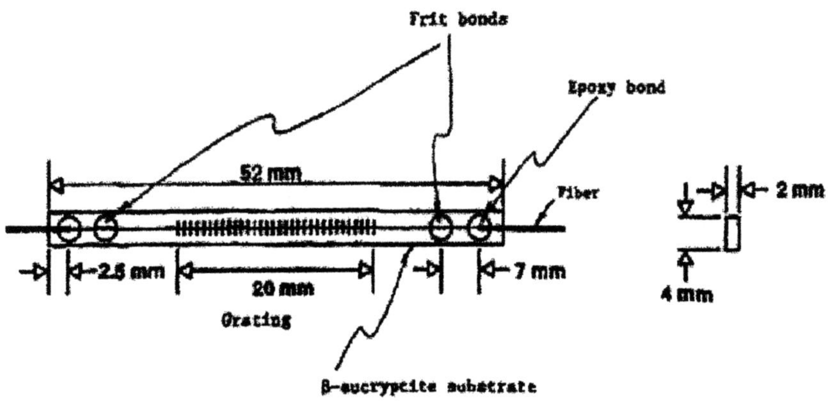

Figure 2.10 Athermalized Bragg grating with B-eucyptite substrate.

There have been a number of studies of the optical reliability of fiber Bragg gratings. However, there exists very little understanding of their mechanical reliability. For instance, Feced et al[31], have shown that when the gratings were produced off-line, substantial degradation occurred. On the other hand, Askins et al.[32] have shown that FBGs can be produced in-line at the same wavelength (248 nm) and flux level (1 J/cm^2) without any strength degradation as illustrated in Fig. 2.11. There is clearly the opportunity for mechanical 'handling' damage during the stripping and irradiation processes when the gratings are prepared off-line. This issue requires further study.

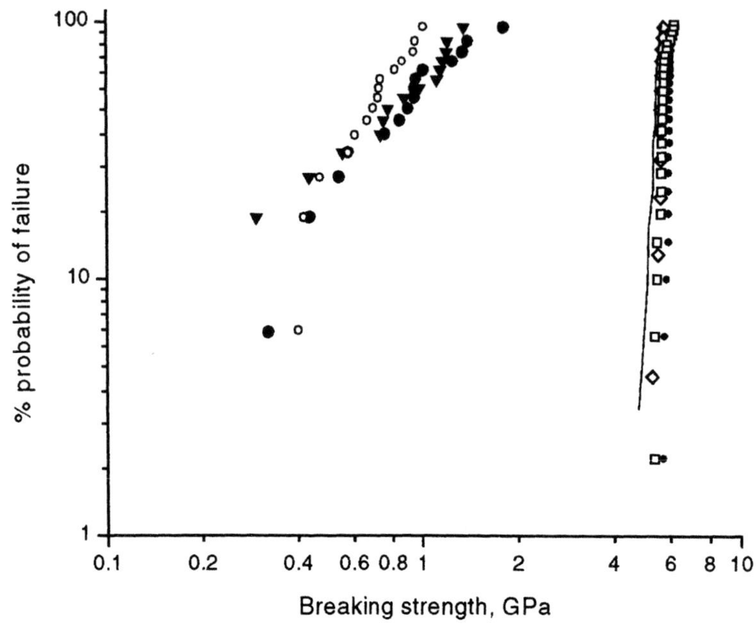

Figure 2.11 Strength distribution of $GeO_2 - B_2O_3$ co-doped fiber. Fiber Bragg Grating was written in-line[32].

2.7 Metallization

Lightguide fibers and devices often require effective protection against moisture ingress. Also, device packaging may require the use of metallic solder rather than organic or inorganic adhesive and/or sealants. In this case, in addition to the removal of the coating, the application of a metallic layer is necessary in order to accomplish satisfactory soldering. There are a number of techniques which can be used for metallization.[33] The best technique for the application of a metal coating is one which involves drawing the bare fiber, in-line, through a molten bath of the desired metal. While this results in a fiber which shows no fatigue or aging, strength in the range 10-12 GPa, and quite high pull-out loads in soldered packages as shown in Figs 2.12 and 2.13,[34] not all metals can be applied in this way. Other techniques such as evaporation and sputtering, as well as electrode and electroless techniques have been in use. Although each technique may have definite areas of application, there exists little fundamental understanding of these processes and of the reliability of metallized fiber and packaging produced.

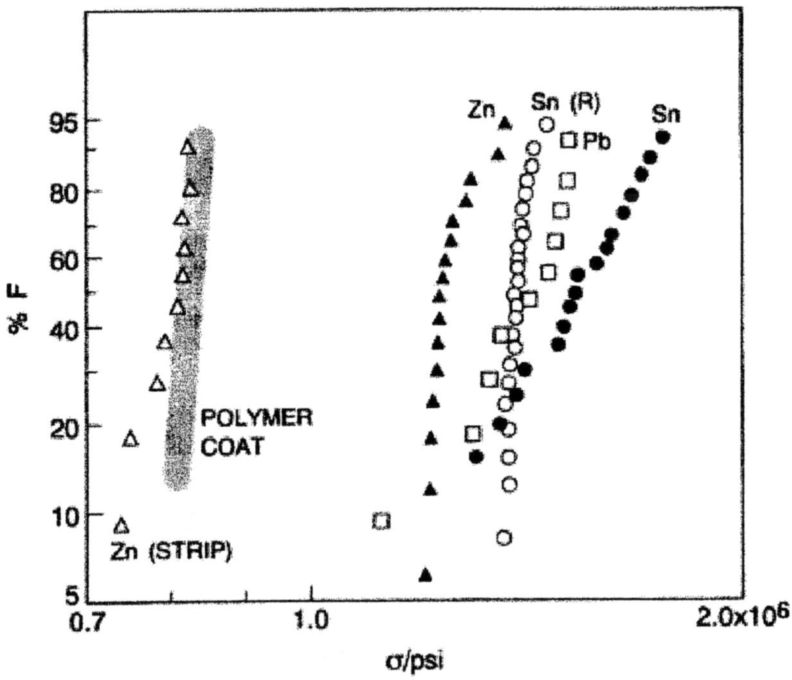

Figure 2.12 Strength distribution for metallized fibers.[34]

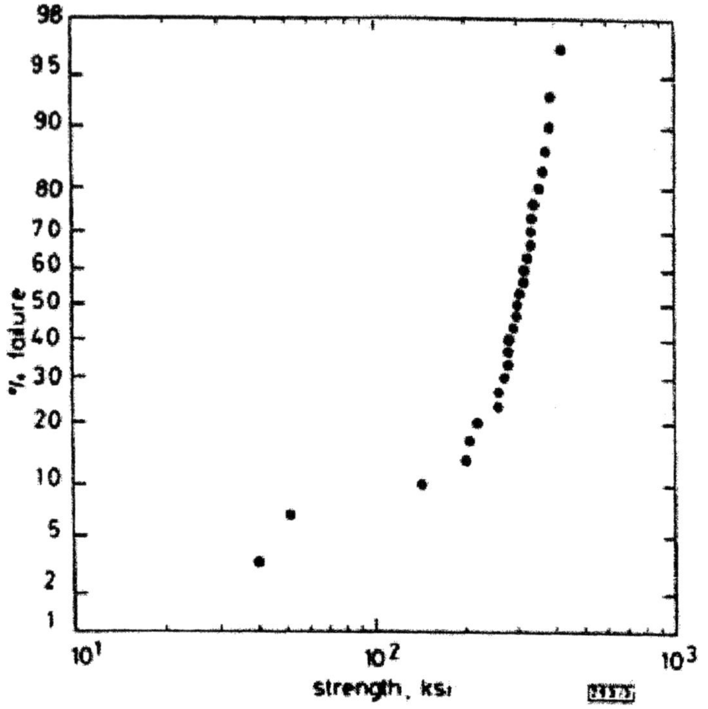

Figure 2.13 Solder joint strength of Ni-plated Al fibers[34].

3. FIBER OPTIC CABLES

Fiber optic cables provide the physical infrastructure through which high-bandwidth and high-capacity advantages of optical telecommunications are realized. While earlier cable designs were aimed at protection of a loose bundle of optical fibers from the harsh environment of telecommunications outside plant, modern fiber optic cable designs have been developed to achieve additional goals such as high fiber counts for increased transmission capacity and high installation/maintenance productivity in the field.

Modern telecommunications fiber optic cables are categorized according to a) the arrangement of fibers; whether fibers are loosely bundled or placed in parallel arrays to form fiber ribbons, b) the location and size of protective plastic tubes that house fibers/ribbons; loose tube cables and core tube cables, c) the type of water-blocking material used; gel-filled cables versus dry-block cables where water-swellable tapes or yarn are used instead of cable filling compounds, and d) the location of cable deployment; indoor/intra-building cables versus outdoor/outside plant cables. Some recently introduced cable products belong to all of these categories since they combine defining features of these four classes. For instance, a high-fiber count RILT (Ribbon-in-the-Loose Tube) cable consists of fiber ribbon stacks embedded in cable filling gels in loose tubes, and it is designed for outside plant applications. Among fiber optic telecommunications cables, filled loose tube cables still represent the largest category although the deployment of core tube cables and dry-block loose tube cables has been expanding at an increasing rate.

In a filled loose tube cable, as schematically depicted in Figure 3.1, optical fibers are embedded in a hydrocarbon gel that fills the void space around the fibers in a protective plastic tube called a buffer tube. In each buffer tube, radial clearance is provided between the optical fibers and the tube wall to minimize tensile and/or lateral loading of the fibers during installation and use over a broad temperature range. Since a loose tube cable consists of cable components with different thermomechanical response, the principal goal of a loose tube cable designer is to structure a cable 1) to provide sufficient radial clearance between the fibers and the buffer tube wall and 2) to maintain this design clearance over the service life of the cable by taking into account a large differential thermal expansion/contraction between cable components and the fiber. For instance, while cable jackets, made from MDPE (medium-density polyethylene) or LDPE (low-density polyethylene) and buffer tubes, made from either polyesters or polyolefins, are of large CTE (Coefficient of Thermal Expansion), metallic shield and CSM (Central Strength Member) are low-CTE components. The lowest-CTE member in the composite structure of a fiber optic cable is the optical fiber. A cable manufacturer's challenge is, then, to incorporate these different components into a cable structure that can maintain optical fibers in an essentially load-free state under widely varying thermal conditions of the telecommunications service environment. In the design and manufacture of fiber optic cables, therefore, the dimensional / environmental stability and compatibility of materials that make up cable components are of critical importance in assuring final product quality, long-term performance and reliability. For product performance and reliability assurance, outside plant cables and premises cables are often tested against generic requirements set forth in Telcordia documents GR-20-CORE[35] and GR-409-CORE[36], respectively. Among the key materials issues that are addressed in these documents are the compatibility of cable filling gels with fiber coatings[37] and buffer tube/core tube materials, hydrolytic stability of buffer tubes/core tubes,[38,39] and dimensional stability of buffer tube materials and its impact on the low-temperature cable transmission performance.[40]

Figure 3.1 Schematic cross-sections of two loose tube cables

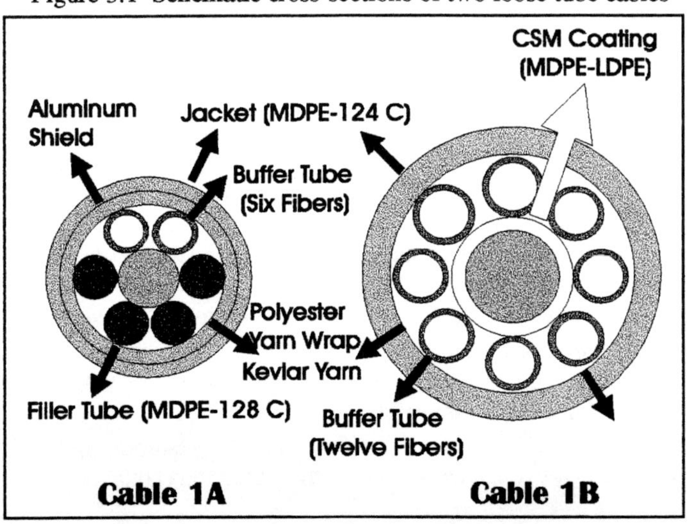

Recent service-affecting transmission degradation incidents underscored the critical importance of compatibility and dimensional stability of cable elements in the long-term performance and reliability of fiber optic cables in the telecommunications service environments where temperature could drop to -40°C in winter months and rise to 35°C during summer months. Section 3.1 below presents an in-depth discussion of the complex materials and cable structure issues that impact transmission performance at low temperature. From this discussion, however, temperature dependence[41] of the rheological (deformation and flow) behavior of buffer tube gels (or cable filling gels) is excluded for clarity since the role of buffer tube gels (or cable filling gels) in the low-temperature cable transmission performance is not well understood, and, therefore, still under investigation. A series of transmission loss incidents also occurred at elevated temperatures and involved fiber ribbon core-tube cables that were installed within innerduct on aerial transmission lines. These incidents resulted from the retraction of fiber ribbons into the cable on aerial installations with large (>150 feet) span length (pole-to-pole distance) and total distance between the splice closures (from 4000 feet to 6000 feet). Since these failures have not been investigated to the extent the low-temperature failures were examined, we will defer a detailed discussion on this subject to a future report.

3.1 Low-Temperature Transmission Performance of Loose Tube Cables

A series of service-affecting transmission loss incidents over a broad temperature range underscored the challenge of meeting the product quality, long-term performance and reliability goals in loose tube fiber optic cable design.[42,43,44] These incidents occurred initially in aerial fiber optic transmission lines operated at 1550 nm. However, subsequent failures took place in 1310-nm transmission facilities. In all cases, the optical loss abated when the ambient temperature returned to moderate levels.

This temperature-dependent transmission loss phenomenon (we also call it TICL, Temperature-Induced Cable Loss) has shown the following characteristics:

- The loss can reach 10 dB or higher and occur over a broad temperature range from 15°C to -40 °C.
- The loss exhibits characteristic wavelength dependence of microbending loss in singlemode optical fibers. It is higher at 1550 nm than at 1310 nm by an order of magnitude.
- The loss occurs within 10 to 50 feet of a cable termination.
- Earlier incidents involved cables of mid-to late-1980s vintage. Most recent failures, however, occurred in cables of 1986 through 1993 vintage.
- Affected cables exhibit central member (CM) protrusion in a splice closure. Proper termination by rigid coupling of the cable sheath and buffer tubes to the central member (CM) suppresses the TICL.
- Laboratory thermal-cycling experiments (between -40°C and 80°C) revealed that significant optical loss appears only upon cooling after conditioning of the cable at a higher temperature such as 80°C. Thus, cables that were not exposed to high temperature would not exhibit transmission loss.
- Optical cable spans that exhibited TICL could be restored to service only by removing the affected cable length and re-terminating the cable.

A series of key laboratory thermal cycling experiments[45,46,47] indicated that the thermal contraction of the buffer tubes was the major contributor to TICL. This finding was very important since the cable jacket, due to its large cross-sectional area and thermal expansivity, had previously been considered to be the major driver of bending-induced loss in loose tube cables. These laboratory experiments combined with extensive analyses of cable materials and the mechanical coupling between components further revealed that a tight cable structure that couples the cable jacket and buffer tubes to the CM could prevent the occurrence of TICL. Furthermore, these investigations uncovered key materials issues that stem from the time- and temperature-dependence of buffer tube dimensional stability.

3.2 Physical Origins of Low-Temperature Transmission Loss

A series of experiments were conducted to measure optical loss both at 1310 and 1550 nm during thermal cycling between -40°C and 80°C in order to identify the physical origin of the TICL phenomenon and the cable component(s) that contribute to it. Seven cable samples from four manufacturers and four samples of cables that failed in the field were installed, one at a time, in an environmental chamber. One end of a 21-m (70-ft) long cable was kept in the chamber while the other end was terminated outside the chamber and maintained at 23°C. Inside the chamber, the cable end was left free to simulate a worst-case installation condition, and fibers from two separate buffer tubes were concatenated for an effective fiber length of 43 meters (141 feet). A typical response of a cable exhibiting temperature-induced cable loss due to thermal cycling consists of optical loss growth through a half a dozen thermal cycles between -40°C and 80° to a steady loss value of 10 dB or higher. For all cable samples tested to date, optical loss disappeared at 80°C and grew to a steady-state value at -40°C as the number of cycles increased.

The laboratory samples and field cables (nine samples from five cable manufacturers) tested were all made of similar materials as shown for Cables 1A and 1B (two cables from one manufacturer) in Fig. 3.1. Our optical measurements to date have indicated that variations in cable components such as the number of jacket layers, the presence of a metallic shield, the buffer tube thickness, the number of filler tubes, and the number of fibers per tube do not correlate with the temperature-dependent optical performance.

Having determined that no single cable characteristic correlates with the cable's temperature-dependent performance, thermal cycling experiments (between -40°C and 80°C) were conducted on test cables that were progressively disassembled to determine the role of major cable components in TICL. In one series of experiments, optical loss was measured on test cables before and after complete removal of the jacket and at subsequent stages during the disassembly of the cable core. Results of these measurements, shown in Fig. 3.2, pointed out that the cable jacket is not a major contributor to TICL, and that it can, in fact, restrain the thermally-induced buffer tube shrinkage from going far enough to cause fiber axis perturbations.

In the second series of optical loss measurements, a cable that had exhibited service-affecting transmission loss in the field was progressively disassembled. The cable disassembly process involved progressive removal of cable components starting with the

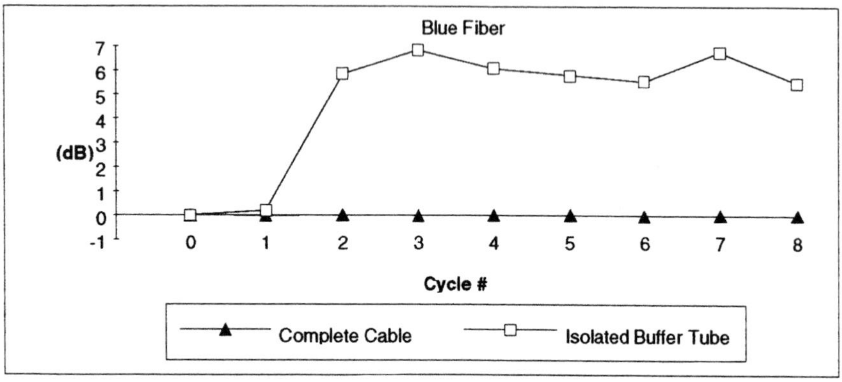

Figure 3.2 Optical loss at -40°C for progressively disassembled Cable 1B

cable jacket over some cable length from the terminated end of the cable inside the environmental chamber, and it continued until the fiber bundles were exposed. As such this process was different from the disassembly of test cables 1A and 1B since these cables were disassembled over their entire length.

3.3 Cable Materials and Structure Issues

3.3.1 Buffer Tube Materials

The optical loss measurements described in the preceding paragraphs indicated that, although test cables were made of similar materials, they behaved very differently at low temperature. Therefore, we focused on the role of cable structure, particularly, cable structure tightness in the control of temperature-dependent performance. In earlier reports[44,45,46,48] we showed that a tightly structured cable with strong mechanical coupling of the cable jacket and core components to the cable CM gives minimal optical loss. From our cable mechanical model, we defined a finite cable length over which cable component slip past each other. This relative movement is a consequence of the large differential thermal expansion/contraction of some cable components; it occurs where cables are terminated at splice closures and the cable ends are poorly constrained. Table 3.1 compares the coefficients of thermal expansion (CTE) of typical cable elements at 25°C.

Table 3.1 CTE of Typical Cable Elements at 25°C

	Cable Jacket	Polyester Buffer Tube	Secondary Coating	Primary Coating	Central Member (CM)	Glass Fiber
CTE $(°C)^{-1}$	2.5×10^{-4}	1.3×10^{-4}	1.0×10^{-4}	$1\text{-}3 \times 10^{-4}$	1.0×10^{-6}	1.0×10^{-7}

Particularly noteworthy in Table 3.1 is that the CTE of the buffer tube is three orders of magnitude greater than the glass fiber CTE and two orders of magnitude greater than the cable central member CTE. In addition, the buffer tube CTE does not have a single value over the telecommunications service temperature range while the Cites for the glass fiber and the CM are independent of temperature. As all thermoplastic polymeric materials do, buffer tubes exhibit three distinct states in their thermal and mechanical behavior: the glassy state, the glass transition region and the rubbery state. One parameter that determines the state of the buffer tube thermomechanical response is its glass transition temperature T_g. As illustrated in the modulus-temperature profile for a PBT (Polybutylene terephthalate) buffer tube in Fig. 3.3, the region labeled with T_g separates the glassy state where buffer tube is rigid and glass-like from the rubbery state where it is softer and more flexible. The buffer tube modulus is significantly lower at temperatures above its glass transition temperature. Thus, the PBT buffer tube for which Fig. 3.3 shows its modulus-temperature profile is rigid and glassy at an ambient temperature of 25°C since its T_g is 60°C. However, this material will turn rubbery soft when its temperature rises above 60°C. Fig. 3.3 also shows the modulus-temperature profile of a recently introduced flexible buffer tube made of PP (Polypropylene) and/or its copolymers with PE (Polyethylene). The modulus of this material at 25°C is an order of magnitude lower than the modulus of the PBT buffer tube material; it turns glassy rigid at temperatures below 0 °C. This flexible material does not exhibit as distinct a glass transition region as the PBT buffer tube material. It softens over a broad temperature range from 0°C to well above 60°C. One order of magnitude lower modulus at 25°C for this material translates into the perception of flexibility and craft-friendliness in field installations of loose tube cables.

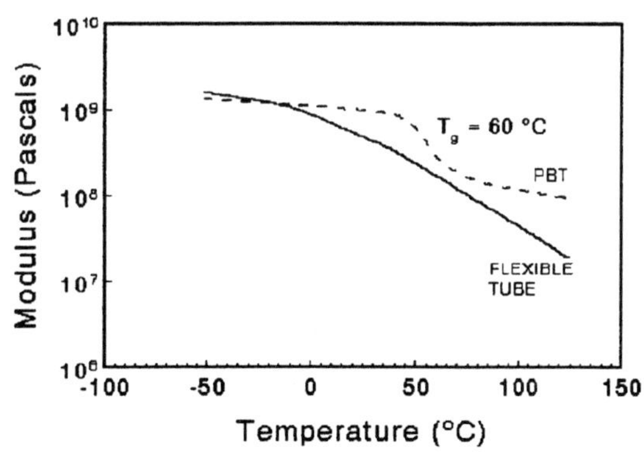

Figure 3.3 A comparison of modulus -temperature profiles of two buffer tube materials

Thermal expansion measurements by high-resolution dilatometry indicate that the PBT buffer tube thermal response is also strongly temperature-dependent and particularly complex above its T_g as shown by the thermal expansion profiles of Fig. 3.4. In its simplest case, a thermoplastic material exhibits a glassy and a rubbery thermal expansion behavior. In each state, thermal expansion is described by a single coefficient. However, for PBT buffer tube materials, the CTE is not only temperature-dependent in the glassy and rubbery states, but it is also time-dependent since it changes with thermal cycling.

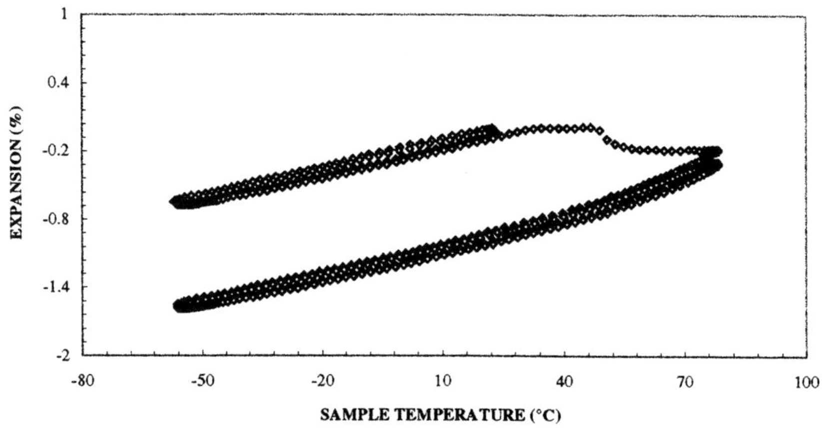

Figure 3.4 Expansion/contraction profiles for PBT buffer tubes during thermal cycling

Typically, the rubbery CTE is three to four times larger than the glassy CTE. Hence, thermally-induced strain in a PBT buffer tube increases suddenly upon a temperature rise through its T_g.

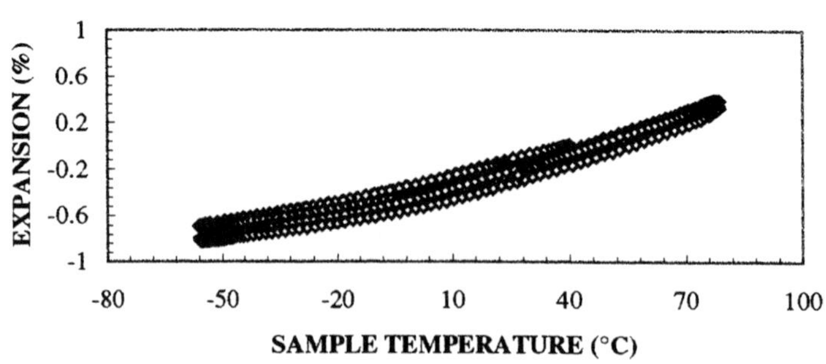

Figure 3.5 Thermal expansion/contraction profile of a PP buffer tube during thermal cycling.

The PP buffer tube, however, already exhibits a large rubbery CTE. Consequently, it will undergo uniformly large thermally-induced strain as the temperature of the service environment changes. Therefore, a loose tube fiber optic cable designer must take this potentially large displacement of flexible buffer tubes relative to the glass fiber and the CM into consideration for minimizing the risk of temperature-induced loss. The thermal expansion/contraction profiles indicate that the thermal response of PBT buffer tubes involves significant displacements that are associated with post-processing (annealing) shrinkage. Although it generates about 0.2 to 0.5 % thermal strain, the post-processing shrinkage is a major part of the total thermally-induced buffer tube strain, and as such it should affect the magnitude of thermally-induced buffer tube displacements that lead to microbending loss. This post-processing dimensional instability in buffer tubes provides an important clue to our understanding of a field and laboratory observation on the need for high-temperature aging prior to the appearance of the optical loss at lower temperature. Simply stated, a PBT buffer tube's dimensional change is much larger when its temperature rises above its T_g. For the PP buffer tube, the dimensional change associated with the post-processing relaxation of residual stresses should be very small since this material remains rubbery soft and, perhaps, never becomes glassy to contain frozen-in processing-induced stresses. Our high-resolution thermal expansion measurements on PP buffer tube materials confirm that net shrinkage at the conclusion of the thermal cycling program, given in Table 3.2, is large in PBT buffer tubes as shown in Fig. 3.4 compared to PP buffer tubes as shown in Fig. 3.5.

Table 3.2 Thermal Cycling Program in the Dilatometry of Buffer Tubes

Step	1	2	3	4	5	6	7	8	9
Furnace Temperature (°C)	-60	25	-60	80	-60	80	-60	80	25

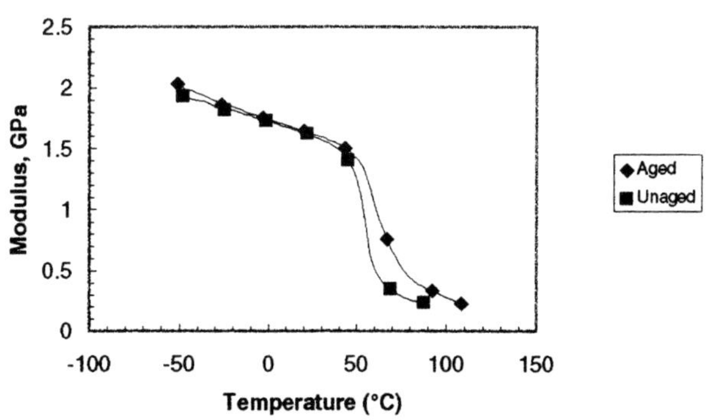

Figure 3.6a Effect of aging @ 85°C in a cable gel on storage modulus (E′) profiles for PBT.

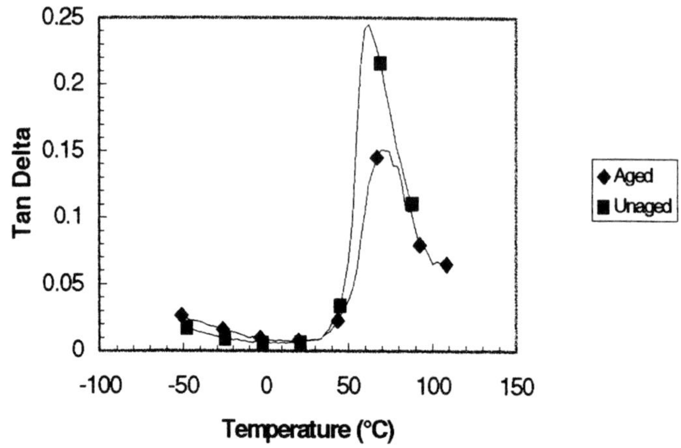

Figure 3.6b Effect of aging @ 85°C in a cable gel on loss tangent profiles for PBT.

Although net thermal cycling shrinkage is small in PP compared to the net shrinkage of PBT, our detailed laboratory analyses point out that long-term exposure to some buffer tube gels used in fiber optic cables may adversely affect buffer tube net thermal cycling shrinkage. This, in turn, may lead to a heightened risk of low-temperature-transmission loss in loose tube cables. Figures 3.6a and 3.6b show the effects of aging in an ETPR (Extended Thermoplastic Rubber)-based cable gel on the dynamic mechanical properties of PBT buffer tubes.

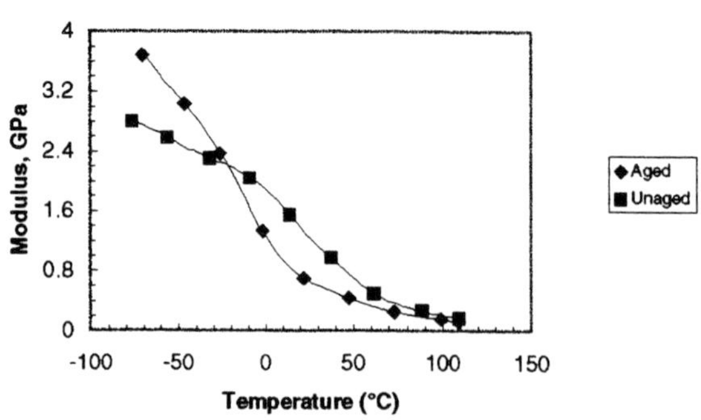

Figure 3.7a Effect of aging @ 85°C in a cable gel on storage modulus (E′) profiles for PP.

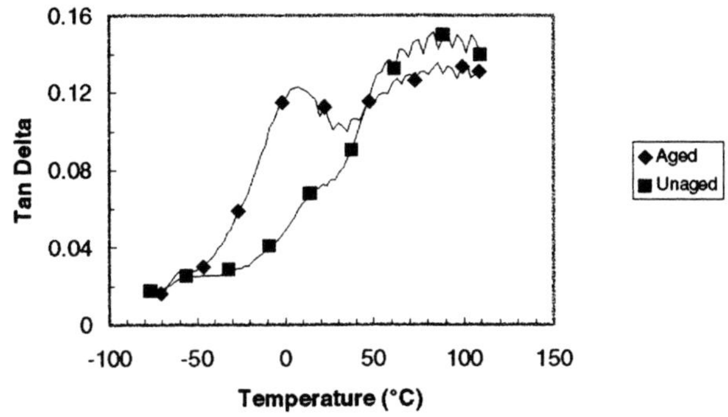

Figure 3.7b Effect of aging @ 85°C in a cable gel on Loss Tangent profiles for PP.

While two-week exposure to the cable gel raises the T_G by only a few degrees and broadens the glass transition region for PBT, it leads to a much more dramatic effect on the properties of PP buffer tubes as shown by the dynamic mechanical spectra of Figs. 3.7a and 3.7b. As storage modulus profiles of Fig. 3.7a and loss tangent profiles of Fig. 3.7b show, exposure of PP to the cable gel lowers the T_G and results in the plasticization of the buffer tube material. Table 3.3 presents a summary of the effects of aging under three different conditions on the mechanical properties of PBT and PP buffer tubes.

Table 3.3 T_g(°C) & Tan δ Peak Height @ 1 Hz

	PBT	**PP**
Before Aging	62 & 0.21	21 & 0.072
85°C/85% RH	71 & 0.13	17 & 0.072
85°C/Uncont. Humidity	71 & 0.13	15 & 0.075
85°C in Gel	69 & 0.14	7 & 0.13

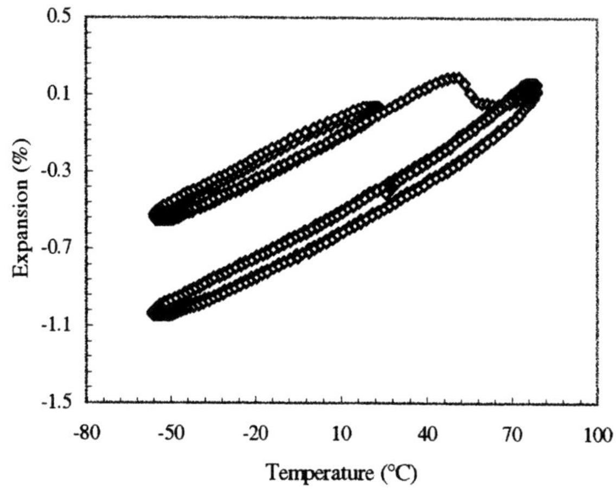

Figure 3.8a Thermal expansion profile of PBT after 30-day aging @ 85°C in a cable gel.

The plasticization effect suggested by the changes observed in the dynamic mechanical properties of
PP is confirmed by the thermal expansion/contraction profiles of Figs. 3.8a and 3.8b. While 30-day exposure to the cable gel does not result in any appreciable change in the overall thermal cycling response and net shrinkage of PBT as shown in Fig. 3.4, it does significantly affect the net shrinkage of PP as indicated by the thermal expansion profile of Fig. 3.8b. Table 3.4 compares net shrinkage for PBT and PP for three different aging conditions and shows clearly that PP becomes dimensionally unstable in general and in the cable gel, in particular. However, temperature cycling experiments in progress with loose tube cables that are made of PP buffer tubes have not shown yet any service-affecting transmission loss. This observation underscores the importance of cable structure as the second factor in the control of low-temperature performance of loose tube cables.

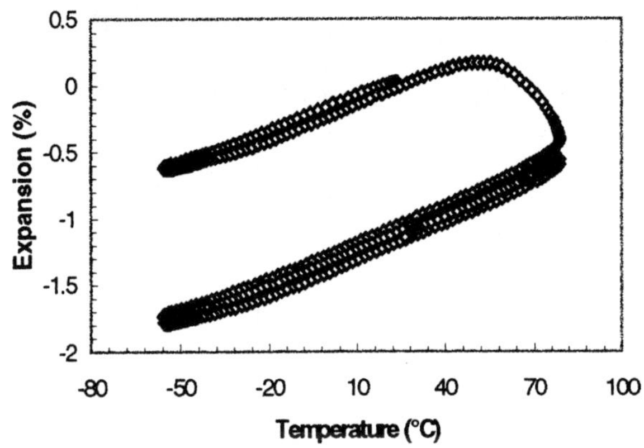

Figure 3.8b Thermal expansion profile of PP after 30-day aging @ 85°C in a cable gel.

Table 3.4 Net Shrinkage (%) @ 25°C & -60°C

	PBT	PP
Before Aging	-0.94 & -1.58	-0.25 & -0.82
85°C/85% RH	-0.07 & -0.72	-0.44 & -0.98
85°C/Uncont. Humidity	-0.10 & -0.76	-0.48 & -1.04
85°C in Gel	-0.35 & -0.99	-0.91 & -1.56

3.3.2 Cable Structure

In our earlier reports, we presented a cable mechanical model to determine the length of "a buffer tube slip region" through which thermally-induced buffer tube displacements occur from the cable ends. This model was based on a (temperature-independent) single value of the glassy PBT coefficient of thermal expansion, and, therefore, it represented a lower bound case for buffer tube slippage within a cable. This analysis provided "slip region length (SRL)" (also called coupling length in earlier reports) as an important parameter of cable structure tightness. The SRL decreases as the cable structure becomes tighter. The SRL parameter is calculated from our cable mechanical coupling model by inputting an experimentally determined pull-out force for the cable CM. Table 3.5 presents a comparison of calculated SRL values against measured 1550-nm optical loss at -40°C for the test cables.

Table 3.5 Optical Loss Performance vs. Cable Parameters

Cable	1A	1B	2A	2B	3A	4A	4B	5A	5B
Slip Region Length (m)	7.2	1.5	2.2	0.3	5.9	1.5	1.8	8.8	16.8
EL (% mm^{-1})	0.4	0.2	0.3	0.1	0.4	0.1	0.1	0.4	0.8
Loss dB (1550 nm -40°C)	11	0.2	7.8	0.03	26	0.03	0.1	Field Failure	Field Failure

This table shows that cables with a SRL value greater than 2 meters exhibit high optical loss. Cables 5A and 5B that failed in the field exhibited the highest SRL values determined to date. Including all the sample cables and field cables analyzed to date, the threshold SRL for TICL to appear is about 3.7 meters or 12 feet. Tightly-structured cables with low SRL values give low optical loss at low temperature. In Table 3.5, we also show the measured values of a second important cable parameter; this is the excess fiber length-to-buffer tube inner diameter (ID) ratio, EL, for each cable. We determine EL by directly measuring the fiber length in 5-foot cable sections.[48]

4. FIBER-BASED COMPONENTS

The fiber-based components considered in this section include optical connectors, fusion splices, and fiber components using a fused-tapered fiber design (such as couplers/splitters/WDMs). The optical fiber used in these components is particularly vulnerable to damage and potential long-term reliability problems because of the actions necessary to fabricate the components: stripping, cleaning, cleaving, and fusing, along with other handling activities. Minimization of mechanical damage is critical during the fabrication process, yet some fiber strength reduction of the fiber is inevitable. Because of this, long-term component reliability is ultimately dependent on the packaging. Well-designed packaging will protect the fiber from stress, so that opportunities for crack growth and subsequent fiber breaks can be avoided. On the other hand, poorly-designed packaging can add additional stresses to fibers, resulting in premature fiber failures. In addition to protecting the fibers from mechanical and environmental stresses, the packaging must also provide submicron alignment over the service life of the component in the case of connectors. Finally, in addition to packaging, there are some additional materials issues to be considered, such as the susceptibility of fibers in couplers to water diffusion.

In spite of all of these issues, the reliability of fiber-based components has in general been excellent, and continues to improve. However, the demands for system performance also continue to increase. For example, high-performance analogue video systems require reflectances lower than –55 dB, DWDM systems will require better wavelength stability than ever before, and optically amplified systems will typically result in higher optical powers being transmitted than in the past, with unknown effects on the long-term fiber component reliability. In addition, the growth of new fiber optic applications accelerates the deployment of the components into outside plant environments, where the stresses that the packages must withstand are greatly increased. Certainly, materials reliability will remain a challenge into the foreseeable future.

4.1 Connectors

A number of material characteristics are critical to optical connector performance and reliability. Figure 4.1 shows a "generic" connector, and illustrates some of the key features common to most connectors that make up today's embedded base in the telecommunications network. These include ST™, FC, and SC type connectors. Each type has a different housing design and latching mechanism, but the basic design and alignment method are the same for all. All of these types consist of an optical fiber that is fixed into a ceramic (usually zirconia) ferrule using an adhesive (usually a two-part thermally-cured epoxy). The ferrule is fixed to a metal or plastic base that is spring-loaded into the outer housing, with a typical spring force of about 0.9 kg (2 lb). The tip of the ferrule is polished to give a radiussed surface, with the fiber core ideally centered on the dome of the polished surface. Two issues which impact the stability and reliability of the package have to do with the effects of the adhesive (glue) and any crimping action on the surface flaws present in the original fiber, or developed during the stripping and/or other handling operations. Figure 4.2 presents an illustration of tensile stresses that develop in the fiber (the 'glue spall' stresses) as first described by Gulati and Hagy.[49] These stresses are the result of the high-CTE adhesive acting on the cracks in the low expansion silica glass. Another potential reliability problem may develop if a'crimp' or crimping style component is used. This will result in the development of Hertzian contact stresses as illustrated in Fig. 4.3.[50] In both of these cases, the stresses may be estimated. In the case of the glue spall stresses, the epoxy plays a central role in the development of the stresses while Hertzian contact stresses may be reduced by the presence of a suitable polymer layer.[51]

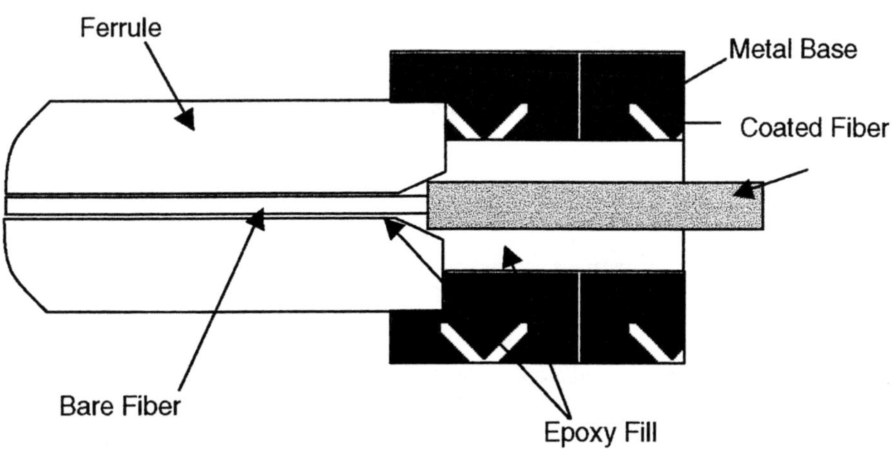

Figure 4.1 Generic connector. The ferrule and base are filled with adhesive, and the fiber is stripped, cleaned, and inserted into the ferrule. The adhesive is cured and then the fiber/ferrule tip is polished to give a radiussed surface.

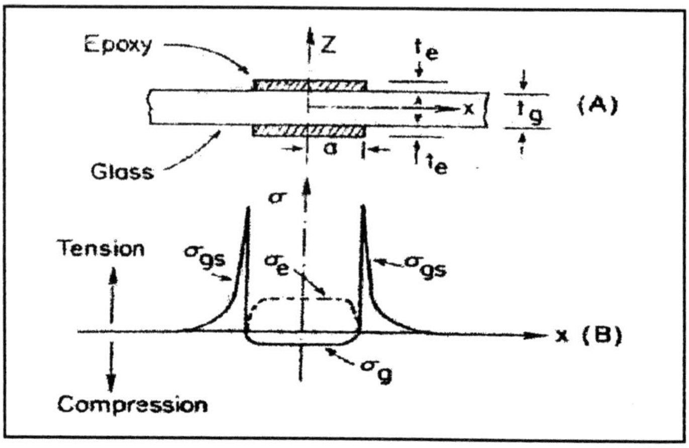

Figure 4.2 "Glue Spall" stresses in an adhesive joint. A) Seal geometry and coordinate system, B) Schematic of stress distribution.

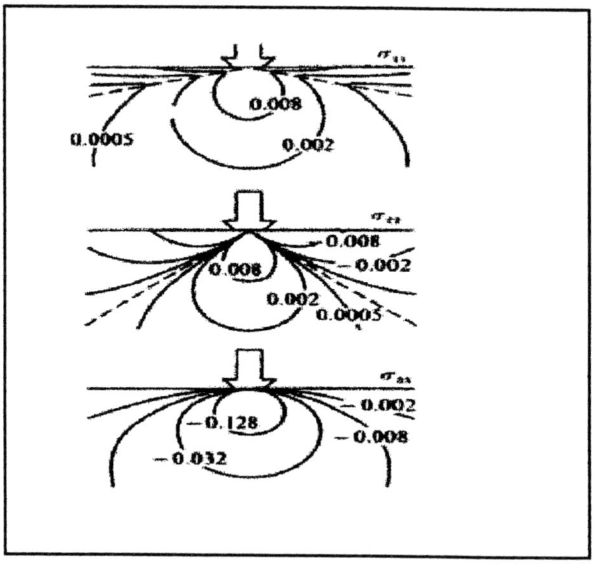

Figure 4.3 Cross-section of normal contact stresses

4.1.1 Structural Support and Strain Relief

The adhesive is critical to connector performance and reliability to provide structural support and strain relief for the stripped section of the fiber. It is important that the adhesive fill be complete and that the fiber be inserted into the ferrule so that as little of the bare fiber be exposed behind the ferrule as possible. If the bare fiber extends out behind the ferrule, and particularly if the adhesive fill is not complete, the fiber can experience bending or buckling stresses in this region that have been shown to result in fiber breaks.[52,53,54,55,56,57]

4.1.2 Optical Performance

The details of the geometry of the polished surface are critical to the optical performance of the connector, and materials stability over the lifetime of the connector is key to the reliability. Figure 4.4 shows the details of the connector endface geometry. The endface geometry is defined by three parameters: the radius of curvature (measured on the ferrule), the apex offset (the offset of the center of the spherically-polished surface relative to the optical axis), and the fiber protrusion/recess (the position of the fiber relative to the spherical surface defined by the ferrule).

Ideally, the fiber is polished flush to the spherical surface, and the spherical surface is centered on the optical axis. However, the more realistic situation is shown in Fig. 4.4, where there is some apex offset and fiber undercut, which result in a finite gap when two connector plugs are mated.[58,59] Due to the spring-loading force and the compliance of the fiber/ferrule materials, some deformation of the endface occurs when the connectors are mated, so that a finite gap may be closed.[60,61] The intent is that the optical fibers make good physical contact, and that the contact is maintained over the service life of the connector. The impact of the loss of physical contact on the optical performance is shown in Figures 4.5 (a) and (b). If a gap opens up at the interface between the two connecting fibers, interference effects due to multiple reflections of light in the gap can cause an increase in the insertion loss and reflectance. As shown in this figure, increases in the insertion loss of up to 0.6 dB worst case, and reflectances as large as –8 dB are possible, for gap sizes of only a few tenths of a micron.[62]

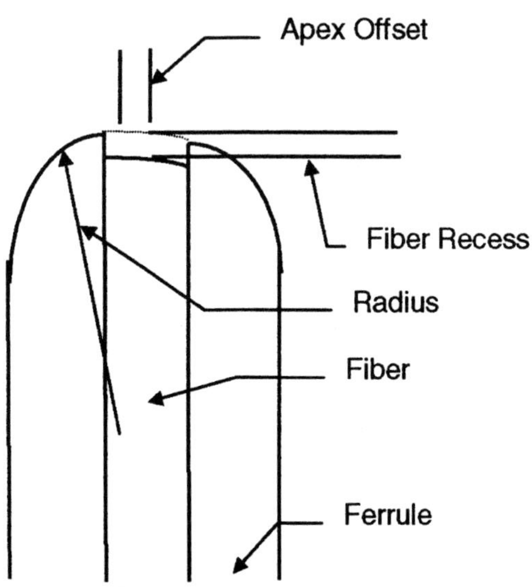

Figure 4.4 Connector endface geometry. apex offset, radius of curvature, and fiber protrusion/recess.

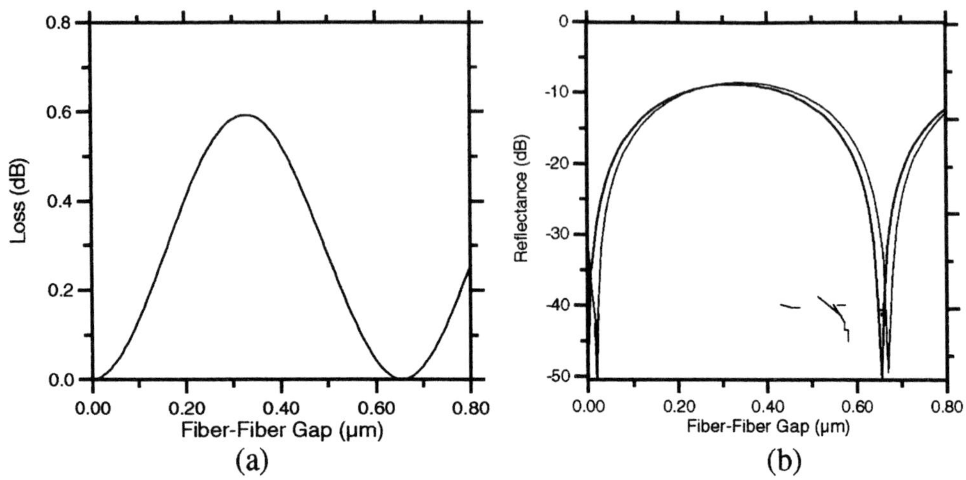

Figure 4.5 (a) Loss and (b) reflectance across a connector interface as a function of the gap size at 1310 nm.

Material dimensional stability is critical to maintaining physical contact. There are three types of material degradation that have been shown to affect the dimensional stability in optical connectors.[63] These include epoxy creep,[64] permanent ferrule deformation,[65,66] and fiber dissolution.

When connectors are mated under normal spring-loading forces, the endface deformation results in a sharing of the load across the fiber/ferrule/adhesive interface. Figure 4.6 shows the typical loading force of the fiber relative to the total loading force for different radii of curvature.[67] The loading force stresses the adhesive interface. Adhesives will creep under load, particularly at elevated temperatures, and this can result in fiber pushback into the ferrule over time. Figure 4.7 illustrates fiber pushback into the ferrule when connectors are aged under load. Long-term tests have shown that most fiber pushback occurs over the first few hours after the application of the load.

Temporary fiber withdrawal can also occur due to fiber pistoning with temperature. This effect is normally attributed to the differences in the coefficients of thermal expansion between the ceramic and fiber, and theoretical analyses based on these assumptions have typically given good agreement with the experimental measurements.[68] However, an alternative theory, based on radial compressive forces has recently been suggested.[69] The initial endface geometry parameters must take into account all of the sources of fiber motion during the connector lifetime and provide enough safety margin in the design to ensure physical contact over the full operating range and take into account some limited amount of fiber pushback.

The permanent ferrule degradation is believed to occur when the zirconia ceramic degrades from a tetragonal crystalline structure to a monoclinic structure.[70,71] Zirconia ferrules are typically stabilized with yttria to optimize strength and fracture toughness. However, this structure can destabilize at high temperatures and humidities. This is believed to result in both ferrule deformation (flattening) across the contact area at the tip of the ferrule and surface roughening. The ferrule degradation can result in increases in the frictional forces between the ferrule and the alignment sleeve inside the adapter. In the worst case, this can affect the ability of the connector plugs to make physical contact when they are inserted into the adapter. Small grain size of the ceramic is one of the keys to minimizing the degradation.[72]

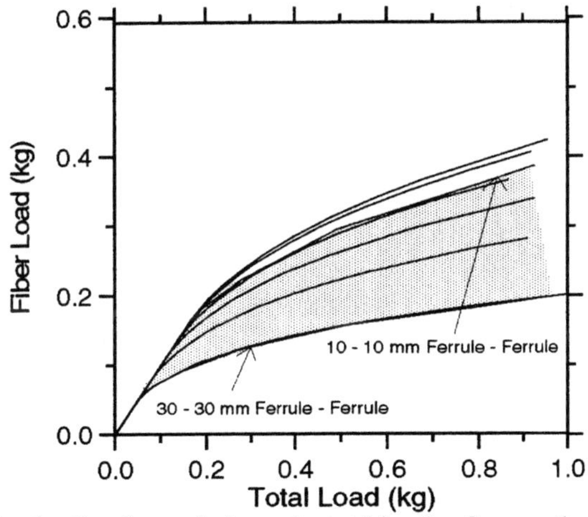

Figure 4.6 Fiber loading force relative to the total loading force on the ferrule endface.

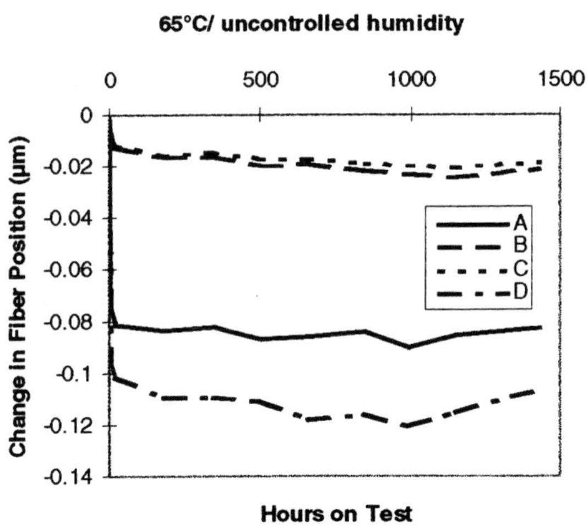

Figure 4.7 Fiber pushback for four epoxies, for a 0.9-kg radiussed load applied at 65°.

Figure 4.8 shows the rate of degradation of different zirconia ceramic ferrule materials.[73] The degradation factor is the difference between the final and initial radius of curvature (before and after testing) divided by the product of the final and initial radii of curvature. Most products tested have been found to be quite stable, but it is important to be aware of the potential problem, particularly for outside plant applications, so proper materials selection can be made.

Some new ferrule materials are appearing on the horizon. For example, glass-ceramic ferrules have recently become commercially available. Glass ceramic ferrules should in principle not suffer the same environmental degradation as zirconia ferrules, and in addition, they should be significantly less expensive, and be easier to polish.[74] It should be noted that the current endface geometry specifications commonly accepted by the industry are all based on a zirconia ceramic ferrule. If different materials are used for ferrules, the requirements for the endface geometry parameters would need to be determined for that particular material. In addition, intermateability issues, such as differences in the loading forces experienced by the fibers inside the ferrule, should be investigated.

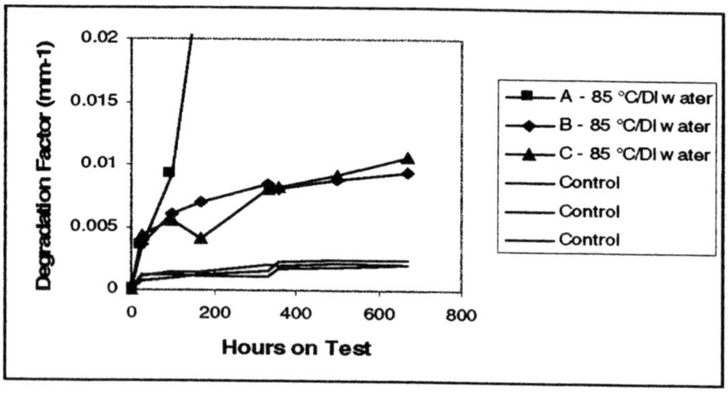

Figure 4.8 Degradation rate for different zirconia ferrules.

Finally, fiber dissolution has been observed as a source of connector failure. Figure 4.9 illustrates the effect. It is a scan of the fiber surface taken using atomic force microscopy (AFM), and shows that the core of the fiber (in this case, the fiber has a pure silica cladding and a germanium-doped core), is preferentially dissolved during aging in 85°C de-ionized water. Although the fiber core is dissolved at a higher rate, it is believed that the cladding is also being dissolved.[75]

The circumstances under which the fiber dissolution occurs are still not fully understood. It is known that connector plugs that are soaked in water with certain metal housings all show severe damage after less than a week of soaking at 85°C. The evidence points to the possibility of certain metal contaminants as the source of fiber degradation. X-ray analyses of metal constituents of the optical connectors have shown a variety of metals to be present. However, the data currently available do not confirm that any particular metal is the cause of the dissolution. At this point no clear conclusions as to the mechanism can be given. Although this effect is not expected to occur often in the field, the severity of the effect when it does occur gives rise for concern, and a need to understand the underlying mechanism.

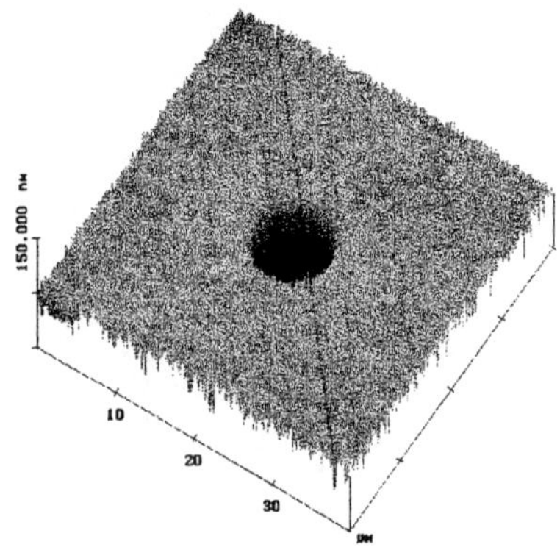

Figure 4.9 Fiber dissolution of the fiber core in an optical connector after aging.

4.2 Fused-Tapered Fiber Components (Couplers/Splitters/WDMs)

The reliability of branching components in upcoming and rapidly growing systems applications such as DWDM may be even more critical than for connectors and splices because of the multiple data streams travelling through a single component. Not only are the components more expensive to replace than single connectors or splices if failures occur, but two or more fibers must be taken out of service in order to implement the repair. Perhaps due to the critical nature of these components, couplers are one of the few types of passive optical components for which some accelerated testing and reliability estimates have been completed.[76,77,78] The laboratory testing undertaken for these studies showed that the estimated FIT rates for couplers were typically 10 to 100 times higher than the targeted FIT rates of 50.[79]

Most of the laboratory failures were due to materials-related problems, including fiber breaks, and adhesion failure. Typically, the couplers/WDMs are mounted to a substrate using an adhesive, and then the substrate is mounted in a package, and the fibers are covered by protective tubing and connectorized. The fused-tapered components themselves are fabricated by stripping and cleaning two or more fibers, wrapping them around one another, and then drawing them out and fusing them at the same time under controlled conditions until the desired optical characteristics are achieved.[80] It is important to avoid applying mechanical stresses to the couplers/WDMs because such stresses can change the coupling characteristics.[81] Figure 4.10 shows the performance of one coupler during temperature cycling in laboratory testing. In this case, the protective tubing used to cover the fiber pigtails was contracting at low temperatures and causing bending-induced losses in the fibers. Packaging problems are the most common cause of failures for these types of components.

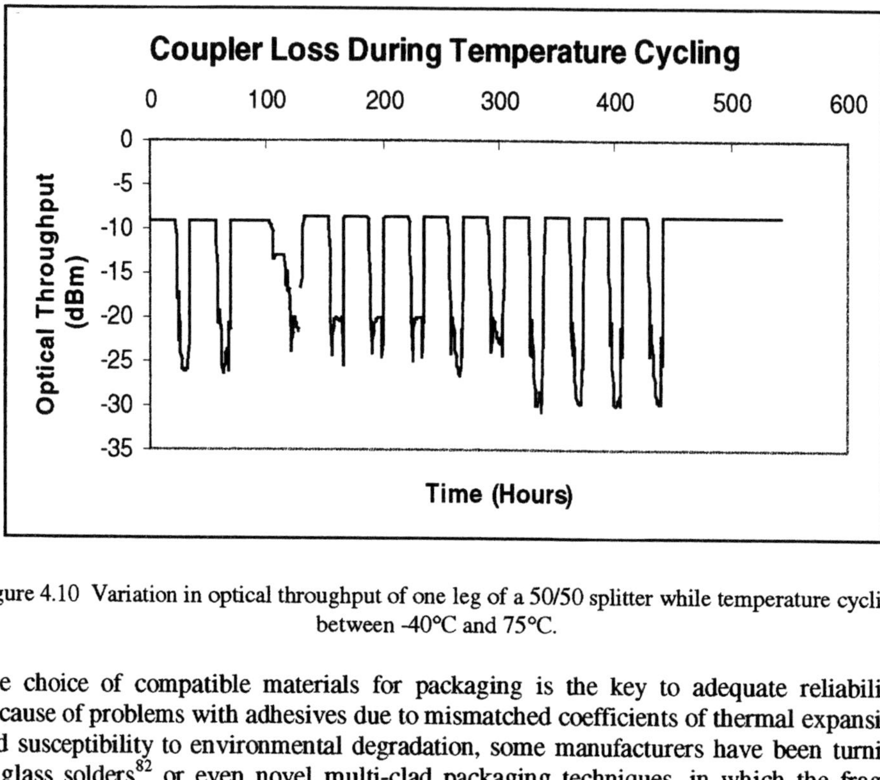

Figure 4.10 Variation in optical throughput of one leg of a 50/50 splitter while temperature cycling between -40°C and 75°C.

The choice of compatible materials for packaging is the key to adequate reliability. Because of problems with adhesives due to mismatched coefficients of thermal expansion and susceptibility to environmental degradation, some manufacturers have been turning to glass solders[82] or even novel multi-clad packaging techniques, in which the fragile coupler is completely encased in a composite wrap that utilizes a glass ceramic.[83] The use of these new materials and processes should enhance product performance and reliability.

Although most reliability issues related to couplers are extrinsic, one potential failure mechanism has been identified that is inherent to the silica material itself. This is the diffusion of water into the silica and subsequent changes in the coupling ratio.[84,85] Couplers were aged in 85% relative humidity environments at various temperatures. When a failure was defined as a 0.5 dB change in the insertion loss of the coupler, the predicted mean time to failure was over 1500 years. Therefore, this failure mechanism, although of theoretical interest, is not expected to be significant.

SUMMARY & FUTURE DIRECTIONS

Since the first demonstration of optical fiber telecommunications by R. Maurer, D. Keck and P. Schultz (Corning, Inc.) in 1970, tremendous progress has been made in the development and use of fiber optic materials, components and systems to meet the rapidly growing demand for high-speed and and high-capacity communications worldwide. While the silica optical fiber manufacturing technology has had continuing refinements for producing low-loss optical fibers at very high (exceeding 10 meters /sec) fiber drawing speeds, many cabling and interconnection technologies have been developed to take advantage of the transparency of glass optical fibers to 1310-nm and 1550-nm lightwave signals for high-bandwidth telecommunications applications. Although many types of optical fibers and cables have been developed to meet the needs of communications service providers for long-term performance and reliable operation, the brittle nature, aging and fatigue of silica optical fiber remains to be the key materials issues that need to be dealt with in order to accelerate the deployment of lightguide fibers in telecommunications networks. Although the developments in manufacturing technologies have resulted in significant price reductions for optical fiber, cable and components, the cost of installation and maintenance of fiber optic networks is still not fully competitive with other alternative technologies. In local area networks, the development of plastic optical fiber with the promise of greater than 10 GB/s transmission rates over 100 meters offers labor cost advantages with simpler and more robust installation over both copper wire and multimedia silica optical fiber. In fiber optic cabling technologies, the compatibility and dimensional stability of cable materials will remain to be the key issues for achieving cost reductions in product manufacturing and field installation/maintenance while maximizing long-term performance and reliability. In fiber-based components, the challenges of the brittleness of silica glass, compatibility and dimensional stability of connector ferrule materials, adhesives, and other packaging components need to be dealt with simultaneously. Although we may consider today's fiber optic technologies to be mature, we should bear in mind the fact that all materials issues discussed in this review article will remain to be of central importance in maintaining the explosive growth of high-density, high-capacity broadband fiber optic communications networks into the future.

ACKNOWLEDGMENTS

The authors gratefully acknowledge the contributions of I.M. Plitz, G.T. Aich, P.B. Grimado, M. Guthrie, R.M. Kanen, G.D. Kiss, M.Koza, M.J. Kozlowski, J.D. Mann, M. Ozgur, R. Schubert, D. Unkel, C.J. Wieczorek, and M.J. Zammit. Many helpful discussions with R.A. Frantz and F. P. Kapron are also appreciated.

REFERENCES

[1] "Advances in Fiber Design and Processing", D.J. DiGiovanni, D.P. Jablonowski, M.F. Yan, Chap. 4 in "Optical Fiber Telecommunications – IIIA" Ed. I.P. Kaminow and T.L. Koch, Academic Press, San Diego, p 63, 1997.

[2] "Advances in Cable Design", K.W. Jackson, T.D.Mathis, P.D. Patel, M.R. Santana, P.M. Thomas, Chap. 5 in "Optical Fiber Telecommunications – IIIA" Ed. I.P. Kaminow and T.L. Koch, Academic Press, San Diego, p 92, 1997.

[3] C.R. Kurkjian and M.J. Matthewson, Elec.Letts., **32**, p 761, 1996.

[4] C. R.Kurkjian, J.T. Krause and U.C. Paek, J. Phys. Colloq., **43**, C9-585, 1982.

[5] C.R.Kurkjian, R.A. Frantz, Optics and Photonics News, **7,** pp 35-39, 1997.

[6] M.J. Matthewson, H.H. Yuce, SPIE Proc., **2290**, 204, 1994.

[7] J.L. Armstrong, M.J. Matthewson, C.R. Kurkjian and C.Y. Chou, IWCS Proc., **46**, 902, 1997.

[8] O.S. Gebizlioglu, C.R. Kurkjian, MRS Symposium Proc., **531**, p 301, 1998.

[9] 5M.J. Matthewson, C.R. Kurkjian and S.T. Gulati, J. Am. Ceram. Soc., **69**, 815, 1986.

[10] O.S. Gebizlioglu, I.M.Plitz, P.B. Grimado, and R.A. Frantz, SPIE Proceedings, **2290,** 7 1994.

[11] O.S. Gebizlioglu, I.M. Plitz, "In-Situ Characterization of Fiber Coatings for Post-Mortem Analyses and Reliability Investigations", NFOEC Proceedings, **Vol.2**, 113, 1996.

[12] C.J. Koester and E. Snitzer, *Appl. Phys. Lett,* **3**, 1182, 1964.

[13] R.J. Mears, L.Reekie, I.M. Jauncie and D.N. Payne, El. Lett. **23**, 1 26, 1987.

[14] E. Desuvire, J.R. Simpson and P.C. Becker, Op. Lett, **12,** 888, 1987.

[15] M. Nakaawa, Y. Kimura and K. Suzuki, Appl. Phys. Lett, **54**, 295, 1989.

[16] P.C. Becker, N.A. Olson, J.R. Simpson, "Erbium-Doped Fiber Amplifiers – Fundamentals and Technology", Academic Press, San Diego, 1999.

[17] W.R. Wagner, "Failure Analysis of Fiber Optic Connectors" in "Advances in Ceramics", **22**, American Ceramic Society, Columbus, Ohio, 1988.

[18] S.L Semjonov, G.S.Glaesemann, C.R.Kurkjian, M.M. Bubnov, "Modeling of Proof Test Level Flaws using Cube Corner Indents", Proc. IWCS, p 928, 1998.

[19] B. Lin and M.J. Matthewson, "Inert Strength of Subthreshold and Post-threshold Vickers Indentations on Fused Silica Optical Fibers", Phil, Mag. A., **74**, 1235, 1996.

[20] E.H. Urruti, J.W. Botelho, M.A. Botelho, 11th NFOEC Proc., **Vol.1,** p 225, 1995.

[21] A. Dwivedi, J.L. Smith, IWCS Proc., p 381, 1996.

[22] J.D. Mann, M.R. Ozgur, O.S. Gebizlioglu, C.R. Kurkjian, 14th NFOEC Proc., **Vol.1,** p 271, 1998.

[23] B.G. LeFevre, "Reliability of Passive Optical Components", Critical Reviews of Optical Science & Technology, **CR50**, SPIE Optical Engineering Press, Bellingham, Washington, p 105, 1993.

[24] G.D. Kiss, "Reliability of Optical Splices", Critical Reviews of Optical Science & Technology, **CR50**, SPIE Optical Engineering Press, Bellingham, Washington, p 118, 1993.

[25] J.T. Krause, C.R. Kurkjian, "Fiber Splices with Perfect Fiber Strengths of 5.5 GPa, $\nu <$ 0.01", Elec. Letts., **21**, 533, 1985.

[26] C.R.Kurkjian, J.D. Mann, O.S. Gebizlioglu, SPIE Proc. **Vol. 3848**, in press, 1999.

[27] L.A. Reith, H.H. Yuce, P.B. Grimado, Proc. of the Intern. Symp, Fiber Optic Networks and Video Comm., p. 316, 1993.

[28] J.T. Krause and D. Stroumbakis, SPIE Proc. on Opt. Ntwk. Eng. and Integrity, **Vol. 2611**, p. 98, 1996.

[29] G. D. Kiss and E. Vogel, 12th NFOEC Proc., **Vol.3**, p. 819, 1996.

[30] G.H. Beall, K.Chung and J.E. Pierson, "Negative CTE β-Eucryptite Glass-Ceramics for Fiber Bragg Grating Athermalization", 18th International Conference on Glass, San Francisco, 1998.

[31] R. Feced, M.P. Roe-Edwards, S.E. Kanellopoulos, N.H. Taylor and V.A. Handerek, Elect. Letts, **33(2)**, 157, 1997.

[32] C.G. Askins, M.A. Putnam, H.J. Patrick and E.J. Friebele, Elect. Letts, **33(5)**, 1333, 1997.

[33] R. Filas, "Metallization of Silica Optical Fibers", MRS Symposium Proc., **Vol. 531**, p. 263, 1998.

[34] P. Simpkins, C.R. Kurkjian, C.M. Schroeder, "Aluminum-coated silica fibers: strength and solderability", *Electronics Lett.*, **31**, 747, 1995.

[35] GR-20-CORE, *"Generic Requirements for Optical Fiber and Fiber Optic Cable"*, Telcordia Technologies, Inc., Morristown, New Jersey, Issue 2, July 1, 1998.

[36] GR-409-CORE, *"Generic Requirements for Premises Fiber Optic Cable"*, Telcordia Technologies, Inc., Morristown, New Jersey, Issue 1, May 1, 1994.

[37] O.S.Gebizlioglu, I.M. Plitz, "Self-Stripping of Optical Fiber Coatings in Hydrocarbon Liquids and Cable Filling Compounds," Optical Engineering, **Vol. 30 (6)**, 749, 1991.

[38] O.S. Gebizlioglu, I.M. Plitz, "Monitoring Accelerated Aging of Polyester Buffer Tubes in Fiber Optic Cables," IWCS Proceedings, **Vol.** 42, p 509, 1993.

[39] B.G. Risch, J. Auton, O. Tatat, "Advanced Impact Polypropylene Buffering Material For Superior Long-Term Hydrolytic and Thermo-Oxidative Stability," IWCS Proc., **Vol. 47**, 698, 1998.

[40] O.S. Gebizlioglu, "Time- and Temperature-Dependent Material Behavior and Its Impact on Low-Temperature Performance of Fiber Optic Cables", MRS Symposium Proc., **Vol. 531**, p 333, 1998.

[41] O.S. Gebizlioglu, I.M. Plitz, M.J. Zammit, "Temperature-Dependence of Buffer Tube Gel Performance in Loose Tube Fiber Optic Cables," 11th NFOEC Proc., **Vol. 2**, 357, 1995.

[42] G.D. Kiss, "Self-Healing Failures in the Aerial Plant," SPIE Proceedings, (1993).

[43] G.D. Kiss, M. Zammit, O.S. Gebizlioglu, P.B. Grimado, M.J. Hlavaty, C.J. Wieczorek, "Low-Temperature Reversible Splice Loss Accompanied by Cable CM Protrusion," NFOEC Proceedings, **Vol. 2**, 411, 1994.

[44] O.S. Gebizlioglu, G.D. Kiss, M.J. Zammit, P.B. Grimado, C. Cascio, G. Karl, "An Investigation of Temperature-Induced Cable Loss," NFOEC Proceedings, **Vol. 2**, 479, 1995.

[45] P.B. Grimado, O.S. Gebizlioglu, M.J. Zammit, G.D. Kiss, "Low-Temperature Transmission Loss in Loose Tube Fiber Optic Cables," SPIE Proceedings, **Vol. 2290**, 29, 1994.

[46] M.J. Zammit, O.S. Gebizlioglu, P.B. Grimado, G.D. Kiss, "Properties of Loose Tube Optical Cables Contributing to Low-Temperature Optical Loss," IWCS Proceedings, **Vol. 43**, 538, 1994.

[47] O.S. Gebizlioglu, P.B. Grimado, G.D. Kiss, M.J. Zammit, "Low-Temperature Performance of Loose Tube Fiber Optic Cables," SPIE Proc., **Vol. 2611**, 189, 1995.

[48] O.S. Gebizlioglu, "Time- and Temperature-Dependent Material Behavior and Its Impact on Low-Temperature Performance of Fiber Optic Cables", MRS Symposium Proc., **Vol. 531**, p 333, 1998.

[49] S.T. Gulati and H.E. Hagy, J. Am. Ceram. Soc., **65**, p. 1, 1982.

[50] B.R. Lawn, "Fracture of Solids", Cambridge Univ. Press, Cambridge, 1993.

[51] M.J. Matthewson, "Adhesion Measurements of Thin Films by Indentation", Appl. Phys. Lett. **49**, 1426, 1986.

[52] M. Kinney, "Optical Fiber Stress Within a Biconic Connector", *Proc. 7th Biennial Conference on Failure Prevention and Reliability*, Boston, MA, **DE-Vol. 9**, p. 75, 1987.

[53] W.R. Wagner, "Failure Analysis of Fiber Optic Connectors", *Advances in Ceramics,* **Vol. 22**, p. 389, 1988.

[54] J.M. Anderson, A.G. Hardee, W.W. King, and B.G. LeFevre, "Design Considerations Relating to Thermally Aggravated Fiber Stress in a Connector Cavity", *Materials Developments in Microelectronic Packaging Conference Proceedings*, Montreal, Quebec, Canada, p. 225, August 1991.

[55] E. Suhir, "Predicted Curvatures and Stresses in a Fiber-Optic Interconnect Subjected to Bending", *J. Lightwave Tech.,* **Vol. 14**, p. 144, 1996.

[56] W.W. King, B.G. LeFevre, and D.L. Stephenson, "Structural Mechanics Aspects of Fractures of Optical Fibers in Connectors", IWCS Proc., **Vol. 45**, p. 228, 1996.

[57] P. Singh, G. T. Galyon, G.M. Jordhamo, and G.W. Peterson, "Fiber Optic Duplex Connector Design Evolution and Analysis", *Proc. 7th Electronic Materials and Processing Congress*, Cambridge, MA, p.169, August 1992.

[58] T. Shintaku, E. Sugita and R. Nagase "Highly Stable Physical-Contact Optical Fiber Connectors with Spherical Convex Ends", J. Lightwave Tech., **Vol. 11**, p. 241, 1993.

[59] T. Shintaku, R. Nagase, and E. Sugita, "Connection Mechanism of Physical-Contact Optical Fiber Connectors with Spherical Convex Polished Ends", Appl. Optics, **Vol. 30**, p. 5260, December 1991.

[60] E. Deeg, "Effect of Elastic Properties of Ferrule Materials on Fiber Optic Physical Contact (PC) Connections", AMP J. Tech., **Vol. 1**, p. 25, November 1991.

[61] E. Deeg and T. Bolhaar, "Contact Zone and Hertzian Stress in Fiber-Optic Connections with Spherical or Ellipsoidal Fiber Endfaces", AMP J. Tech., **Vol. 2**, p. 29, November 1992.

[62] Special Report "Single-Mode Connector Technology", Telcordia, SR-ARH-002744, Issue 1, August 1994.

[63] L.A. Reith, O.S. Gebizlioglu, M. Koza, J. Mann, M. Ozgur, and T. Bowmer, "The Dimensional Stability of Adhesives, Zirconia, and Silica in Optical Connector Ferrules and their Impact on Optical Performance", MRS Symp. Proc., **Vol. 531**, 65, 1998.

[64] R. Ziebol, H. Roberts, and B. Daniel, "Permanent Fiber Withdrawal in Fiber Optic Connectors", 8th NFOEC Proc., **Vol. 1**, p. 211, 1992.

[65] L.A. Reith, R.A. Frantz, P.B. Grimado, I.M. Plitz, O.S. Gebizlioglu, E.M. Vogel, F. DeRosa, and E. Beebe, "Optical Connector Reliability During High Humidity Temperature Cycling", 11th NFOEC Proc., **Vol. 3**, p. 647, 1995.

[66] "Optical Connector Reliability," S. Bianco et al, COST 246 Workshop, October, 1996.

[67] L.A. Reith, P.B. Grimado, R.A. Frantz, I.M. Plitz, W.W. Wood, and D.A. Dolinoy, "Effects of Fiber Pushback in Ceramic-Ferrule Connectors on Connector Intermateability", 10th NFOEC Proc., p. 225, 1994.

[68] I.M. Plitz, O.S. Gebizlioglu, L.A. Reith, and P.B. Grimado, "Reliability Issues Concerning Thermally-Cured Adhesives in Fiber Optic Connectors", IWCS Proc., **Vol. 44**, p. 519, 1995.

[69] M. Takahashi, "Generating Mechanism of Maintaining Force for Optical Fiber Installed in Ferrule Hole", J. Lightwave Tech., **Vol. 16 (4)**, p. 549, 1998.

[70] L.A. Reith, E.M. Vogel, D. Biswas, F. DeRosa, E. Beebe, and J. Mann, "Reliability of Optical Connectors in Hot, Humid Environments", 12th NFOEC Proc., **Vol. 3**, p. 675, 1996.

[71] E. Makrides-Saravanos, P.J. Huspeni, C. Saravanos, "Ceramic Ferrule Performance under High Temperature and High Humidity Conditions", 13th NFOEC Proc., **Vol. 2**, p. 215, 1997.

[72] E. Makrides-Saravanos, P.H. Huspeni, and C. Saravanos, "Creep Deformation of Zirconia Ceramic: Effect of Composition and Sintering Temperature", 14th NFOEC Proc., **Vol. 2**, p. 291, 1998.

[73] L.A. Reith, M. Koza, J. Mann, R. Nickel, M. Ozgur, and D. Rader, "Ceramic and Epoxy Reliability in Connectors Exposed to Heat and Humidity", 14th NFOEC Proc., **Vol. 2**, p. 279, 1998.

[74] Y. Takeuchi and S. Mitachi, "High-Strength Glass-Ceramic Ferrule for SC-Type Single-Mode Optical Fiber Connector", IEEE Phot. Tech. Lett., **Vol. 9**, p. 1502, November 1997.

[75] L.A. Reith, J. Mann, O.S. Gebizlioglu, C.R. Kurkjian, T. Bowmer, E. Scerbo, and J. Armstrong, "Connector Materials Reliability in Hot, Humid Environments", 13th NFOEC Proc., **Vol. 2**, p. 193, September 1997.

[76] M. Gadonna, M. Redstall, and S. Gundersen, "Reliability for PON Power Splitters", SPIE Proc. on FO Materials and Comp., **Vol. 2290**, p. 170, 1994.

[77] N. Fabricius, "Reliability of Passive Fibre Optic Branching Devices" IEE Colloquim on FO Cable Systems, No. 1994/023, 1994.

[78] D.J. O'Brien, K.W. Murphy, S.M. Robinson, J-L Malinge, W.R. Willson, and F. Jean, "Quantifying the Reliability of Fiber-Optic Branching Devices", 10th NFOEC Proc., p. 285, 1994.

[79] M. Gadonna, T. Draycott, and S. Gundersen, "Results of Accelerated Lifetests on Passive Optical Splitters", Proc. 13th Annual Conf. on Eur. Fibre Optic. Commun. And Networks, p. 65, 1995.

[80] A.J. Stevenson and S.A. Barr, "Fibre-Optic Coupler Fabrication at AOFR", International J. Optoelectron., **Vol. 6**, p. 127, 1991.

[81] K.P. Oakley and M. Corke, "High Reliability Packages for Fused Tapered Components", 11th NFOEC Proc., p. 843, 1995.

[82] H.S. Daniel, D.R. Moore, and V.J. Tekippe "A Glass Solder Process for Packaging Fiber Optic Components", 10th NFOEC Proc., p. 71, 1994.

[83] D.R. Young, A.J. Morrow, K. Gadkaree, D.E. Quinn, "Packaging of High Reliability Couplers", Proc. 44th Elect. Comp. & Tech. Conf., p. 1004, 1994.

[84] C.V. Cryan, J.R. Curley, F.J. Gilham, D.R. Maack, B. Porter, and D.W. Stowe, "Long Term Splitting Ratio Drifts in Singlemode Fused Fiber Optic Splitters", 11th NFOEC Proc., p. 746, 1995.

[85] D.R. Maack, D.W. Stowe, and F.J. Gilham, "Confirmation of a Water Diffusion Model for Splitter Coupling Ratio Drift Using Long Term Reliability Data", 12th NFOEC Proc., p. 559, 1996.

Passive Optical Components

Reliability of fiber optic connectors and splices

Bruce G. LeFevre and Stephen C. Mettler

Lucent Technologies/Bell Labs, Norcross, GA

Interconnection Products Group

ABSTRACT

The understanding of fiber optic connector and splice reliability has progressed in several areas in recent years. Material properties, failure mechanisms, environmental conditions and design parameters critical to performance are generally well known and are being addressed in the technical literature and in standards. Still lacking however, are reliability models for measuring and predicting failure rates. Acceptance tests are frequently used to approve a product for service, but there are no standard life-stress models for predicting failure rates over time. This and related problems are under study by several industry organizations. These topics are addressed in this review.

Keywords: Fiber optic, connectors, splices, reliability, performance, standards, design, materials

1. INTRODUCTION

Fiber optic connector and splice products are sold and installed worldwide in volumes 10 to 100 million units per year. Both technologies have evolved over the last several years to accommodate network trends toward increasing traffic density and migration of fiber closer to the workplace and home. Along with other passive components, they have undergone sufficient changes in application, design and installation practice to raise new reliability questions and cause existing ones to be re-examined. The authors' intent is to provide a timely review of the subject of reliability of these components. Within that context, the following topics will be addressed.

- Technology trends
- Applications
- Designs and materials
- Performance requirements and standards
- Failure mechanisms and analysis
- Reliability models (measurement and prediction)

Although connectors and splices share the common function of joining fibers, segregation of application and design makes it both convenient and instructive to treat the subjects independently.

2. CONNECTORS

2.1 General

Connector assemblies are single or multi- fiber devices consisting of fiber-terminating plugs and adapters (or receptacles). Their function is to provide remakeable connections between fiber ends or between fibers and packaged passive or active devices (e.g couplers, splitters, wavelength division multiplexers, transmitters, receivers) to facilitate network reconfiguration. Since fiber terminating processes require carefully controlled polishing, tuning and measuring operations, they are generally carried out in a manufacturing facility. The resulting products are pre-connectorized cable assemblies (e.g. patch cords, preterminated building cables or preterminated cable panels) or packaged devices with adapters. Connectors are also sold as components which can be assembled in the field with appropriate tool kits and procedures.

Connectors are used in all types of multimode (MM)) and single mode (SM) fiber networks under operating conditions ranging from climate controlled building environments to uncontrolled outside plant conditions[1,2,3]. Some typical functions are:

- Connecting equipment to network in a central office or commercial building
- Cross connects in a central office or commercial building telecommunication closet
- Connecting fibers to outside plant electronics equipment
- Connecting patch panels to hub or switch in commercial building
- Connections in outside plant patch panels, cabinets, pedestals and closures (above and below ground).
- Connecting outside cable plant to central office or premise network in a building basement

Within this range of environments, connectors must provide stable and repeatable connections with respect to physical function, insertion loss (IL or attenuation) and reflectance (or return loss, RL). Except for cases where components are designed with fixed levels of attenuation for purposes of balancing power (attenuators), loss and reflectance are kept as low as is practical. Maintenance of function and performance within acceptable limits in the above applications and environments is the basic reliability requirement for connectors.

2.2 Designs and Technology trends

Except for a very limited number of special applications, fiber optic connector designs are based on butt-joint fiber alignment. Most are single- or duplex- fiber devices using cylindrical ferrules (ceramic, metal or some combination) in the plug assembly and split resilient alignment sleeves (ceramic, metal or plastic) in the adapter assembly. Multifiber connectors accommodating arrays of 2 to 12 are used as well, but in much smaller numbers. Array styles are typically based on precision machining (or silicon v-groove) and polymer molding technology. Regardless of the style, the technical and manufacturing issues are non-trivial because submicron tolerances are generally required.

Applications of connectors, until fairly recently, have been limited primarily to controlled building or partially protected environments, but as noted below, that trend is changing.

Industry standards covering designs, applications and performance have been in existence or under development for many years, but have not fully evolved to a consistent and useful form with respect to defining, measuring and reporting *reliability*.

In recent years, the following trends have evolved;

- A move toward smaller size (Small Form Factor, SFF) designs
- An increase in the use of injection molded plastics
- An increase in designs and applications of duplex and multifiber variants
- Increased application in harsher outside environments
- Maturation of industry interface and performance standards
- Development and refinement of specialty or hybrid components (e.g. attenuators)

The relevance of these trends to the current understanding of connector reliability will be addressed in this paper where needed.

2.3 Design examples

Table 1 lists representative examples of available connector designs covering some of the most commonly used styles and variants of cylindrical ferrule and array types. Two of these are shown in figures 1 and 2. The relevant interface standards are developed within the International Electrotechnical Commission (IEC), Subcommittee 86B and the Telecommunications Industry Association (TIA), Subcommittee 6.3 To ensure *mechanical* intermateability between different vendor products made to the same design. It is important to note that conformance to an appropriate interface standard affects, but does not guarantee, performance or reliability. This topic will be covered in section 2.5.

All the designs in table 1 are based on adhesive bonding of fibers inside tightly toleranced ferrule capillaries and precision polishing of the ferrule endface to ensure fiber contact or accurate positioning within a device. Some designs have tuning capability (active fiber core alignment) to minimize loss. Acceptable reflectance performance for most all applications requires a PC (Physical Contact) ferrule finish. Some applications require further reduction of reflectance, which is achieved with an APC (Angled Physical Contact) finish. *Note: In Japan, APC may be used as an acronym for Advanced Physical Polish.*

There are other less common designs using innovative technologies including ferrule-less fiber alignment, index matching membranes, and features permitting termination without adhesive bonding or polishing. Since these alternate technologies have had limited use and performance history, this paper will focus on the ferrule and array designs exemplified in table 1.

2.4 Critical features, parameters and conditions

Alignment and positioning of fibers to achieve optical performance requires maintenance of tight tolerances on several ferrule dimensions and parameters. These are discussed here for the case of cylindrical ferrule designs[4]. Similar requirements either exist or are under development for array styles.

Ferrule Geometry (see figure 3):

Ferrule and fiber geometry are the major factors controlling optical performance. The following ferrule requirements apply primarily to single-mode applications. Multimode requirements are similar but less stringent.

1. Outer diameter, roundness and cylindricity (< 0.5 μm)
2. Hole diameter and eccentricity (< 0.7 μm)
3. Endface radius (10 mm> R < 25 mm for PC and 5 mm> R < 15 mm for APC)
4. Dome eccentricity (< 50μm)
5. Endface angle (nominally, 0 for PC and 8, 9 or 12° for APC)
6. Fiber undercut (± 100 nm)

Parameters 1-3 affect alignment and loss while 4-6 affect fiber contact (and, secondarily, alignment for APC). Fiber contact has a major influence on reflectance. Without contact it is both high and unstable.

Endface Polish

Polishing produces a surface layer glass with a different structure, density and refractive index from that of the uniform fiber core[5,6]. The result is a limitation on the minimum achievable reflectance in a PC finish, even with fiber-to-fiber contact. The surface layer produces an intrinsic reflectance which varies with its thickness and structure from approximately -25 to -60 dB, depending on the quality of the finish. With an APC finish, reflectance is independent of the surface layer and can be reduced to less than - 60 dB.

Ferrule spring force:

The axial compression force required to bring a ferrule to its fully mated and latched position must be within a certain range. A minimum value is necessary to compress the ferrule ends and ensure fiber endface contact with allowance for frictional resistance from the alignment sleeve. A maximum is necessary to prevent fiber/ferrule damage.

Component material environmental stability:

To maintain optical performance throughout the intended product life, the above critical parameters must remain within certain limits. This in turn requires environmental stability of the component materials in the ferrule-fiber assembly (see figures 4 and 5). The primary concerns are dimensional or phase transformations in the ferrule, particularly if it is a zirconia ceramic, or motion of the fiber due to low creep resistance of the adhesive bonding it to the ferrule. These and other material stability problems will be discussed under *failure mechanism, section 2.6.*

With the above conditions met, and present day typical fiber tolerances, it is possible to maintain performance of loss < 0.8 dB (maximum value of randomly mated product sample) and reflectance between -20 and -60 dB for PC and < - 60 for APC. With tuning, it is possible to achieve a maximum loss of < 0.3 dB. Values for a specific product depend on its precision and quality which in turn are dictated its particular *performance requirements.* Different optical performance grades are specified in some standards (e.g. IEC 61753-1, see section 2.5).

2.5 Performance, quality assurance and reliability requirements

Performance, quality assurance (conformance) and *reliability* are interrelated concepts but distinct from one another in meaning. They are addressed in numerous standards and specifications, which, unfortunately, are often misunderstood and misapplied. The primary International and US passive component standards organizations are IEC/SC86B (Fiber Optic Interconnecting Devices and Passive Components) and TIA-FO 6.3 (Fiber Optic Interconnecting Devices and Passive Products). Each has taken significant steps to structure their connector standards for clarity and utility in these areas to provide a common basis for product acceptance. The organizations and systems are similar in many respects and appear to be converging over time. The IEC system is reviewed here because its is an international body and its document structure is more fully developed.

IEC/SC86B Connector Standards

Tests and Measurements (IEC 61300 series): These specify standard optical, dimensional and other measurement procedures and environmental test methods.

Interface Standards (IEC 61754 series): These provides all the essential information required to manufacture or use products, of a particular design, which will *mechanically* intermate. Connectors and adapters that comply with the appropriate standard will fit together, but will not necessarily meet a required level of optical or environmental performance. *Optical* interface standards are presently under development to define essential parameters of the waveguide interfaces of component interconnections.

Performance Standards (IEC 61753 series): These contain a series of mechanical and environmental tests and optical measurements with clearly defined conditions and pass-fail criteria. They are intended to prove fitness of design and manufacturing technology for a particular application and severity of environment. They are performed on a "once-of " basis. Compliance means that the product has passed *initial qualification*, but says nothing about on-going quality or service reliability over time.

Reliability Standards (IEC 62005 series): A seven part series of documents dealing with various aspects of reliability of connectors and other passive components. These standards are intended to ensure that a component can meet the performance requirements for a stated *time* period. They are discussed in more detail in section 2.7.

Quality Assurance/Conformance Standards (IEC QC 001002 set): These deal with procedures for ensuring and certifying that product components and assemblies, as manufactured, remain fixed over time. The requirements are referenced in the *Generic Specification* (see below)

IEC/SC86B Connector Specifications

A *Detail Specification* is a product procurement document. It references various interlinking standards, and lists other requirements, to develop a detailed set of acceptance criteria. It is the useful end result of a three - level document structure for which the *Generic Specification* provides the basis. The generic describes the scope, structure and interlinking of all other documents in the system.

A detail specification is written for a particular application and environment. In addition to other necessary information, including quality acceptance details, it references the requirements of the relevant interface, performance and reliability standards, as needed. Taken together the set of requirements provides the manufacturer and user with a product which (1) fits the physical interface (*interface standard*), (2) has a proven fitness for the intended application at the time of production (*performance standard*), (3) has a manufacturing quality that will remain at the initial level (*quality conformance standard and requirements*) and, finally, (4) will meet the performance requirements for a stated period of time at a maximum failure rate (***Reliabilty*** *standard*).

Table 2 gives a list of required tests from *IEC 61753-2-1: Fiber Optic Interconnecting Devices and Passive Components Performance Standard - Part 2-1: Single mode connectors for category U - Uncontrolled environment.* Category U represents the operating service environment that may occur in an outside plant pedestal cabinet or building wall enclosure and is one of three defined in the General and Guidance section (Part 1) as follows:

- E = An extreme environment with a temperature range of -40 to +85°C, liable to water immersion.
- U = A semi-protected environment with a temperature range of -40 to +70°C with no humidity limitations.
- C = A controlled environment with a temperature range of 0 to +55°C with a maximum relative humidity of 85%

The tests are performed on a fixed sample selected at random according to test and measurement methods from the 61300 series. They are grouped and sequenced in a defined way to test the fitness of the as-manufactured product for the category U environment. They compose a set of *qualification* requirements that verify *initial* fitness but do not give direct quantitative information on reliability parameters such as failure rate over time or cumulative failure fraction over the lifetime. A common mistake is the inference of a high a high level of reliability from conformance of a product to such a set of tests. This is a misapplication of their purpose for reasons discussed in section 2.7: *Reliability models.*

The IEC *Reliability Standards* address the problem of failure as a function of *time* in a service environment. Series IEC 62005-x, for passive components, comes in seven parts, in various states of development. They are:

- Part 1: Introductory guide and definitions
- Part 2: Example (s) of quantitative assessment based on accelerated aging tests
  Part 2A:Temperature and humidity, steady state
  (Note: This example deals with branching devices)
- Part 3: Guide to relevant tests for evaluating failure modes and failure mechanisms and failure mechanisms for passive optical components
- Part 4: Product screening
- Part 5: Relating accelerated tests to standardized service environments
- Part 6: The use of field data to determine, specify and improve passive component reliability

- Part 7: Life-stress models

These documents are applicable but not specific to connectors. The normative portions are quite general, but can be built on to provide specific examples of methods for measuring and estimating reliability from field and test data. This will be discussed in more detail in section 2.7: *Reliability models.*

Because many people, especially in the US, are more familiar with the structure of Bellcore (TelCordia) requirements for connectors in GR-326-CORE, it is important to understand the similarities and differences. between it and the IEC document set. GR-326 is, by name, a *Generic Specification.* In the IEC system it would be more properly regarded as a *Detail Specification* because it contains all the elements of *standards* requirements in one form or another including, interface (references TIA-604-x standards), test and measurements (references TIA FOTP's), performance (service life tests), quality assurance (referred to as reliability assurance) and ***reliability*** (extended service life tests). The application of these tests to the problem of quantifying failure probability is discussed in section 2.7.

2.6 Failure mechanisms

In a previous review of this subject[7], it was pointed out that the problem of developing reliability models for connectors is complicated by two factors: (1) connectors are multi-part assemblies consisting of several different materials, and (2) they are exposed to a variety of stresses, the sequence and severity of which depends on the application. The result is a mix of competing failure mechanisms for which there is no single accelerating test condition. Since the technology and applications have not changed in any major way, neither has the problem, although some progress toward a solution has been made. Much of it has been in the area of understanding failure mechanisms. The major ones are discussed in this section.

Part 3 of the IEC reliability standard distinguishes between failure *modes, mechanisms and effects.* The major ones are identified and described as follows:

- Failure mode: Optical or mechanical degradation
  - Increased insertion loss
  - Increased reflectance
  - Loss of mechanical failure

- Failure mechanism: The primary (physical/chemical) process producing failure
  - Deterioration of adhesive
  - Fiber microbending
  - Fiber endface mechanical damage (wear)
  - Particle contamination of fiber endface (wear-generated or external debris)
  - Mechanical (or thermal) damage to housing assembly or latching mechanism
  - Alignment sleeve cracking

- Failure effect: Consequence (manifestation) of the failure to the function of the part or system (or observed effect under failure mode analysis)
  - Loss of physical contact (fiber endface)

- Fiber failure
- Fiber endface damage
- Loss of mechanical mateability

The following is a brief summary of the present understanding of these phenomena and their significance with respect to overall connector reliability. They are discussed from the standpoint of ceramic-ferrule style connectors, the most common type, however, many of the effects apply with modifications to other types as well.

Fiber Breaks:

This topic has recently been a comprehensive review of all probable causes of fiber breaks[8] in ferrule style connectors. In well-designed and manufactured products, breaks occur as random, low rate events in the backbone-ferrule assembly as a result of a tensile stress component. Tensile stress may arise from one or more of several sources including (i) thermally induced buckling, (ii) bending from transverse misalignment, (iii) debris or geometric occlusions (iv) thermal expansion of the adhesive, or (iv) externally applied loads to the cable/fiber. The probability of failure is greatly affected by (i) the amount of stripping and handling damage to the fiber during termination, (ii) the materials and construction of the fiber and terminated cable elements, (iii) the method of termination and (iv) the stability of the adhesive with respect to creep resistance and bonding strength. Because the failure probability is extremely craft sensitive, field terminations are susceptible to fiber breaks than are factory terminations..

Deterioration of adhesive:

This topic has been reviewed in a series of papers by Reith and coworkers, the latest of which is reference 9. Adhesives are subject to creep or loss of bonding strength in hot humid environments. The potential consequences of these effects are: (i) a reduction of load isolation of the fiber which protects against breaks and (ii) fiber motion within the ferrule which can result in loss of physical contact between mated fibers. Loss of contact, as previously noted, causes high and unstable reflectance and is of particular concern for connectors in the outside plant which are subject to widely varying temperature and uncontrolled humidity. The most commonly used adhesives are thermally cured epoxies which generally perform well in service extremes. In laboratory tests, those with the highest glass transition temperatures tend to perform best.

Ferrule degradation:

As noted in section 2.4, ferrule endface geometry is critical to connector optical performance. Any significant change in radius of curvature or surface condition of the ferrule may affect fiber contact. The most commonly used connector ferrule material is yttria-tetragonally-stabilized (TZP) zirconia with which has been shown to undergo surface roughening, swelling and creep at high temperature and humidity[9,10]. The first two of these effects are attributed to the transformation of the metastable tetragonal zirconia phase to a monoclinic structure. Creep has been shown to be dependent on grain size and processing parameters[11]. Roughening occurs on a scale that can have a significant effect on endface geometry and ferrule/sleeve friction. Creep of mated connectors causes flattening of the ferrule endface radius. These changes have the potential to cause failures by loss of fiber contact, but, as yet, no correlation has been made between them and optical performance of connectors.

Ferrule degradation is a legitimate concern for plastic array type connectors as well, however, there is no body of information in the literature documenting its frequency of occurrence nor the dominant mechanisms. Certainly creep is a likely one. For the case of metal ferrules, one has to be concerned about oxidation, corrosion and creep in very extreme environments, however there is no evidence that these occur to any great extent in most applications.

Particle contamination of endface (debris), sleeve cracking and wear

These topics are related because they involve material choices and cost trade-offs. Most ferrule type connectors use resilient ceramic (typically zirconia) or metal (typically phosphor bronze) alignment sleeves. The latter is cheaper. Both are subject to wear and generation of debris during repeated matings. It is generally conceded that wear and debris are more prominent with metal sleeves but that cracking is more likely with the ceramic. In a previous review of connector reliability by one of the authors,[6] the view was stated that both mechanisms have low probability of causing failure. Information and experience accumulated in the intervening years have not changed this view. Both type failures should be regarded as low rate random events.

Mechanical failures

Mechanical failures include such things as damage to the connector-cable/(fiber) interface (e.g. cable jacket and strength members, bend limiter) and failure of the latching mechanism, housing or mounting features of the adapter. These can occur as a flexing, twisting, impact or vibration. They are mechanical handling issues which can be designed for, and should not generally be of great concern unless the terminated fiber has no reinforcement from the cable strengthening elements as in some ribbon element terminations with multifiber connectors or single fiber outside plant terminations. In both cases the weak link in the assembly is the fiber which does not have good stress isolation.

With the increased use of injection molded plastic components, creep rupture of housing assemblies and latch mechanisms becomes an issue. Creep rupture is one of the easily overlooked failure mechanisms in plastic connector components even after many years of experience in other fields. This is sometimes a case of new applications re-discovering old problems. These type failures can normally be detected the design phase with the use Highly Accelerated Life Test (HALT methods).

The use fiber amplifiers in systems raises concern about the effects of high power density on fiber endfaces. At least two potential failure modes have been identified (i) fusion of debris particulates and (ii) fiber pitting. Both are the result from the combination high power and poor contact. The long-term degradation effects of high power on materials used in attenuator connections are also of concern, but there is limited information on the subject at present.

Other organizations, in addition to IEC/SC86B, are addressing the problem of connector reliability. One of these is COST (European **Co**operation in the field of **S**cientific and **T**echnical research), an organization of 28 countries within the European Commission.

COST 246 was five-year program on the study of "Materials and Reliability of Optical Fibers and Components", the results of which have just been published (see reference 3). For connectors, there were several interesting findings and conclusions:

Failure Modes:

Environmental and mechanical test results on ceramic ferrule style connectors from several participating laboratories showed failures by all the modes discussed above, excluding those relating to high power. The surprising result was the number of mechanical failures of fibers, latching mechanism and adapter housings. These were the most common events. Other failures involved changes in loss or reflectance beyond performance limits. The exact causes were not identified, but ferrule degradation in the form of roughening and endface flattening and fiber motion due to epoxy creep were also observed.

Field reports of various styles of ferrule connectors in the COST 246 study indicated that many "failures" were the result of poor installation and handling practice, a recurrent theme in any discussion of connector reliability. Sleeve cracking and endface debris problems were also reported. The debris problems were not correlated with the sleeve material type.

Conclusions of COST 246:

- Due to the multiplicity of designs and applications, reliability questions must focus on specific, rather than general circumstances.

- Reliability models based on test data are not sufficiently accurate at this time. Reliability predictions with reasonable confidence intervals should, for the present, be based on field data.

COST is initiating a new five-year program on reliability of passive components which will focus on improving life-testing methods and reliability estimation models. It will include connectors with emphasis on *new technology*.

TIA Working Group FO 6.3.8 studies connector reliability. Its objective is to develop guidelines for test methods to accelerate service failures and analytical methods for predicting reliability. Focusing on specific applications, the group is currently looking at failure modes under thermal cycling. In a recent report of an ongoing test program involving ceramic ferrule style connectors[12], the following observations were made.

- Insertion loss changes show trends which may be extrapolated to end of life by "failure" to meet optical requirement limits (an important consideration for determining failure lifetime statistics).

- Insertion loss and reflectance changes do not correlate well - an expected result because loss of fiber contact affects reflectance much more than it does insertion loss.

- Comparison of two commercial grade epoxies showed better performance correlating with the greater T_g.

The work of FO 6.3.8 is ongoing. The group maintains unofficial liaison with IEC/SCB/WG5.

2.7 Reliability models

The objective of a reliability model is to predict failure rate as a function of time (FIT's) or cumulative lifetime failure probability. To the authors' knowledge there are no published models for estimating connector (or splice) reliability. The problem for connectors is especially difficult because of the numerous designs, materials, service environments and stresses the components may see in service. The current draft (under review) of Part 7 of IEC 62005-x deals with reliability (life stress) models for passive components by giving general requirements in the *normative* section and *informative* worked examples in the appendix.

The *normative* requirements of IEC 62005-7 simple state that:

1) The argument for reliability must be consistent with existing physical knowledge of the failure modes stimulated by the device under specific stresses (described in Part 3 or in the examples presented in Part 7 of the standard).

2) The argument for reliability must be *statistically* consistent with all the data gathered in experimental study of the device in question. Elimination of statistical outliers must be justified by reasonable arguments based on an understanding of experimental procedure. Statistical confidence bounds must be placed on estimates of reliability.

3) The following 3 approaches are legitimate ways of supporting arguments that a device is reliable *with respect to a specific stress*:

   A) *Acceptance testing*: where devices survive a period of time at accelerated test conditions. The time duration and stress of the acceptance test must be justifiable based on an argument using a model which bounds failures during life by a large plausible class of failure modes.

   B) *Life-stress models*: based on degradation or failure under accelerated testing.

   C) *Previous qualification:* to the standard of a similar material system, where a detailed report of the method is available, may be used as a basis either for a failure/degradation model or an acceptance test.

The *informative* examples in IEC 62005-7 deal with acceptance tests and life-stress models.

Acceptance tests are run to determine if there is a reasonable confidence level of high product reliability. A sample of the product is aged to the equivalent of end of life. A highly reliable product will show *no failures* and minimal degradation. A lower bound estimate of lifetime reliability can be obtained from the data. This is the approach taken in Bellcore (TelCordia) GR-326 with respect to the reliability tests. Since binomial statistics apply, the difficulty with determining accurate reliability estimates is readily apparent from the following argument. For an acceptance test in which there are no failures, 95% confidence that there is less than 1% probability of failure during life requires a sample of size of 298. For a sample size of 28, typical of qualification or performance certification tests, the upper-bound failure probability increases to 10% at 95% confidence. Pass-fail type tests on practical sample sizes, whether described as service life, extended service life, or qualification tests, do not give tight confidence intervals on cumulative failure at end of life unless additional assumptions are made. Furthermore, they give no information on failure rate as a function of time.

Equivalent lifetime reliability tests, such as those in GR-326, are generally purely on the basis of engineering judgement. One can use a strategy developed by LuValle and coworkers[13] to design acceptance tests for thermally activated failure mechanisms from physical/chemical kinetics principles. The method uses approximations of material parameters in an Arrhenius rate model to determine time-temperature equivalence in the form of *Demarcation Maps*. These tests can be combined with measurements of degradation and assumptions about the component population distribution to tighten reliability estimates for a given sample size. It is possible to reduce the sample size of 298 to about 13 and still achieve 95% confidence of 99% reliability, again assuming no failures.

Life-stress models are based on accelerated tests which produce failures in the sample lot. From this type data, a life distribution (failure probability as a function of time) can be derived. The general methodology is described in reference 14. Examples involving mechanical failures in connectors are given in reference 15.

Application of reliability test methods to connectors has not gone beyond straight-forward acceptance testing based on engineering judgement. The need for more predictive life-stress models has been recognized, but they have not been developed for connectors as yet. In the authors' view, this is as much the result of engineering priorities as it is of technical difficulty.

2.8 Connector summary

The understanding of connector reliability has progressed in several areas in recent years. Material properties, failure mechanisms, environmental conditions and design parameters critical to performance are better known and are being addressed in the technical literature and in standards. Still lacking however, are methods for measuring and predicting failure rates. This problem is under study by at least three organizations.

3. SPLICES

3.1 General

A splice is a permanent joint or connection between two optical fibers designed to minimize the loss of light (signal) transmitted through the fibers while ensuring the long-term physical integrity of the splice. Reliability of splices is based on their ability to remain intact and functioning over the design life of the system under all expected environmental conditions. Unlike connectors, most splices are intended for connecting cables in the outside plant and are not designed to be re-entered and remade, although some mechanical splices may allow a few remakes.

Optical fiber splicing technology is divided between mechanical and fusion techniques. Fusion splices use heat to fuse the glass into one continuous optical path. Mechanical splices use other materials to align, support and protect the fibers. The dominant trend in recent years has been toward fusion. Mechanical splices are still deployed in some new spans, but are now used primarily for restoration.

The international generic specification on "Splices for Optical Fibers and Cables" is IEC 1073-1 prepared in 1994. It describes hardware and accessories, quality assessment procedures, and measurement and environmental test procedures. This document references the relevant IEC test procedures discussed in section 2.5. listed in table 2. Because splices are expected to survive in the outside plant, only the most severe environmental testing criteria are usually applied to them (category E in section 2.5 above).

Most early optical fiber cables had less than 100 fibers per cable and single-fiber splicing was generally performed. However, the number of fibers in cables is growing rapidly because of the increase in data traffic over the telecommunications network. With hundreds of fibers in a cable and many splice points, cable designs using ribboned fibers have evolved and mass splicing techniques can greatly reduce installation time and costs. Continuing improvements in fiber geometry and in splicing methods have greatly reduced both single-fiber and mass splice loss. This process has become more complicated with the evolution of new fiber types for 1550 nm systems. These G.655 fibers require changes and improvements in splicing techniques to achieve the same field splice loss distributions normally expected in standard (1310 nm) fiber.

3.2 Fusion splices

At this time, electric arc-fusion is the most widely used method to make reliable single or mass optical splices in the field. The fusion process uses fast, efficient splicing machines from a number of manufacturers, which have been continuously improved over the last two decades.

To make a fusion splice the protective coatings are stripped from the fibers which are cleaved and then positioned and aligned between two electrodes in the splicing machine. An electric arc heats the silica glass until the "melting" or softening point is reached while the fiber ends are being pressed together so that a continuous splice is obtained. The splice is then proof-tested for strength and a suitable protection device is applied to the splice to protect the bare fiber and to allow handling and storage without adversely affecting the physical integrity of the splice.

The fiber core alignment in fusion machines can be passive (v-groove alignment) or active (light injection and detection (LID) system or core/cladding profile monitoring and alignment (PAS) system). Active alignment machines use three-dimensional movement to actively align the cores or the outside diameters (ODs) of the two fibers being spliced. The splicing machine minimises the splice attenuation by either focusing on the core or cladding of the fibres with its vision system to directly align them or by optimizing the transmitted light through the fibres. Any fiber core eccentricity present can produce transverse offset (misalignment of the fiber cores) when the surface tension of the molten glass pulls the fibers into alignment on their (ODs) during fusion. This movement of the fibers due to surface tension can cause core distortion (or core deformation) which causes additional splice loss.

In mass fusion splicing, individual fiber alignment is not currently available and initial alignment is controlled by the positions of the fibers in v-grooves. Passive alignment machines use only fiber longitudinal movement so accurate core alignment requires good fibre geometry. Core alignment depends on the combination of fiber core eccentricities, fiber OD difference, ribbon geometry and the alignment and cleanliness of the v-grooves in the splicing machine. The cleave quality and the intensity and the duration of the arc as well as any differences between the two fibers being spliced determine the splice loss. The final splice loss, for each fiber pair, also depends on core distortion and core misalignment caused by the surface tension effects and how well the machine maintains fiber alignment during splicing. Surface tension effects can be more important for mass fusion because the longer arc times (15 to 30 seconds for mass fusion vs. 1 to 2 seconds for single-fiber fusion) allow more time for self-centering. In addition, variations in ribbon structure may perturb the position of the fibers in the heating zone, producing higher splice loss. Ideally, the ribbon should have all the coated fibers in contact with the adjacent fibers and in one plane. As fiber geometry and ribbon structure continue to improve, the movement caused by surface tension decreases and the splice loss caused by both core distortion and final core misalignment are reduced.

Proof-testing is recommended after the splice is completed to check for minimum strength. It is very important to establish a defined level of mechanical strength for the splice that is related to its expected lifetime. The splice is subjected to a tensile proof-stress for a short period of time by most splicing machines. Some splicing machines perform this test with the spliced fibres in the splicing chucks and some perform it after placing the spliced fibres in the holders for heat shrink protector application.

After proof-test, the splice point is protected. The "protector" is a mechanical device or restored coating, that provides both mechanical and environmental protection to the single or multiple splice. It must not affect either the attenuation of the splice or its functional properties. All bare fibre must be covered. Designs may include heat-shrink sleeve, "clam-shell", fiber re-coating and encapsulating protectors.

Protectors for single-fibre fusion splices should be capable of accepting either 250 μm (nominal) diameter coated fibres, 900μm (nominal) diameter buffered fibres, or 250μ m/900μm combinations. They usually require tools or equipment to assemble and install. The most commonly used protectors are heat shrink tubes with a rigid member inside to keep the splice point straight. If tube-type protection devices are used, they must be placed over one end of the fibres or ribbons to be spliced before splicing. The user must make sure that any heat shrink protector used is compatible with the heating oven on the splicing machine being used.

Handling of the fibers during coating removal, cleaving, splicing, and splice protection contribute to the long-term mechanical reliability in the field. Any contact with the bare glass during these steps can cause flaws that reduce the strength of the completed splice.

In particular, mechanical stripping of the coating reduces strength significantly. High strength splicing processes for undersea cable use chemical stripping using hot sulfuric acid to avoid this problem. They also make sure that nothing touches the bare fiber during cleaving, splicing and splice protection[16].

3.3 Mechanical splices

Mechanical splices have the following features:

- A surface for aligning fiber ends
- A retaining mechanism to keep the fibers in alignment
- An index-of-refraction matching material (gel, grease or adhesive) between the fiber ends to reduce loss and reflectance.

Both single fiber and ribbon fiber designs are in existence. Some utilize factory pre-termination for faster connection in the field.

Mechanical splicing is relatively simple and inexpensive since it does not require the use of a special machine, as does fusion splicing. Installation tools are used to fix the fibers in a protective housing, generally without the need of electrical power. Some allow tuning by hand for minimum loss. Fiber end preparation is accomplished either by cleaving or grinding and polishing. The integral housing of the assembly provides mechanical and environmental protection. Separate protectors are not required.

Index-matching material is used between the ends of the fibers in mechanical splicing to reduce both loss and reflectance. These materials are chosen to match the index-of-refraction of the fiber core. The most common materials are silicone gels and greases. UV- curable adhesives and epoxies are sometimes used. The gels, UV- adhesives and epoxies are cured and cross-linked materials. The greases are resin suspensions in oil. Most pre-index-matched mechanical splices use index-matching grease (though some manufacturers call it gel).

Cured materials are generally more stable than the greases. Properly cured silicon gels retain their physical properties over a wide range of environmental conditions and have some degree of resilience and resistance to flow under stress. They are mixed and applied and cured at the time of splicing. Since batch and sample variations are possible, it is good practice to place a "witness sample" of the material at the splice point so that its stability can monitored.

Some of the greases have been shown to separate into an oil and a resin under adverse conditions. The long-term environmental stability of index-matching greases should be confirmed by HALT testing before use under severe or unusual environmental conditions. The supplier of the index-matching material should provide complete information about its behaviour at different temperatures (especially the extremes) and its estimated lifetime, in terms of maintaining the initial optical performance. Proof tests are usually not performed on mechanical splices.

3.4 Splice loss estimation and measurement

Most active alignment and some passive alignment machines also provide an estimate of the splice attenuation after the splice is complete by observing the core alignment before and after splicing (PAS systems) or by measuring the transmitted light through the splice (LID systems). For ribbon cables however, all of the current mass-fusion machines estimate splice attenuation using PAS systems. This number is truly an estimation based on measurements made by the fusion machine predicting how the fiber cores align at the point where the fusion actually occurred. The estimation is usually optimistic and should not be used as actual splice loss. This estimation is only useful for determining if a splice is "generally good" or "generally poor".

Field measurement of splice loss is a difficult task. During construction of a fiber route, it is limited to using inexact techniques. The two most commonly used techniques are 1) splice loss estimation from the fusion splicing machine, and 2) one way OTDR readings of each splice. Both of these techniques are admittedly poor at producing quantitative splice loss results but, can be used to evaluate gross high splice losses so that the splice may be remade if necessary. The only method to determine the actual splice loss in the field is by using two-way OTDR readings and averaging the results (bidirectional OTDR).

3.5 Factors affecting splice loss

Singlemode splice loss depends on the degree of alignment of the fiber cores and on the amount of core deformation caused by the fusion process. In singlemode fiber, splice loss sensitivity to transverse offset is inversely proportional to the MFD, a larger MFD reduces splice sensitivity to transverse offset. Transverse offset is the primary contributor to fusion splice loss. Difference in MFD itself between fibers in a splice causes very little splice loss; however, it can cause significant one-way OTDR measurement errors[17].

Fiber manufacturers cannot recommend a maximum splice loss with any fiber type because of the numerous variables in the field that can't be controlled by the fiber manufacturer. There is considerable discussion in the international optical fiber community at this time concerning the economic sense of remaking many splices to achieve individual low loss splices. With the advent of higher bit rate transmission systems, most optical fiber systems are dispersion limited rather than loss limited, especially in the 1550 nm window where Er–Doped amplifiers are available. In addition, splice loss is statistical in nature and any system with more than a few splice points will meet the end-to-end loss requirements because the mean splice loss on any single fiber path will tend toward the overall mean splice loss for which the system was designed. Cumulative system loss during construction can be measured, even with one-way OTDR, as each new cable is spliced into the span. If this measurement meets the statistical mean design loss requirements, individual splice losses are less important.

3.6 Failure mechanisms

Because splicing is a craft sensitive operation, most failures take place during installation. Those which occur during service depend on the type of splice. Fusion splice failures almost always involve fiber breaks because the signal cannot be affected by misalignment. Mechanical splice failures may be due to fiber breaks signal changes resulting from lateral misalignment, endface separation, or changes (degradation or contamination) of the index matching material. These are reviewed with comprehensive bibliographies in references 3 (chapter 6) and 7, and will not be covered in any detail here.

Since fusion splicing is dominant, the failure mechanism of greatest concern is fiber fracture. Examination of low-strength breaks shows that the fibers fail in areas adjacent to the splice that have flaws or inclusions from stripping or handling in the splicing process. These are frequently the result of damage occurring during cable end preparation or fiber storage after splicing. In spite of the concern for potential failures, the overall rate is quite low.

3.7 Splicing reliability

As pointed out in section 2.7 above, there are no published reliability models for optical fiber splices. This is also the case with mechanical splices, the splice point itself is reinforced but the fiber near the splice has been handled. Most reported splice failures are found by OTDR. The COST 246 study concluded that the process of establishing failure rates for different mechanisms will have to rely on field data. That information has not yet been assembled in a comprehensive and useful form. The same general principles and IEC standards apply to splices as to connectors, but they have yet to be applied.

3.8 Splicing summary

Both fusion and mechanical splicing methods and equipment have proven to be highly reliability in service. Millions have been made and survived in the outside plant under extremely adverse conditions for many years. Failure rates have been so low that it has been difficult to establish FIT rates or develop reliability models. As fiber extends further into the access network, closer to the customer premises, the emphasis on reliability will increase. Both fiber and splicing equipment manufacturers are continuing to improve quality and reliability.

ACKNOWLEDGEMENTS

Thanks to R. Musk, P. Longhurst, S. Swanson, M. Shahid and W. King for providing useful material or informative discussion on connector designs, applications and standards, to M. LuValle for extensive help in understanding the methods and limitations of reliability models and to S. Sheldon for the illustrations.

REFERENCES

1. *Generic Requirements for Single-Mode Optical Connectors and Jumper Assemblies,* GR-326-CORE, Issue 3, September 1999, Section 2.2: Connector Applications.

2. *Commercial Building Telecommunications Standard,* TIA-EIA-568A, October, 1995, Section 12.4: Connecting Hardware for Optical Fiber Cable.

3. T. Volotinen, W. Griffion, M. Gadonna and H. Limberger (Editors), *Reliability of Optical Fibers and Components:Final Report of COST 246,* Chapter 9: Connecting and

Branching Components, p 289, Springer, Berlin, 1999.

4. R. Musk, "Fiber Optics: Some Practical Aspects of an Optical Link," *Short course* IWCS '97 Short Course, *46th International Wire and Cable Symposium,* Philadelphia, September, 1997.

5. M. Kihara, S. Nagasawa and T. Tanifuji "Return Loss Characteristics of Optical Fiber Connectors", *Journal of Lightwave Technology,* ***14****, no.9,* September 1996, p. 1986.

6. K. Kanayama, Y. Ando, R. Nagase, S. Iwano and K. Matsunaga and S. Iwano, "Advanced Physical Contact Technology for Optical Connectors" *IEEE Photonics and Technology Letters,* **4,** pp 1284-1287.

7. B. G. LeFevre, "Reliability of Passive Optical Components," *FiberOptics Reliabilty and Testing: Critical Reviews of Optical Science and Technology, SPIE, Vol* ***CR 50,*** D. K. Paul, ed, 1994, p 107

8. B. G. LeFevre, W.W. King, C. K. Sidbury, and D. L. Stephenson, "Analysis of Fiber Fracture in Connectors," *Technical Proceedings, 14th Annual NFOEC, Vol.II ,* Orlando, 1998, p 257.

9. L. A. Reith. M. Koza, J. Mann, R. Nickel, M. Ozgur, and D. Rader, "Ceramic and Epoxy Reliability in Connectors Exposed to Heat and Humidity" *Technical Proceedings, 14th Annual NFOEC, Vol.II ,* Orlando, 1998, p 279.

10. E. Makrides-Saravanos, P. J. Huspeni, C. Saravanos "Ceramic Ferrule Performance Under High Temperature and High Humidity Performance" *Technical Proceedings, 13th annual NFOEC, Vol.II ,* 1997, p 215.

11. E. Makrides-Saravanos, P. J. Huspeni, C. Saravanos, "Creep Deformation of Zirconia Ceramic: Effect of Composition and Sintering Temperature" *Technical Proceedings, 14th Annual NFOEC, Vol.II ,* Orlando, 1998, p 291.

12. G. Brown, "Accelerated Aging of Epoxy Type Fiber Optic Connectors", presentation to TIA FO 6.3.8, June 1999, unpublished.

13. M. J.LuValle, L. R. Copeland, S. LuValle, M.J., Copeland, L.R., Kannan, S. Judkins, J.B., and Lemaire, P. J., "A strategy for extrapolation in accelerated testing," *Bell Labs Technical Journal,* ***3****,* No 3, 1998, p.139.

14. W. Nelson, *Accelerated Testing: Statistical Models, Test Plans and Data Analysis,* John Wiley, New York, 1990.

15. B. G. LeFevre, J. W. Elling, C. S. Perry and R. L. Sweatt, "Life-stress Relationships and Accelerated Testing of Connectors and Splices," *Fiber optic Reliability and Testing: Benign and Adverse Environments,* SPIE Conference Proceedings, **2074**, Boston , September 1993.

16. J. T. Krause & C. R. Kurkjian, "Intrinsic glass strength achieved in fiber splices," OFC '84, New Orleans, LA, paper WI7.

17. S. C. Mettler, "Monte-Carlo Analysis of the Effect of Mode Field Diameter Mismatch on Single-Mode Fiber Splices," *NFOEC '92*, Washington, D.C., p. 647.

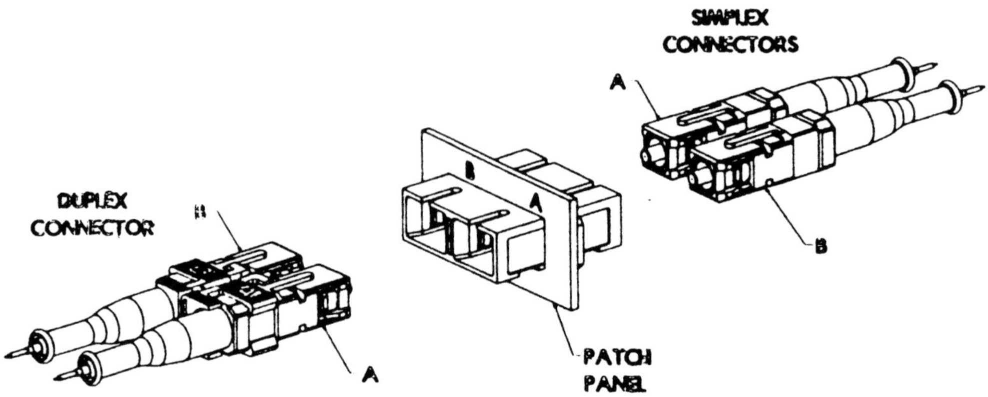

Figure 1. Cylindrical ferrule type connector, SC style

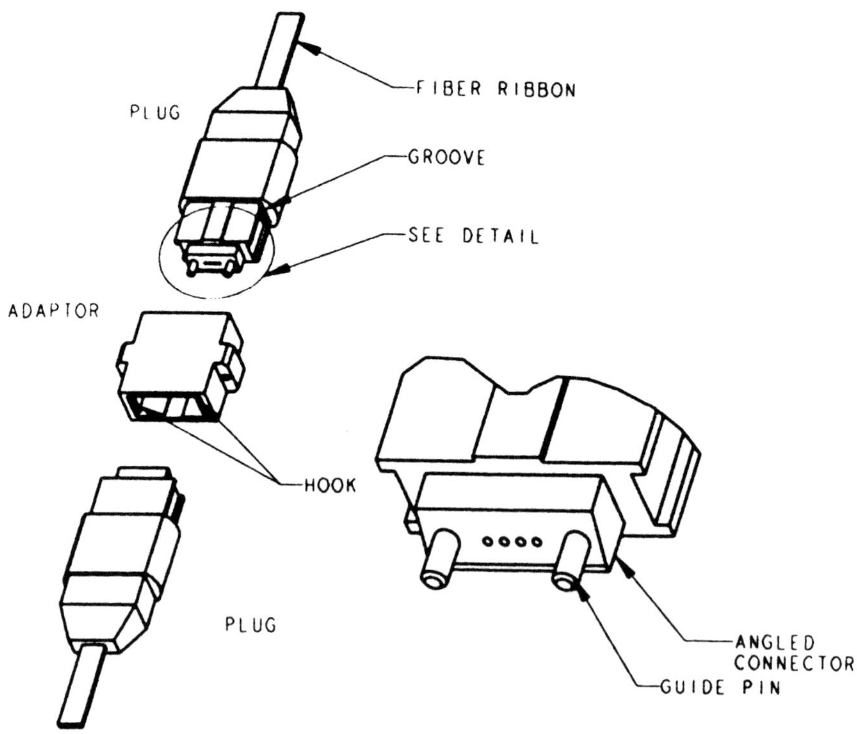

Figure 2. Multi-fiber array type connector, MPO style

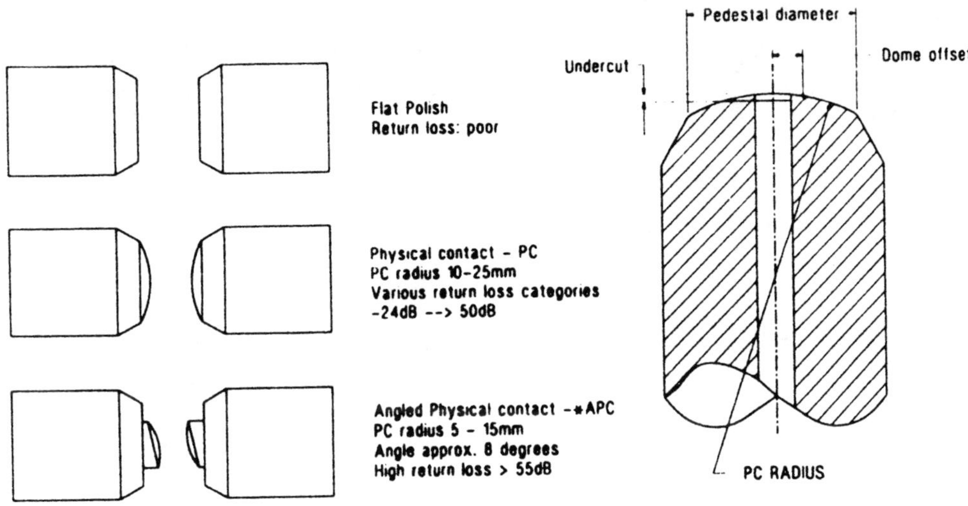

Figure 3. Ferrule/fiber geometry for PC and APC connectors (courtesy of R. Musk)

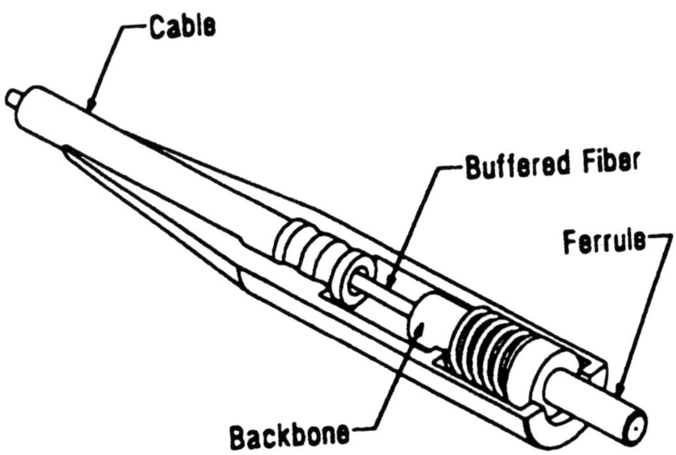

Figure 4. Typical connector assembly construction on patch cord cable

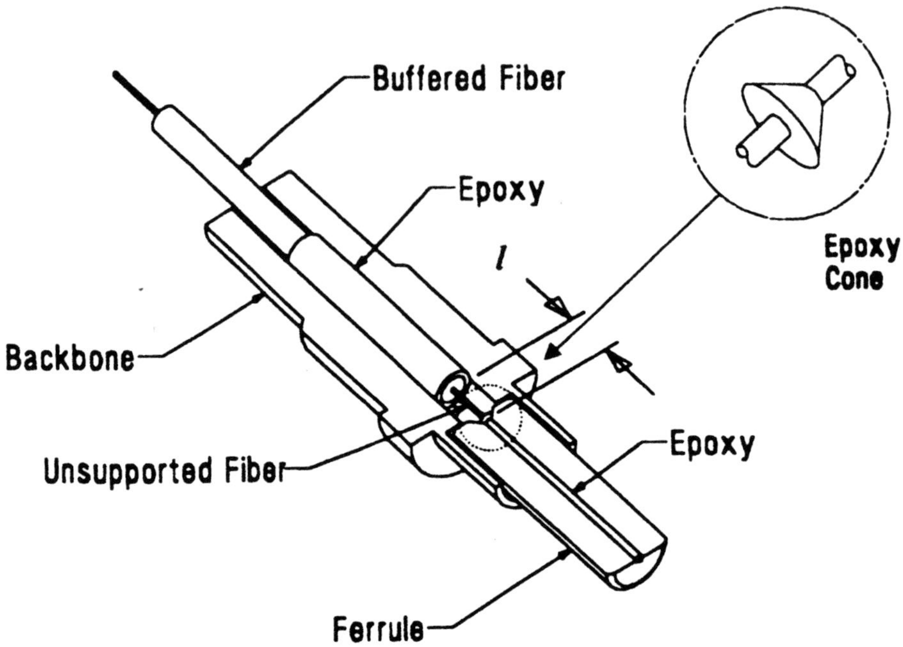

Figure 5. Typical construction of connector backbone/ferrule assembly on terminated buffered fiber

TABLE 1: SOME COMMON CONNECTOR STYLES

Designation	Type/Fiber Count	Interface Standard
FC	cylindrical ferrule-sleeve (2.5 mm) / simplex	IEC 61754 -13
ST (BFOC 2.5)	cylindrical ferrule-sleeve (2.5 mm) / simplex	IEC 61754 -2
SC	cylindrical ferrule-sleeve (2.5 mm)/ simplex - duplex	IEC 61754 -4
MU	cylindrical ferrule-sleeve (1.25 mm, SFF)/ simplex - duplex	IEC 61754 -6
LC	cylindrical ferrule-sleeve, (1.25 mm, SFF)/ simplex - duplex	TIA/EIA-604-10
MPO (MTP)	Multifiber array-pin alignment/2-12	IEC 61754 -7
Mini MPO	Multifiber array-pin alignment (SFF)/2-4	IEC 61754 -10

TABLE 2: PERFORMANCE REQUIREMENTS FOR SINGLE MODE CONNECTORS IN AN UNCONTOLLED ENVIRONMENT FROM IEC STANDARD 61753-2-1

TEST	REQUIREMENTS	TEST METHOD
Insertion Loss (reference)	< 0.5 dB	IEC 61300-3-4 (Method B)
Return Loss	Class S ≥ 25dB Class T ≥ 35dB Class U ≥ 40dB	IEC 61300-3-6 (Method 1)
Return Loss	Class S ≥ 50dB Class T ≥ 55dB Class U ≥ 70dB	IEC 61300-3-6 (Method 2)
Attenuation (random mate)	Grade P ≤ 0.35dB Max 1.00dB for ≥ 97% matings Grade Q ≤ 0.20dB Max 0.60dB for ≥ 99% matings	IEC 61300-3-34
Vibration (sinusoidal)	Allowable loss variation: ≤ 0.20dB	IEC 61300-2-1
Bending moment	Allowable loss variation: ≤ 0.20dB @ 1550 ± 30 nm	IEC 61300-2-7
Fiber cable retention	Allowable loss variation: ≤ 0.20dB @ 1550 ± 30 nm	IEC 61300-2-4
Torsion	Allowable loss variation: ≤ 0.20dB @ 1310 ± 30 nm	IEC 61300-2-5
Tensile strength of coupling mechanism	Allowable loss variation: ≤ 0.20dB @ 1310 ± 30 nm	IEC 61300-2-6
Impact	Allowable loss variation: ≤ 0.20dB @ 1550 ± 30 nm	IEC 61300-2-12, Method A
Static side load	Allowable loss variation: ≤ 0.20dB @ 1550 ± 30 nm	IEC 61300-2-42
Mating durability	Allowable loss variation: ≤ 0.20dB @ 1550 ± 30 nm	IEC 61300-2-2
Cold	Allowable loss variation: ≤ 0.20dB @ 1550 ± 30 nm	IEC 61300-2-17
High temperature endurance	Allowable loss variation: ≤ 0.20dB @ 1550 ± 30 nm	IEC 61300-2-18
Damp heat (Steady State)	Allowable loss variation: ≤ 0.20dB @ 1550 ± 30 nm	IEC 61300-2-19
Change of temperature	Allowable loss variation: ≤ 0.20dB @ 1550 ± 30 nm	IEC 61300-2-22
Dust	Allowable loss variation: ≤ 0.20dB @ 1310 ± 30 nm	IEC 61300-2-27
Condensation	Allowable loss variation: ≤ 0.20dB @ 1550 ± 30 nm	IEC 61300-2-21

Active Optical Components

Future Directions In 980nm Pump Lasers: Submarine Deployment To Low Cost Watt Class Terrestrial Pumps

V. Gulgazov, G. Jackson, K. Lascola, J. Major, Jr., R. Parke, T. Richard, V. Rossin, and K. Zhang

SDL, Inc.,

80 Rose Orchard Way

San Jose, CA 95134

OVERVIEW

The demands of global bandwidth and distribution are rising rapidly as Internet usage grows. This fundamentally means that more photons are flowing within optical cables. While transmitting sources launches some optical power, the majority of the optical power that is present within modern telecommunication systems originates from optical amplifiers. In addition, modern optical amplifiers offer flat optical gain over broad wavelength bands, thus making possible dense wavelength de-multiplexing (DWDM) systems. Optical amplifier performance, and by extension the performance of the laser pumps that drive them, is central to the future growth of both optical transmission and distribution systems.

Erbium-doped amplifiers currently dominate optical amplifier usage. These amplifiers absorb pump light at 980 nm and/or 1480 nm, and achieve gain at wavelengths around 1550 nm. 980 nm pumps achieve better noise figures and are therefore used for the amplification of small signals. Due to the quantum defect, 1480 nm lasers deliver more signal photon per incident photon. In addition, 1480 nm lasers are less expensive than 980 nm lasers. Thus, 1480nm pump lasers are used for amplification in situations where noise is not critical. The combination of these traits leads to the situation where many amplifiers contain 980 nm lasers to pump the input section of the Er-doped fiber with 1480 nm lasers being used to pump the latter section of Er fiber. This can be thought of as using 980 nm lasers to power an optical pre-amplifier with the power amplification function being pump with 1480 nm radiation.

This paper will focus on 980 nm pump lasers and the impact that advances in 980 nm pump technology will have on optical amplification systems. Currently, 980 nm technology is rapidly advancing in two areas, power and reliability. Improving reliability is becoming increasingly important as amplifiers move towards employing more pump lasers and using these pump lasers without redundancy. Since the failure rate allowable for an amplifier is not a function of the number of pumps employed in the amplifier, the allowable failure rate of an individual pump laser is decreasing for next-generation amplifiers. This will lead to specifications for terrestrial pumps well below 1000 FIT, and may lead to the case where high power amplifiers need laser pump reliability to approach 100 FIT. In addition, 980 nm laser diodes are now being deployed in submarine systems where failure rates lower than 100 FIT are commonly specified. It is obvious that both terrestrial and submarine markets are pushing allowable failure rates for pumps for optical amplifiers to continually decrease.

A second push for improvement is in the output power of 980 nm pump modules. There exist a number of motivations for increasing the output power of pump lasers. First, each additional channel in a DWDM system requires additional power. To first order, a doubling in channel count implies a doubling in pump power. Second, larger amplifiers require multiple pumps. Higher output power from pump modules allows for fewer pumps, less complicated control systems and smaller size amplifiers.

The discussion of this paper will focus on how current development progress of 980 nm laser diodes addresses these issues: better reliability and higher output powers.

HISTORICAL BACKGROUND

The first generation of 980 nm lasers was developed in the early 1990's [1, 2]. The first chips sold in appreciable volume had operational powers of 150 mW with kink-free power 200 mW. The p-up version of this chip, the 6520 was qualified in 1995. The multicell lifetest used to qualify the 6520 was composed of 161 diodes populating 4 cells with two operation powers: 150 mW, 200 mW, and two heatsink temperatures: 50°C, 85°C. With 555 device-khrs accumulated this test yielded the value of failure rate of 450 FIT at an operational power of 150 mW at 25°C. Since then, a significant number of devices has been deployed, thus accumulating field hours. Assuming that 85% of shipped diodes were deployed 6 months following diode shipment, the total number of field hours of the 6520 chip is estimated to be 297 million. It is assumed that the laser diode has been operated CW with a facet output power of 150 mW at a temperature of 25°C. Further, it is assumed that every field failure is reported to the factory. With 6 reported field failures, this gives a failure rate of 35 FIT with a 90% confidence level. It is worth noting that none of the failures have been due to facet catastrophic optical damage (COD). This failure mode has been completely eliminated by design. The dominant root cause for random failures is the propagation of epitaxial dislocations into the laser diode active region during operation. The propagating dislocations create absorptive regions that result in optically opaque material. This damage is internal to the chip and does not reach the facet.

The continuous demand to increase chip power resulted in the operation of the 6520 chip at powers higher than nominal 150 mW. Chip performance and kink-free power did allow for modest increases in output power. However, the kink power and reliability of the 6520 were not sufficient to meet increasing demand. To achieve the 250-300 mW kink levels currently required by the amplifier community, with high-reliability, a substantial chip redesign was necessary that included optimization of epitaxial layers, device design and packaging processes.

The incorporation of these changes dramatically improved the performance of the 980nm laser diode and resulted in the present device, the 6530. To aid in a comparison between the two chips, Table 1 lists important performance and reliability metrics for both the 6520 and 6530 980nm laser diodes.

Table 1. Comparison of 6520 and 6530 chip parameters.

Specification	6520	6530	6530	6530
P_{op} (mW)	150	150	200	250
P_{kink} (mW)	200	230	250	300
I_{op} (mA)	200	200	260	330
V_{op} (V)	2	1.8	2	2.25
Dissipated heat density (mW/ m)	0.33	0.21	0.3	0.5
Aspect ratio	2.9	2.6	2.6	2.6

Within the next section of this paper the failure modes and reliability of the 6530 will be discussed, including many aspects of the successful qualification program for submarine deployment of these laser diodes.[3]

RELIABILITY

To qualify the 6530 for terrestrial and submarine applications a multi-cell lifetest was started in December 1997. This test included 240 diodes selected from 5 epitaxial wafers distributed among 6 cells. After identification of a gradual degradation failure mode, to be discussed later within this paper, an additional high current cell and high temperature cell were started with 40 diodes in each cell selected from the same 5 epitaxial wafers used in the original 6 cells. Thus, a total of 5 epitaxial growths, 8 operating conditions, and 320 diodes form the core of the reliability testing program for the 6530.

The diodes in the multi-cell test have been operated at constant current and temperature. The optical power used in modeling of these devices is defined as the average optical power at the operating conditions for each respective cell. The diode p-n junction temperature is calculated using the thermally dissipated power and a diode package thermal resistance of 41.7^{0}C/Watt. It should be noted that several of the cells are driven at extreme overstress conditions with current and junction temperatures as high as 600 mA and 169^{0}C, respectively. These conditions are significantly harsher than reported previously [2,4-7], and are specifically designed to derive a failure acceleration model for the 6530 and to demonstrate the overall robustness of the design.

In previous lifetests, the primary failure mode had been identified as random sudden failure [7]. In this case, the output power of the laser diode falls suddenly without any precursory indication. The development success of the 6530 has resulted in a random sudden failure rate so low it has become difficult to model. Since failure occurs randomly, a large number of device operating hours have been required to establish an acceleration model for random failures rate with high confidence. These harsh multi-cell test conditions are intended to maximize the number of accelerated device hours and increase the corresponding number of observed defects.

In addition to the 320 laser diodes operating at eight distinct operational points, 2861 diodes have been randomly sampled over the past year from the 6530 production line. This population is operated at approximately 450 mA and 70^{0}C heatsink temperature, corresponding to ~235 mW. In total, this population has accumulated more than 3 million actual device hours. Fig.1 shows a Weibull plot of surveillance sudden failures.

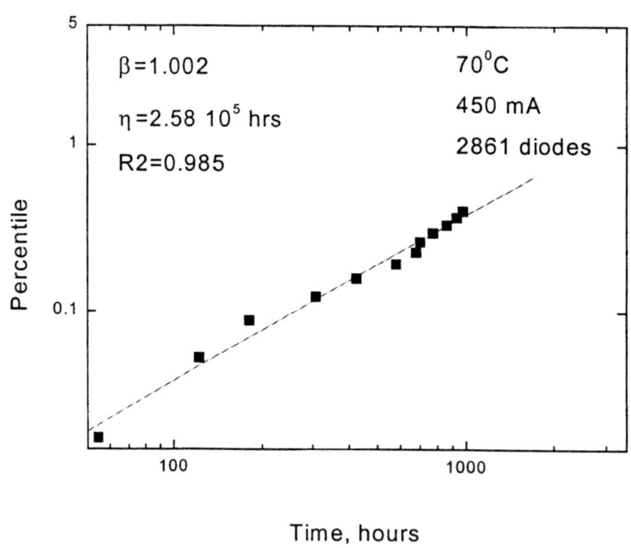

Fig.1 Weibull plot of sudden failures on surveillance test.

Fig. 1 clearly demonstrates a constant failure rate (=1) for random sudden failure with MTTF=29.5 years at the operation conditions of 450 mA (~235 mW) and 70^{0}C. For statistical certainty of temperature acceleration of random sudden failures, 240 diodes were put onto testing at 450 mA with a heatsink temperature of 107^{0}C.

Thus, the reliability testing of the 6530 is, at the time of the writing of this paper, composed of ten distinct operating conditions and 3421 devices. Five of the cells, with total 200 laser diodes, have operated more than 13,000 hours. This set of data is sufficient for modeling of the reliability of the 6530. In order to define an acceleration model consistent with the failure rate observed in all of the cells and surveillance tests a maximum likelihood calculation for power-law current acceleration and Arrhenius temperature acceleration has been performed. A thermal activation energy of E_{act}=0.64eV

$$FailureRate \quad \propto \quad I \bullet \exp\left(-\frac{0.64eV}{k_B T_j} \right)$$

and power of current acceleration m=1.2 provide best fit to experimental data. To yield a conservative estimate, a linear current acceleration model has been used as shown explicitly below

T_j is the temperature of the active region of the device. The active region temperature is given as

$$T_j \quad = \quad T_h + R_T(I \cdot V - P)$$

Where T_h is a heatsink temperature, R_T the thermal resistance, I the drive current, V the forward voltage, and P the output power. Accelerated hours in both tables are calculated using this acceleration model. With this model a random failure rate of 49 FIT is calculated (90% confidence level), at a 25 C heatsink temperature and a 200 mW output power. FIT calculations based on independently on multicell or surveillance testing give very close FIT values (64 and 46, respectively) pointing to consistency of the acceleration model.

In previous generations of 980 nm devices the sudden random failure mode dominated other failure modes. As the sudden random failure has been nearly eliminated in the 6530, a gradual wear-out failure mode, similar to that observed in InP devices, has also been observed in the 6530 multi-cell test. This type of failure mode has been detected only in the harshest cells. The appearance of this failure mode is a direct result of the extreme overstress conditions experienced by diodes in these cells. The onset of wear-out is also accompanied by an increase in the sudden failure rate. Fig. 2 shows the lognormal distribution of sudden failures vs. time normalized to the wear-out time of individual diodes to 90% of initial power. It includes diodes from all 4 cells exhibiting wear-out. With more than 13,000 hours of testing, power degradation has only been observed for laser diodes simultaneously having drive currents exceeding 500 mA and junction temperatures exceeding 110°C.

Fig. 2 shows that most of the sudden failures lie along the line pointing to correlation of sudden failures and wear-out. The first 7 failures (open squares) deviate from the line and are considered as sudden random failures and are therefore included into the sudden random failure count. They all happen at normalized time less than 1. A practical criteria used throughout this work is that a random failure is any failure occurring at an output

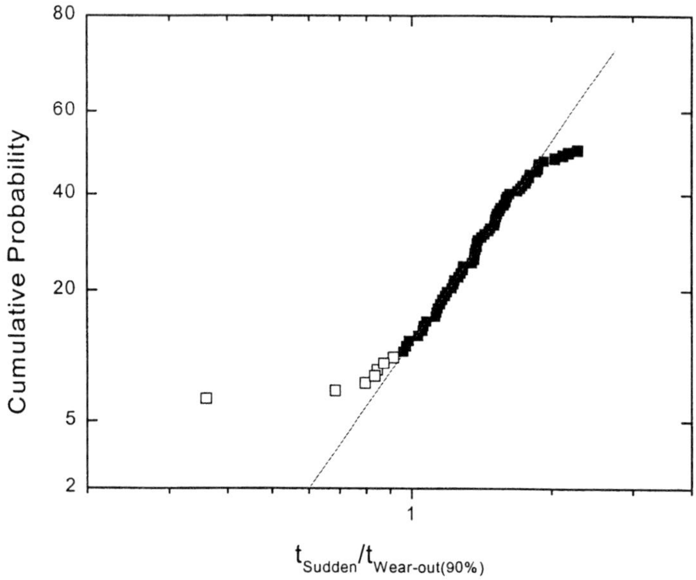

power greater than 90% of initial power, since the sudden random failure rate rises rapidly once the onset of wearout is observed.

Fig. 2 Weibull plot of all failures normalized to the mean wear out time.

This definition is an important distinction in the interpretation of lifetest results. For all calculations here, any failure prior to the power degrading below 90% output power is considered a sudden random failure. In the extremely overdriven cells where degradation does exist, the diode failure is defined as the time, or extrapolate time, at which the diode produces 90% of initial output power. To model the degradation, a quadratic function is fit to the data. Linear modeling of degradation does not hold here, in that it is observed that the degradation rate of InGaAs laser diodes increases gradually over time, thus required a polynomial function to achieve a reasonable match to the data. In most cases observed in this work, diodes that have degraded more than 10% exhibit accelerated sudden random failure rates. For the InGaAs laser diodes studied within this work, the use of linear degradation over time to an output power of 50% of initial power and an assumption of no increase in failure rates with increasing degradation [6], do not appear valid. The usage of such assumptions would lead to a substantial underestimation of failure rates.

Several of the extreme cells have different junction temperatures, 169^{0}C, 145^{0}C and 113^{0}C, and a common operating current, 500 mA. Analysis of the end-of-life times calculated as the time of 90% of initial power of diodes in these three cells allows the derivation of a thermal activation energy for wear-out. The average value for the wear-out activation energy for all 5 wafers is 0.64 eV. This value is larger than the 0.2-0.3 eV reported for ridge-waveguide [7] or self-aligned buried –ridge-waveguide 980 nm laser structures [6]. Based on lifetest results a wear-out failure mode with small activation energy is not observed in the 6530 laser diode. The wafer with shortest life in the multi-cell test has an end-of-life time (90% output power) of 48 years when extrapolated to use conditions of 200 mW and 25^0C heatsink temperature. Comparison of end-of-life times

$$\textit{Time to } 90\% \textit{ power} \quad \propto \quad \exp\left(-\frac{0.64eV}{k_B T_j}\right)$$

of the diodes operating at the same junction temperature, but at different currents, did not result in any current dependence. Therefore, the time to wear-out is primarily driven by temperature and can be expressed simply as

Recall that the temperature listed above is the junction temperature and as such does rise considerably above the heatsink temperature with increasing drive current.

The ultimate check of reliability is field data. The 6530 chip has accumulated 52.6 million device hours in field when estimated as discussed previously. No failures have been reported. This corresponds to a failure rate of less than 44 FIT with 90% confidence level at operation conditions (200mW, 25^0C).

NEXT GENERATION 980 LASER DIODES

Several trends within the communications industry have provided continued pressure for improvement in 980 nm pump module technology. The first is simple economics. The most straightforward way of quantifying this is the price of the pump module divided by the output fiber power. As a note, because of the quantum defect difference between 1480 nm pumps, the cost, in $/mW, of a 1480 nm pump is approximately 1.5 times less than an equivalent 980 nm pump. Thus, for economic equivalence between a 980 nm pump and a 1480 nm pump, the 980 nm pump would have to cost a factor of 1.5 less than the 1480 pump (or produce 1.5 times more output power). The price normalized to unit power is an important motivator for the development of higher power laser diodes.

The second trend is that the industry is packing more components into smaller volumes. This is important for two reasons, first, it raises size as an important constraint, and secondly, the thermal dissipation of the module becomes critical due to the higher packing density of active components. This makes the development of pump lasers capable of substantially higher total efficiency very critical to minimize thermoelectric power consumption and dissipation. 980 nm laser diodes have substantially lower operating currents and thermal dissipation than 1480 nm pumps and therefore require less expensive drivers and heat removal systems.

The discussion of the next generation of 980 nm pump laser diodes will focus on these three areas; power, efficiency, and reliability. The development goals for the 6540, the next generation 980 nm pump, are a substantial increase in performance in each of these areas, thus providing the industry with a revolutionary new product for amplification systems.

The next generation of 980 nm chip – the 6540, is designed to increase optical operating power to well beyond 300 mW. This objective was approached by a complete optimization of the 980 nm chip; working to minimize all drivers of failure while maximizing efficiency and power.

The biggest challenge to overcome was to increase the optical kink power to ensure mode stability beyond the range of all practical operating conditions. A multi-variant optimization technique, focused upon key device parameters, was used to design the device structure for optical performance. The resulting device design removes optical kink from being a practical limitation of device operation. This breakthrough will allow for a significant increase in the operational power of the chip. Fig. 3 shows the L-I curve for a typical 6540 chip. The output power of the chip is measured into a 0.55 N.A. aperture. The parallel far field of the device is shown in Fig. 4 for output powers of 60, 300, and 500 mW. The perpendicular far field of the device measures 23 degrees at the full width half-maximum points. The aspect ratio of the chip is 2.6.

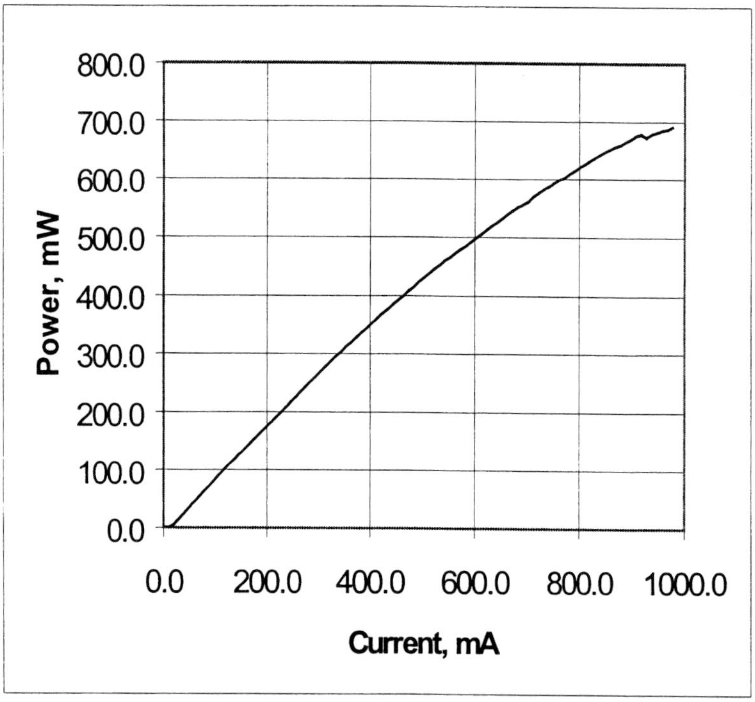

Fig. 3 The output power of the 980 laser diode, as measured into a 0.55 NA aperture. No optical instability is observed to output powers as high as 700 mW CW.

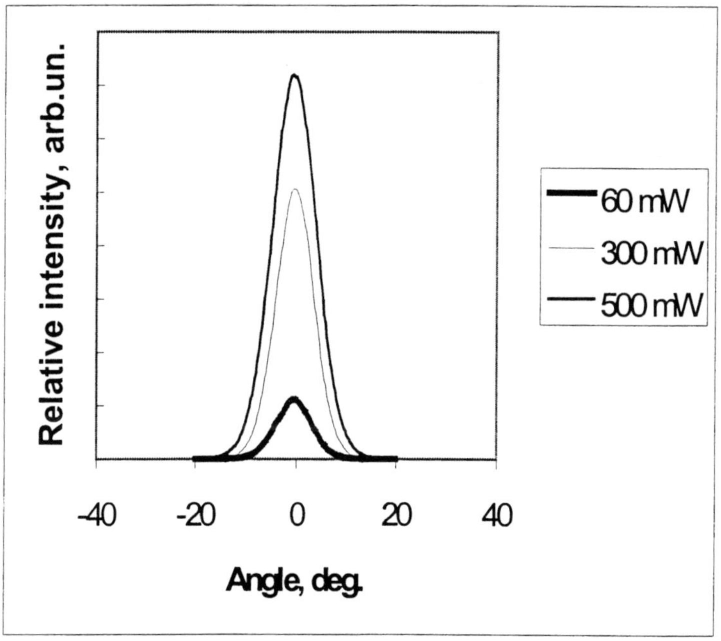

Fig. 4 The parallel far-field of a typical 980 laser diode. The laser diode stays in single mode operation at all power levels. No beam steering is detected. The full width at half maximum of the perpendicular far-field (not shown) is 23 degrees.

Fig. 3 shows that an optical power of 700 mw CW can be reached without observing kink. No beam steering was observed in the whole power range as shown on Fig. 4. The small nonlinearity observed near 700 mW is not an optical mode instability and is actually caused by a change in lasing wavelength. Further increase of the current results in thermal rollover. At thermal rollover, the maximum optical power obtained for the chip is 800 mW. With pulsed operation, the output power reaches 1.4 W at a pulsed current of 2 A (duration 200 ns, repetition rate 10 kHz) with no failure or degradation observed. It is worth noting that unlike high-power decoupled confinement heterostructures [8], the 6540 has a very low threshold of 17 mA.

Laser emission was coupled to lensed fiber with grating with results shown on Fig. 5. Coupling efficiency was 80% and optical power ex-fiber exceeded 320 mW without kink. The experiment has been performed on uncooled set-up, so heating causes a decrease of coupling efficiency at currents higher than 500 mA. In a cooled package, coupled powers of 0.5W CW are expected.

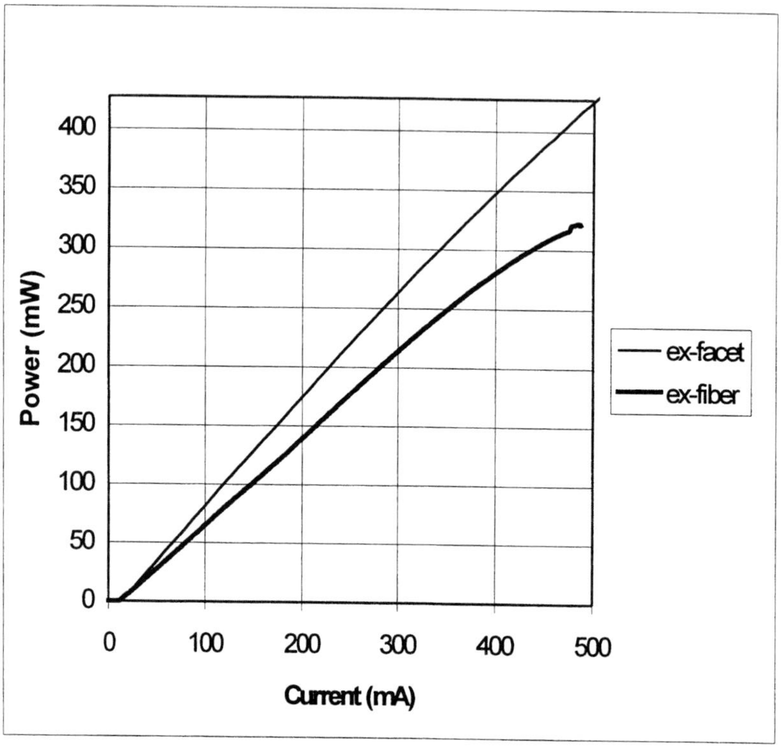

Fig. 5 The fiber coupled output power of the 980 nm laser diode. The uncooled device shows stable output powers exceeding 320 mW.

Table 2 shows several key aspects of operation for the 6540. First, the output power capability of the device is substantially upgraded, with an output operating power of 350 mW, an increase of 40% over the previous capability of the 6530. Second, the device is substantially more efficient at all powers than the 6530. For an operating power of 250 mW, the 6530 has dissipated power

Table 2. Comparison of chip parameters

Operational power, mW	200			250		270	350
Specification	6520	6530	6540	6530	6540	6540	6540
I_{th} (mA)	12	14	17	14	17	17	17
I_{op} (mA)	275	260	235	330	295	320	420
R_s (Ohm)	3.4	2.6	1.2	2.6	1.2	1.2	1.2
Wallplug efficiency	32%	28%	53%	35%	51%	50%	46%
Dissipated power (mW)	420	320	180	470	243	275	415

of 470 mW. The 6540 only dissipates 243 mW for an output power of 250 mW, representing a 50% decrease in dissipated heat. Obviously, this is a dramatic decrease in dissipated power, making the 6540 very well suited for the stringent needs of future optical amplifiers with tight packaging densities. As thermal dissipation becomes a larger issue, the ability of 980 nm laser diodes to have very high output powers with low thermal output may become a large discriminator between various 980 nm pump suppliers, and a potential factor in choosing between 980 nm and 1480 nm technologies. 1480 nm pump lasers typically dissipate very large amounts of heat, limiting their utility in amplification systems with very tight thermal budgets. Finally, with a nominal 75% coupling efficiency, the 6540 laser diode will produce module operating powers greater than 250 mW, a 67% improvement over the 150 mW operating powers of modules now commonly available.

Table 3. Reliability comparison

	6530		6540			
P, mW	200	250	200	250	350	500
FIT	49	100	11	15	26	93

Table 3 compares reliability estimates for the 6540 and 6530 chips. To model the reliability of the 6540, a thermal activation energy 0.64eV is used with a linear dependence upon current density (A/cm^2) for failure rate acceleration. This model is chosen simply to match the 6530 model, since the basic design and operation of the devices are similar. Since the thermal resistance of the 6540 is much less than that of the 6530, the junction temperature T_j is much closer to heatsink temperature T_h for any given current. Combined with the higher efficiency and changes in chip design, this results in a nearly 5-fold decrease of the estimated failure rate of the 6540 as compared to the 6530 at 200 mW operation power. Moreover, at an output power of 500 mW, the failure rate of the 6540 is predicted to be less than that of the 6530 at 250mW and below 100 FIT. Thus, the 6540 could operate reliably and without optical instabilities at 500 mW, giving a fiber coupled output power potentially as high as 350 mW. It should be stressed that full multi-cell testing of the 6540 laser diode has not started and that this is a simple extension of the physics of the 6530 to the 6540 laser. However, in all metrics that drive reliability, the 6540 is dramatically improved over the 6530, leading to the expectation of large increases in reliable output power.

CONCLUSIONS

Within this decade, 980 nm pump laser diodes have progressed from laboratory demonstrations to volume production with exemplary reliability track records. 980 nm laser diodes are now in deployment for submarine systems, with failure rates predicted to be less than 50 FIT at facet output powers of 200 mW. The next generation of 980 nm laser diodes will raise module output powers by over 50% without sacrificing device reliability or increasing module thermal dissipation.

As previously discussed, there are many market forces driving output power of pump modules up. It is useful, in understanding these forces, to define the ultimate goal for 980nm pumps from the perspective of an amplifier designer. The ultimate amplifier is compact, inexpensive and has minimal thermal dissipation. The control of the amplifier is made as simple as possible. One route to such an ideal amplifier is the incorporation of a 980 nm pump laser with power sufficiently high such that only one pump is required to drive an entire broadband amplifier.

Future generations of 980 nm pump lasers will provide such high output powers that many multiple pump amplifiers can be executed with the incorporation of a single high-power 980 nm pump laser. These amplifiers will be compact, with form factors no larger than a small cellular phone.

The authors are grateful for the support of the Manufacturing group at SDL. We wish to thank Josiane Perinet and Didier Sauvage of Alcatel Optronics for discussions of chip reliability. Finally, the contributions of our customers in helping us understand their product needs is gratefully acknowledged.

REFERENCES

1. W.J. Gignac, J.S. Major, Jr., W.E. Plano, D.F. Welch, and D. Scifres, 18th Europe. Conf. Optical Commun., 1992, Mo B2.3, pp.69-72

2. H. Meier, 20th Europ. Conf. Optical Commun., 1994, pp.947-954

3.V.V. Rossin, R. Parke, J.S. Major, Jr., J. Perinet, P. Chazan, M. Biet, D. Laffitte, D. Sauvage, A. Gulisano, N. Archer, and S. Kendrick, 10th Optical Amplifiers and Their Applications Conference, Technical Digest, 1999, pp.175-178.

4. A. Kasukawa, M. Ohkubo, T. Ijichi, and Y. Ikegami, LEOS'98, pp.243-343.

5. H.P. Meier, C. Harder, and A. Oosenburg, OFC'96, ThG2, pp. 229

6. M. Usami, N. Edagawa, Y. Matsushima, H. Horie, T. Fujimori, I. Sakamoto, and H. Gotoh, Optical Fiber Communications Conference, 1999, PD39-1.

7. T. Strite and C. Harder, "Uniphase's 980 nm pump lasers show their reliability", Uniphase Laser Enterprise AG laser reliability report.

8. T. Fujimoto, Y. Yamada, Y. Oeda, A. Okubo, Y. Yamada, K. Muro, Proceedings of SPIE "Fabrication, Testing, and Reliability of semiconductor Lasers III", Vol.3285, 1998, pp.80-87.

Recent progress of fiber Bragg-grating stabilized laser diode modules for EDFA pumping

Akira Mugino
WA-Project Team, R&D Division,
The Furukawa Electric Co., Ltd.
6,Yawata-Kaigandori, Ichihara, Chiba, 290-8555 JAPAN

ABSTRACT

In this article, the recent progress and the performance, applications and reliability for 1480nm and 980nm pump laser module with Fiber Bragg Grating (FBG) wavelength stabilized for EDFA pumping purpose are discussed.

Keywords: Fiber Bragg Grating, 1480nm & 980nm Pumping Laser Diode Module, EDFA, wavelength stabilized laser, InGaAsP & AlGaAs/GaAs, lensed fiber, coupling efficiency, Spatial Hole Burning (SHB), Anti-Reflection (AR) coating, wavelength pulling width.

1. INTRODUCTION

The improvement of optical fiber performances brings up great development for optical communication systems. Aiming at high speed and high capacity optical transmission, high-speed modulation such as Time Division Multiplexing (TDM) is being used. High-speed transmission of 10Gbps is being realized in 1996. The goal is now aimed at 40Gbps. However, due to the rapid growth in the Internet traffic demands, for example, one home user needs 100kbs for general information and 1Mbps will be needed for image data transmission. On the other hand, it is difficult to transmit such huge information data by only improving the modulation speed because of modulation limit of electronic circuits or optical components. To lay more optical fiber may be the solution if without considering the cost. Therefore, the only solution for increasing the transmission capacity is to use Wavelength Division Multiplexing (WDM) together with TDM. At OFC '99 post deadline paper section, 3 Tbps by combining 160Gbs x 19 channel was reported[1]. This means optical transmission is going into the "Tera-bit per second" forefront. By the way, Tbps class transmission can only be realized by using EDFA to amplify the optical signals in order to get long distance transmission. Due to the ITU standard, the wavelength intervals for WDM system are recommended to 100GHz, that is about 0.8nm. For more dense WDM (D-WDM) systems, 50GHz spacing are advisable. However, the increasing in channels for D-WDM means the needs for more gain potential of EDFA. This means that it is necessary to supply more pumping power for EDFA. On the other hand, the fiber output power of pump laser diode module that are classified as the highest output grade are generally about 160mW-180mW in 1480nm wavelength and 120mW-150mW in 980nm wavelength respectively. Even though using polarization-multiplexing technique, only 2 times output power can be achieved by using 2 laser modules.

Correspondence: E-mail: mugino@ch.furukawa.co.jp ;
WWW: http://www.furukawa.co.jp Telephone: +81-436-42-1775; Fax: +81-436-42-1736

Generally, 980nm-pump laser module is famous for its low noise figure and low temperature dependency properties. Also it is more efficiency than 1480nm-pump laser for EDFA pumping. However, the pumping wavelength limits in 980nm is only about 5nm central at 980nm. Moreover, the gain spectra of EDFA are greatly affected by the pump wavelength sensitivity in 980nm region[2-4]. For these reasons, wavelength stabilized 980nm laser diode modules are necessary. On the other hand, fiber amplifier applications such as high power amplification, remote pumping and D-WDM systems require 1480nm pump power levels in excess of 500mW. As stated above, single laser module cannot fulfill those requirements. The only way to increase pumping power is to use wavelength division multiplexing to add all the output power of several 1480nm-pump laser modules together by WDM couplers. Since the pumping wavelength in the 1480nm region for EDFA is about 50nm ranged from 1450nm to 1500nm, it is better to lock the wavelength of the 1480nm pump laser modules to certain nano-meter. If the wavelength stability is within +/-1.5nm, for example, nearly 16 modules with different output wavelengths of 3nm separation can be used to produce high output pumping for EDFA. The problem here is how to lock the wavelengths of those modules. UV-written fiber gratings[5] has resulted in a variety of important technological advances in this purpose. Fiber Bragg gratings technique becomes a key manufacturing technology that provides convenient, low-cost, and reliable solutions to applications such as filtering, optical feedback, wavelength control, and sensing, or even thought as dispersive elements. Moreover, fiber-grating technology has had a particularly strong impact on the design of active devices such as fiber lasers, fiber amplifiers, and laser diodes. There is a great merit that by directly fabrication of low loss grating into the fiber core leading to highly spectrally selective property, despite the coupling loss, extremely low loss resonators with respect to the amplifier materials like semiconductor laser, etc. can be constructed. Also confirmation that fiber Bragg gratings are indeed thermally robust over typical operating temperatures insured its application for optical communication devices.

In this paper, we present an overview of the important role of fiber gratings in the design and performance of laser diodes especially for EDFA pumping purpose focusing on 1480nm and 980nm pumping wavelengths. Recent progress of our research will be discussed in detail concerning about the designs and structures of the laser chips, the comparison of the laser modules with or without a fiber Bragg grating external reflector. Also, module coupling structure, coupling technique and efficiency for 1480nm and 980nm laser chip and module, output power performance, wavelength stability and total reliability will be discussed respectively.

2. FBG STABILIZED PUMP LASER

2.1. Feedback Operation Classification

The effects of reflections, which feed light back into semiconductor laser diodes, have been studied for many years extensively[6-10]. Recently, the use of intra-core UV-written fiber Bragg grating as the optical feedback device for a semiconductor laser is widely studied. Concentrated on the feedback level, there are three types of applications to use FBG for the wavelength locking of the laser. The first one is for single mode operation which anti-reflection (AR) coating of the laser chip front facet is necessary. Generally, very low AR reflectivity (mostly less than 10^{-4}) is strongly recommended such that only the external cavity mode are survival for single-mode laser operation[11-17].

The feed back level is generally higher than –10dB. The second one is for mode-locked purpose[18] by modulating the injection current of the laser at the characteristic frequency of the optical cavity, which is determined by the position of the FBG to produce full control of the pulse width and frequency of the laser. The third one is for what we want to discuss; that is coherence-collapse[6,9-10,18] operation for wavelengths stabilized pump laser[19-28]. This coherence-collapse regime can be obtained if the FBG acts as a partial reflective device, which provides a fairly strong feedback level for a particular wavelength where the FBG position is placed beyond the coherence length of the laser diode. Generally, Fabry-Perot laser diode (FP-LD) type laser chip is used for its commercially availability. In the absence of FBG feedback, the FP-LD operates in a multi-mode oscillating where the wavelength of the modes are dependent to the gain peak position for a particular injection current level. If the "incoherence" feedback from the FBG is presented, the laser enters the coherence-collapse regime, which is characterized by a substantial broadening of the laser diode modes and a strong reduction of the coherence length of the laser. The lasing spectrum then consists of a multitude of external cavity modes whose amplitude is modulated by the mode structure of the diode and the reflection spectrum of the FBG. It is a well-known phenomenon for these kinds of external cavity laser such that the oscillating modes will be forced to operate within the optical bandwidth of the FBG. It is because that the optical loss is the least for such particular wavelength selected by the FBG in the optical resonating cavity formed between the rear facet and the reflecting point of the FBG. Thus stabilization of the laser wavelength was achieved by forcing the lasing mode to fall inside the spectral width of the FBG. Since the FBG parameters such as reflectivity, bandwidth (FWHM) and central reflection wavelength will affect the properties of the external cavity laser; we will make a detail analysis of those parameters relative to laser diode's front facet reflectivity. Moreover, though many references[20, 22] stated that it is not necessary to have an AR-coated semiconductor laser diode for this particular coherence collapse regime, we find that the reflectivity of this AR-coated (or low reflectivity coated) facet influence the stability of the lasing wavelength relative to the limit of the operating current which in turns affecting the FBG wavelength pulling effect. Since there is no 100% coupling between the laser diode and the output pigtail fiber of the laser module where Bragg grating is embedded, the coupling efficiency will give a great influence to the drop of output power for a FBG stabilized pump laser compared with a general one.

3. THEORY OF EXTERNAL CAVITY LASER

3.1. Principle of Effective Reflectivity Model

A schematic simulation model for calculation of an external laser cavity is shown in Fig.1. By considering the effective reflectivity r_{eff} and r_{10}, the threshold condition can be expressed as

$$\left[\frac{r_{12}+C^2 r_{23}\exp[-2j\beta_2 L_2-(\alpha_3+2j\beta_3)(L_3+L_{geff})]}{1-C^2 r_{21} r_{23}\exp[-2j\beta_2 L_2-(\alpha_3+2j\beta_3)(L_3+L_{geff})]}\right]\cdot r_{10}\cdot\exp\{-[\Gamma(\alpha_1-g)+2j\beta_1]L_1\}=1 \quad (1)$$

In equation (1), L1 is the laser chip length, g is the gain coefficient, Γ is the active layer confinement factor. Also, the absorption coefficient and the propagation constant of the laser chip and fiber are $\alpha 1, \alpha 3, \beta 1, \beta 3,$ (air is $\beta 2$) respectively. The power coupling efficiency between chip and fiber is also represented by C where L2 is the distance between chip front facet to fiber front end; L3 is the distance from fiber end to grating head; L_{geff} is the effective coupling length; $\Delta\lambda_{fg}$ is the FWHM; and r_{23} is the reflectivity of the FBG. By using r_{eff}, we can calculate the relationship between the AR coating reflectance, FBG reflectance and the power coupling efficiency C relative to the effective reflectance R_{eff}. Fig.2 shows the computed results for C is set to 75% where Rg is the center wavelength power reflectance of the FBG. Table 1 shows the 1480nm and 980nm band semiconductor laser chip parameters that are used for simulation.

3.2. Simulation Results of Output Power Considering SHB for 1480nm Laser

In our fabrication process, the rear facet reflectance of the 1480nm laser diode is coated to 98% (HR) where the AR reflectance is coated to 6%. Because of the asymmetrical reflectance of the laser chip, the output power in the AR end is very high compare with the HR end-face. This unbalance of output power will introduce the Spatial Hole Burning (SHB) effect, which will make the output power saturate with respect to the decrease of AR reflectance. Therefore in order to maximize the laser output power one must consider the SHB effect[29]. Fig.3 shows the module fiber end output power of 1480nm-semiconductor laser diode module relative to the AR coating reflectance assuming that the coefficient is 75% and the injection current is 500mA. In Fig.3, it is shown that the laser output power started to saturate when the front facet reflectance is near 2% whereas the conventional method (no SHB consideration) shows no saturation. Also if the reflectance is less than 0.5%, the output power will give an undesirable result of power dropping. Next, as looking at Fig.1 carefully, it is obvious that we can use R_{eff} to replace the AR reflectance to evaluate the output power of the FBG external cavity equivalently. It is important to note that the optimal R_{eff} value can be realized by choosing AR-reflectance, FBG reflectance and coupling conditions suitably (see Fig.2) so as to maximize the output power or to minimize the power degradation caused by introducing a FBG external cavity with coupling efficiency less than 100 %. This effect will be discussing further in details for 980nm-band FBG laser diode modules.

3.3. FBG Wavelength Locked 980nm Pumping Laser

3.3.1. Wavelength Pulling Width Concept

Here, an important concept so called "***wavelength pulling width***" is introduced in order to understand the wavelength locking mechanism of the 980nm-band FBG wavelength stabilized external cavity laser. It is different from 1480nm-lasers where InGaAsP/InP semiconductor materials are used; 980nm-lasers are mostly fabricated by using AlGaAs/GaAs. For this reason, the beam of the 980nm-laser is of ellipse shape with x-axis (parallel to the active layer) beam diameter larger than y-axis (perpendicular to the active layer) diameter. This property makes it difficult to couple the elliptical beam into the single mode fiber by symmetrical two-lens scheme and less coupling efficiency than 1480nm-laser are usually expected. For this reason, from equation (1), since the coupling efficiency C will influence the threshold condition so as the lasing wavelength, it is desirable to predict the stability of wavelength locking strength by a good definition concept. Here we express the strength of wavelength locking by means of "***wavelength pulling width***". It is defined as the width difference between the FBG central wavelength λ_{FBG} to the laser chip gain peak wavelength limit $\lambda_{gain\text{-}FP}$, such that the lasing mode of the FBG-laser itself will change suddenly from FBG locked lasing mode to Fabry-Perot lasing mode at wavelengths beyond this particular gain peak wavelength $\lambda_{gain\text{-}FP}$. In Fig.4, the mirror loss and gain profile of the FBG wavelength stabilized 980nm-laser are shown where an idea AR reflectivity (zero) of the laser front facet is assumed. It is shown that if the reflection of the laser chip front end is negligible, the mirror loss will strongly depend on the reflection spectrum profiles of the FBG as shown in Fig.5. This explains why FBG can combine with a Fabry-Perot laser chip to lock the lasing wavelength. In this case, by using the parameter shows in table1, we can calculate the total pulling width of the "FBG-locked laser" respected to the center wavelength of the FBG, λ_{FBG}. In Fig.6, it is shown that the maximum (or idea) wavelength pulling width is about 14nm (28 for both sizes) where Rg is set to 6%.. However, for the real cases, the AR reflectance will be of 1% to 5%. Also, the bandwidth of the FBG must be chosen to fit the longitude mode numbers and laser's FSR. In Fig.7 to Fig.10, the mirror loss profiles and wavelength pulling width for different bandwidth (1nm and 2nm) of 1% AR coating reflectance are shown. The FBG reflectance is set to 10% for calculation. It is obviously that the change in FBG bandwidth will influence the mirror loss shape greatly. By comparing Fig.8 and Fig.10, one can carefully observe that the stability or variation of lasing wavelength relative to the shift of gain peak caused by injection current or temperature will be different respect to FBG bandwidth. However, the change of FBG bandwidth is not important in sense of pulling width influence. In this simulation, change of FBG bandwidth from 1nm to 2nm will just produce a little change of only 0.07nm of pulling width. In next section, the change of AR reflectivity related to wavelength pulling width will be considered together will the coupling efficiency and spatial hole burning effect for optimization of output power and wavelength stability in terms of pulling width.

3.3.2. Trade Off Property between Output Power and Wavelength Pulling Width

In this section, another important result concerning about the optimization of output power relative to wavelength stability will be discussed. For EDFA pumping purposes, output power of the laser diode module will be taken as the top order of priority compare to other properties. There are three factors that will directly impact the output power of FBG-lasers. These are FBG reflectance, AR reflectance; wavelength shift limit for current injection below the End of Life (EOL); and coupling efficiency of laser chip to FBG fiber. Increasing the FBG reflectance will increase the pulling width of the FBG-laser but giving degradation of output power. Increasing the pull width of the FBG-laser will give a secondary effect to enlarge the limit of current injection range. For example, if the kink level of the 980nm FBG-laser is above 500mA, one can operate the current as high as to the kink limit of 500mA for no FBG wavelength stabilized 980nm-lasers because there is no restriction concerning about the wavelength stability. This means that the shift of wavelength to longer wavelength lasing region by large amount of injection current is not important. However in the case of FBG wavelength locked 980nm-laser, the laser must lasing within a stabilized wavelength region which conventionally is less than $\pm$1.0nm relative to the specified center wavelength. On the other hand, the shift of wavelength due to current injection for a conventional 980nm laser is about 0.03nm/mA (greater than 1480nm-lasers). Thus, for a injection current range, for example, range of 250mA or 300mA, will shift the laser gain peak wavelength from about 7.5nm to 9.0nm. This means that it is necessary that the least wavelength pulling width of such 980nm FBG wavelength stabilized laser is greater than 9nm for 300mA ranges current injection. If the pulling width is only about 7nm, the injection current will be limited to less than 250mA ranges despite that the kink level is as high as 500mA. In this case, if the injection current range is set to more than 250mA, the lasing wavelength will jump to long wavelength side suddenly because of failure wavelength locking.

The secondary factor is the AR reflectance. SiN or SiO2 is generally used as the anti-reflection film materials for an AlGaAs/GaAs based semiconductor laser diode. This will make the lower-limit of the AR-coating reflectivity to about 0.5%. Generally, the AR-reflectance will be designed from 1% to 10%. Our simulation shows that the change of AR reflectance from 1%, 5%, and 10% will make the wavelength pulling width become to 10.48nm, 8.74nm, and 8.18nm respectively. This means that the designation of the AR coating, for example, 1% and 5% will give a result of pulling width (one side pulling) difference of (10.48nm-8.74nm)=1.74nm. Since both sides pulling can be used to design the laser chip with respect to a particular FBG wavelength, it will give a total difference of 3.5nm for pulling width. This value means that one can earn 3.5(nm)/0.03(nm/mA) =118mA current injection margin by suitably choosing of the FBG wavelength for particular EDFA systems.

The third factor is the coupling efficiency of the laser chip to pigtail fiber (FBG embedded). Since the power coupling coefficient C can be change dramatically during the assembling process of the module, knowledge of the relationship between power coupling efficiency C relative to AR or FBG reflectance is facilitated. Like 1480nm-lasers, one more factor need to concern is the Spatial Hole Burning. In conventional calculation treatment, the threshold carrier density Nth is treated as a constant. However, because of the AR-coating and HR-coating, the distribution of the carrier will be forced to change during the longitude position of the laser chip as shown in Fig.11, where the change of power relative to laser chip position is computed. Near the HR rear facet, the carrier density is much higher than the AR front facet about 2 times and distribution is of exponential shape. From this results, the relation of AR reflectance (or effective reflectance Reff) to output power for 250mA current injection is shown in Fig.12. Saturation of output power is observed when the AR reflectance is greater than 1%. Together with those simulated results, we demonstrated how to design a 980nm FBG-laser module by using Fig 13. In Fig 13, the left hand side y-axis represents the power degradation percentage of the FBG-laser module with respect to no FBG laser module; the right hand side y-axis, which represent the wavelength pulling width, and the x-axis represents the FBG power reflectance. Also the AR reflectance is set to 1% for this designation. By using this graph, one can optimized the FBG reflectivity by starting with percentage of power degradation or starting with wavelength pulling width, depending on what the priority is. For example, less than 8% power degradation, and greater than 9nm pulling width condition will allow one to find out that less than 6% reflectance of FBG must be used for 1% AR reflectance.

4. EXPERIMENTAL RESULTS

4.1. Experimental Result of 1480nm FBG Stabilized Laser Diode Module

In our experiment, AR reflectance is made to 2%, 4%, and 6% where FBG reflectance of 10%, 4% and 1% are used. Power coupling efficiency C is about 75% and conventional two-lens coupling scheme is used (see Fig.25). These mean the effective reflectance R_{eff} will be equal to 13.4%, 8.13% and 4.6%, etc., for a 2% AR coated laser diode module. The final fiber output power level is measured and the output level is plotted in Fig.14. It can be shown from Fig.14 that those combinations of AR and FBG reflectance are just equal to a single laser chip with AR reflectance coated to 13.4%, 8.13% and 4.6% respectively. Therefore the experimental results are perfectly matched with theoretical prediction for a laser diode with AR coated as shown in Fig.14. Here the power showed in Fig.3 and Fig. 14 are the total output power generated by every wavelength spectrum. One may realize that the important thing for the application of WDM multi-channels pumping technique[30-32] is to keep the percentage of output power concentrated at a particular wavelength window with preset specified width. This concept is illustrated in Fig.15, such that the "***power in band***" percentage is the most important factor that really used to pump an EDFA for a particular chosen or specified wavelength window width. In Fig.16, the "***power in band***" for no FBG 1480nm pump laser diode module (shadow points) is compared with a FBG wavelength stabilized 1480nm pump laser diode module (solid points). It is shown that even the window's width is as narrower as 2nm, the power in band is nearly over 70% of the total output power. Generally, for WDM pumping technique application, wavelength stabilization requirement will be of 3nm for FWHM value. Also, Fig.17 shows the difference of the change of wavelength range relative to injection current between the with/without FBG for 1480nm-band pump laser module. It was shown that the laser wavelength is stabilized within ± 0.5nm centered near to 1485nm where the wavelength unstable (no FBG) module will have 0.02nm/mA variation range relative to injection current. Another example for comparison of with or without FBG output power effect is shown in Fig.18 such that laser chip of 4% reflectance AR coating are used where the reflectance of the center wavelength of the FBG is also 4%, which gave the effective reflectance to be equal to 11.5%. In this case the output power for 75% coupling efficiency of AR reflectance 4% will shift to effective reflectance of 11.5% (see table 2). This means that 145mW-output power will drop to 135mW-output level where the degradation percent of FBG module relative to no FBG module is about 7%. Table 2 shows the relationship between the AR reflectance, the FBG reflectance and the effective reflectance. By the way, it is important for a laser diode to emit stable wavelength even though there is no thermal electrical cooler (TEC). In Fig.19, the TEC temperature of the FBG wavelength stabilized pumping laser diode module is varied from 5 to 65 ℃. It is shown that the lasing wavelength of the module is locked to the FBG center wavelength (1485nm) very stable relative to the 60℃ temperature ranges. It is obviously that FBG locking the laser wavelength perfectly where the fluctuation of the wavelength is inside ± 0.5nm ranges.

4.2.The relationship between the FBG's FWHM and FSR of the Pumping Laser

The longitude mode Free Spectral Range (FSR) of the 1480nm-band pump laser module is about 0.4nm for a laser chip cavity of 0.8mm. It is important to optimize the cavity length of the laser chip with respect to the FWHM of the FBG reflection spectrum. This is because that the rate of change of output power to the change of injection current; that is dL/dI, will perform an oscillating shape due to the unstable mode hoping between the longitude mode confined within the FWHM of the fiber grating. For example if the bandwidth of the FBG is about 1nm, the main power occupied by the main mode will shift into the adjacent mode for increasing or decreasing injection current. In this case, the output power is mainly contributed by three main adjacent modes that will be moved inside the 1nm bandwidth of the FBG. Therefore it is necessary to design the FBG-bandwidth at least to 3 times of the FSR of the laser for safety or stable output power. It is easy to understand that for example, in a particular case, the most high power main mode is not coincide with the central wavelength of the FBG reflectance spectrum. And with a separation of 0.2nm from the FBG central wavelength will means that one of the mode adjacent to this main mode will line outside the FBG-bandwidth. In this case, the total output power inside the bandwidth of the FBG is only counted for 2 modes only. The third high power mode is falling out from the FBG bandwidth, making the total power of this FBG pump laser to become unstable. Fig.20 shows the variation of dL/dI for this particular case such that FBG-bandwidth is designed to 1nm and the length of the laser chip is designed to 0.8mm. The spectrum showing the change of main modes with respect to the marked region (① to ⑩, see Fig.20) and the shift width of the main mode is shown in Fig.21. It is obviously that the FBG bandwidth must be carefully chosen to obtain stable power output status. It is advisable that FBG-bandwidth is better to design in a way such that at least all the 20dB down level sub-modes are being included inside the FBG FWHM-bandwidth. Thus, these kinds of "***oscillating kink phenomena***" can be solved easily by just optimizing the FBG-bandwidth. Fig.22 shows the L-I property of a conventional fabricated FBG 1480nm-pump laser module where the FBG bandwidth was chosen to be 2nm where almost all the 20 dB down sub-modes are included. These results illustrated that optimum designs of the chip longitude mode spacing relative to the FBG-bandwidth can solve this kind of kink problem evidentially.

4.3. Chip and Module Structure and Reliability of 1480nm Laser

Fig.23 (b) shows the schematic diagram the 1480nm-band laser chips. It is constructed by using buried heterostructure (BH) with Muti-Quantum-Well configuration. Structural optimizations are made in terms of high performance and high reliability. This can be achieved by optimized the composition ad thickness of GRIN-SCH structure, together with the optimization of BH structure grown by MOCVD. The aging data for the laser chips is shown in Fig.24[33]. The output power was set to be 80% of the maximum output power at 60℃. The MTTF of more than million hours is predicted at 25℃.

In Fig.25, the configuration of the 1480nm-laser module is shown. A two lens coupling scheme is used to achieve high coupling efficiency over 80% where the beam of the laser is decided to circular shape. In Fig.26, high temperature agent test was shown where the testing time is over 10 thousand hours and no degradation was observed. FBG wavelength stabilized 1480nm-laser module with output power over 200mW with low driving current about 700mA was commercially available[24]. Fig.27 shows the L-I property. Also, Fig.28 shows the output spectrum example where the FBG center wavelength is set to 1463nm with laser bandwidth of 1.4nm. These kinds of pump lasers are used in systems and applications, which were reported in reference [30]-[32]. Excellent laser module and very good pumping properties were achieved for high power and high wavelength stability. As for reliability of FBG wavelength stabilized 980nm-laser module, agent test is under proceeding now and in Fig.29 several hundred hours result is presented. Although there is not enough data for discussing the reliability of the FBG-laser module, from Fig.29, it is shown that no degradation is obtained and we will report this result further.

4.4. Chip Performance of 980nm-Laser Diode

In this section, several important properties of FBG wavelength stabilized 980nm-pump laser are being discussed in detail. Firstly, the laser chip construction of 980nm laser is difference from 1480nm laser. In Fig.23 (a), the ridge waveguide structure of 980nm-laser chip was shown. Developments are concentrated towards high quantum efficiency for low power consumption and narrow beam divergence for high coupling efficiency into single-mode fiber. Also, facet passivation for high reliability is achieved. Catastrophic Optical Mirror Damage (COMD) is commonly observed in 980nm-lasers especially transverse-mode stabilized narrow stripe lasers. The most important item to obtain high reliability in 980nm lasers is facet passivation technology[34]. Fig.30 shows a typical L-I curve of our 980nm laser. No initial COMD is observed, and output power is limited due to the thermal effect. The chip light output power of over 400mW is obtained, with a high kink current of 400mA. The typical output power of 200mW is obtained at a current of 250mA. The FWHM of FFP's parallel and perpendicular to the junction plane are 11deg. and 23 deg., respectively. Also aging characteristics are shown in Fig.30. The condition is 250mW at 60℃ in APC mode. These aging tests reveal that MTTF of half million hours is predicted under 200mW operation at 25 ℃. Also, the agent test of no FBG 980nm-laser module is tested. Fig.31 shows the results of several modules for about 2000 hours. From this data, it is shown that the modules where lensed fiber coupling scheme is used have good agent test responds and degradation is less than 2% for 2000hrs. Also we are processing in the agent performance for FBG wavelength stabilized 980nm laser diode modules and the results will be reported as soon as possible.

4.5. Performance of FBG Wavelength Stabilized 980nm Laser Module

The coupling of laser chip to fiber end of 980nm-laser modules are successful made by using wedge shape lensed fiber. Fig.32 is a SEM picture showing the coupling part of the wedge fiber and the 980nm-laser chip. For the purpose of comparison, a FBG of 978.5nm center wavelength, 6% reflectance and 2nm bandwidth are fused as an external reflector. The L-I property before and after fusing of this FBG is shown in Fig 33 It is shown that the module fiber end power output drop from 167mW to 155mW at 250mA injection current condition showing that the power degradation after the attachment of this FBG is about 7%. Also the wavelength fluctuation is less than 0.3nm as shown in Fig.34 for the injection current range from threshold value (about 35mA) to 300mA. Also the output power relative to coupling efficiency is shown in Fig.35. It can be confirmed that the coupling efficiency is sandwiched between the 70% ~ 80% region by locating the effective reflectance R_{eff} to the point near 6.5% since 1% AR and 6% FBG reflectance is used for this particular module. Also the spectrum output of this module is shown in Fig.36 such that the center wavelength of FBG-stabilized 980nm pump laser module is about 978.7nm at 25℃ with a broadened laser line-width of 0.55nm. From the shape of the spectrum of this module, it can be suggested that this laser module is operating at coherence collapse regime as stated in section 2.1. Fig.37 shows the stability of the locked wavelength compared with no FBG locking. Also, a schematic diagram of "real size module" is shown in Fig 38 where the FBG is location outside the laser package.

5. CONCLUSION

The recent progress of fiber Bragg grating stabilized 1480nm and 980nm-pump laser for EDFA pumping purpose is discussed. Also we have shown how to maximum the output power of a FBG external cavity 980nm and 1480nm pump laser module by considering the effective reflectivity of the external cavity and related it to the output power through spatial hole burning effect theoretically. Also together with the wavelength pulling width concept and injection current range consideration, method to decide the specification of FBG being used are discussed conceptually. Measurements of the fabricated laser modules give good matching between theoretical and experimental results. Excellent laser module performances are realized with less than 10 % power degradation compared with no FBG application for conventional 1480nm and 980nm laser modules.

6. FIGURE CAPTIONS

Figure 1 : the schematic simulation model to derive the effective reflectivity of an external cavity laser.
Figure 2 : the relationship between AR, FBG and effective reflectance for 75% coupling efficiency of 1480nm FBG pump laser diode module.
Figure 3: simulation results showing the relationship of laser front facet reflectivity to laser output power with respect to with/without Spatial Hole Burning effect consideration.
Figure 4: simulated threshold property showing the mirror loss and laser gain curve shape where AR reflectance is set to zero percentage, FBG reflectance is set to 6% and coupling efficiency is 75%. Also gain peak wavelength is set equal to FBG central wavelength in this case.
Figure 5: an example of fiber Bragg grating (FBG) spectrum.
Figure 6: simulated wavelength pulling width value with respected to Fig.4 parameters.
Figure 7: simulated threshold property showing the mirror loss and laser gain curve shape for AR =1%, FBG bandwidth of 1nm and C=75%.
Figure 8: simulated wavelength pulling width value for AR=1 %, FBG bandwidth=1nm and C=75%.
Figure 9: simulated threshold property showing the mirror loss and laser gain curve shape for AR =1% and FBG bandwidth of 2nm, C=75%.
Figure 10: simulated wavelength pulling width value for AR=1 %, FBG bandwidth=2nm, C=75%.
Figure 11: schematic image and graph showing the theoretical prediction of laser chip internal carrier and power distribution by SHB consideration
Figure 12: theoretical output power simulation result by considering spatial hole burning effect.
Figure 13: wavelength pulling width and power degradation percentage graph with respect to FBG reflectance where AR is set to 1%.
Figure 14: comparison of experimental results and simulated predictions of FBG pump laser diode module with AR facet power reflectivity=2%, FBG central wavelength power reflectivity=1%, 4%,10% showing that theoretical and experimental results matched perfectly.
Figure 15: the schematic diagram illustrating the power in band concept.
Figure 16: comparison of power in band percentage relative to window width w for FBG and no FBG laser modules.
Figure 17: comparison of wavelength shift range for without FBG and with FBG laser modules.
Figure 18: comparison of I-L curve properties and spectrum properties for without FBG and with FBG laser modules.
Figure 19: output spectra of FBG stabilized 1480nm laser modules with respect to operating temperature as parameter.
Figure 20: L-I curve property showing the kink phenomena where 1nm FBG bandwidth is used for a 0.8mm cavity length 1480nm laser chip.
Figure 21: spectra of the kink points relative to the marks ① ~⑩ showing that the main mode shifting is about 0.5nm. This is why the reason that the "oscillating kink" occurred.
Figure 22: L-I property of FBG stabilized 1480nm laser module with FBG bandwidth of 2nm.
Figure 23: schematic construction diagram of 1480nm (a) and 980nm (b) laser diode chip devices.
Figure 24: agent test example results of 1480nm laser chips.
Figure 25: structure of 1480nm laser module using two lens coupling scheme.
Figure 26: high temperature agent test result of 1480nm laser modules.
Figure 27: L-I curve property of ultra high output power FBG wavelength stabilized 1480nm laser module.
Figure 28: output power spectrum of FBG stabilized 1480nm laser module with FBG center wavelength of 1463nm where the laser linewidth is about 1.4nm.

Figure 29: The agent test results showing no degradation during 380hours for FBG wavelength stabilized 1480nm laser modules.
Figure 30: 980nm laser chip L-I and FPP properties together with agent test result for 5000 hours.
Figure 31: 980nm lensed fiber coupled laser module agent test results for 1800 hours showing that no degradation of the laser modules.
Figure 32: SEM picture showing the coupling status of lensed fiber head with a 980nm laser chip.
Figure 33: L-I curve for comparison of without and with FBG stabilized 980nm laser modules' output power and degradation percentage.
Figure 34: relationship between the injection current and the output wavelength of FBG locked laser module showing that the variation of wavelength is less than 0.3nm.
Figure 35: output power properties with respect to effective reflectance for different coupling efficiency of 980nm FBG wavelength stabilized laser modules where injection current is set to 250mA.
Figure 36: wavelength spectrum of fabricated FBG stabilized 980nm laser module.
Figure 37: I-λ properties showing the wavelength stability for the FBG locking 980nm laser modules compared to no FBG locked modules.
Figure 38: schematic diagram showing the "real size" of the FBG stabilized 980nm laser module.
Table 1: Parameters of 1480nm and 980nm FBG pump laser diode module for simulation of effective reflectivity Reff and other static properties.
Table 2: values showing the relationships between AR,FBG reflectance and effective reflectance.

7. ACKNOWLEDGMENTS

The author would like to thank Dr. A.Kasukawa, Mr. S.Koyanagi, Mr.T.Kimura, T. Takase, Mr. Seki, Dr. Y.Irie and for their fruitfully discussion. Also, the author would like to acknowledge all members of the pump laser team at Furukawa Electric.

8. REFERENCE

1. S.Kawanishi, H.Takara, K.Uchiyama, I. Shake, and K. Mori, "3 Tbit/s (160Gbit/s x 19 ch) OTDM/WDM Transmission Experiment", OFC/IOOC '99 Postdeadline paper, PD1, pp.PD1-1-PD1-3, Feb., 1999.
2. K.W. Bennett, F.Davis, P.A.Jakobson, N.Jolley, R.Keys, M.A. Newhouse, S.Sheih, M.J.Yadlowsky, Optical Amplifiers and their Applications, Vol.16 of OSA Trends in Optics and Photonics Series (Optical Society of America, Washington,D.C., 1997), paper PD4-1-PD4-4.
3. Michael J, Yadlowsky, Leslie J.Button, "Pump-mediated inhomogeneous effects in EDFAs and their impact on gain spectral modeling", OFC'98, TuG5, pp35-36, 1998.
4. NE Jolley and F Davis, "Measurement of pump mediated inhomogeneity in an EDFA at liquid nitrogen temperatures", OAA, TuD6-1, pp.139-142, July 1998.
5. G.Meltz, W.W.Morey, and W.H.Glenn, "Formation of Bragg gratings in optical fibers by a transverse holographic method", Opt. Lett., Vol.14, pp.823-825, 1989.
6. R.W. Tkach and A.R.Chraplyvy, "Regimes of feedback effects in 1.5-mm distributed feedback laser," J. Lightwave Technol., vol. 4, pp.1655-1661, 1986.
7. G.P. Agrawal, IEEE J. Quantum Electron., QE-20, 468 (1984)
8. J.Helms, C.Kurtzke, and K. Petermann, IEEE J. Lightwave Technol., LT-10, 1137 (1992)
9. J.Wang and K.Petermann, IEEE J. Quantum Electron., QE-27,3 (1991)
10. J.Helms and K.Petermann, IEEE J. Quantum Electron., QE-26, 833 (1990)

11. D.M.Bird, J.R.Armitage, R.Kashyap, R.M.A., Fatah, and K.H.Cameron, "Narrow line semiconductor laser using fiber grating," Electron. Lett., vol. 27, pp.1115-1116, 1991.
12. P.A.Morton, V,Mizrahi, T.Tanbun-Ek, R.A.Logan, P.J.Lemaire, and H.M.Presby, "Stable single mode hybrid laser with high power and narrow linewidth," Appl.Phys. Lett., vol.64, pp.2634-2636, 1994.
13. R.Kashyap,"Wavelength uncommitted lasers," Electron. Lett., vol. 30, pp.1065-1066,1994.
14. W.H.Loh, R.I.Laming, M.N.Zervas, M.C.Farries, and U.Koren,"Single frequency erbium fiber external cavity semiconductor laser," Appl. Phys. Lett., vol.66, pp.3422-3424, 1995.
15. F.N.Timofeev, P.Bayvel, L.Reekie, J.Tucknott, J.E.Midwinter, and D.N.Payne, "Spectral characteristics of a reduced cavity single-mode semiconductor fiber grating laser for applications in dense WDM systems," in Proc. 21st European Conf. Optic. Commun. (ECOC'95), Brussels, Belgium, 1995, paper Tu.P.26, pp.477-480.
16. R.J.Campbell, J.R.Armitage, G.Sherlock, D.L.Williams, R.Payne, M.Robertson, and R.Wyatt, "Wavelength stable uncooled fiber grating semiconductor laser for use in an all optical WDM access network," Electron. Lett., vol.32, pp.119-120, 1996.
17. M.Ziari, J.-M. Verdiell, J.-L Archambault, A.Mathur, H.Jeon, R.-C.Yu, and T.L.Koch, "High speed fiber grating coupled semiconductor WDM laser," in Proc. Conf. Lasers Electro-Opt. (CLEO'97), Baltimore, MD, 1997, paper CMG1.
18. E.C.Vail, M.Ziari, V.Agrawal, A.Mathur, J.Major, Y.Zou, J.P.Zhang, P.Freeman, S.MacCormack, I.Booth, and J.Walmsley, "Laser source with high stimulated Brillouin scattering threshold for dynamic gain control in fiber amplifiers," OAA, ThD16, pp.199-202, 1999.
19. C.R.Giles, T.Erdogan, and V.Mizrahi,"Simultaneous wavelength stabilization of 980-nm pump lasers," IEEE Photon. Technol. Lett., vol.6, pp.907-909, 1994.
20. B.F.Ventrudo, G.A.Rogers, G.S.Lick, D.Hargreaves, and T. M. Demaya, " Wavelength and intensity stabilization of 980nm diode lasers coupled to fiber Bragg gratings," Electron. Lett., vol.30, pp.2147-2148, 1994.
21. D.Hargreaves,G.S.Lick, and B.F.Ventrudo, "High-power 980nm pump module operating without a thermoelectric cooler," in Proc. Opt. Fiber Commun. Conf. (OFC'96), San Jose, CA, 1996, paper ThG3, pp.229-230.
22. Jean-Luc Archambault and Stephen G.Grubb, "Fiber Gratings in Lasers and Amplifiers", J.of Lighwave Technol. vol.15, No.8, Aug., 1997, pp.1378-1390.
23. Atsushi Hamakawa, Takashi Kato, and Goro Sasaki, "1480nm pump fiber-grating external-cavity-laser with two fiber gratings,"OECC'97, 9D2-5, pp.224-225, 1997.
24. S.Koyanagi, A.Mugino, T.Aikiyo and y.Ikegami, "The Ultra High-Power 1480nm Pump Laser Diode Module with Fiber Bragg Grating," OAA '98, MC2, pp.35-38, 1998.
25. Atul Mathur, Mehrdad Ziari and Mats Hagberg, "High power grating stabilized 1480nm flared semiconductor pump laser," OAA'99, FB5, pp.263-266, 1999.
26. Akira Mugino, Toshio Kimura, Yuichiro Irie, Takeo Shimizu, "Realization of high power and wavelength stabilized 980nm-pump laser diode module with Fiber-Bragg –Grating by optimizing the effective reflectivity with Spatial-Hole-Burning effect," IOOC/OFC'99 ,TuC4, pp.29-31.
27. M.Usami, N.Edagawa, Y.Matsushima, H.Horie, T.Fujimori, I.Sakamoto, and H.Gotoh, "First Undersea-Qualified 980nm Pump Laser Diode Module evaluated

with Massive Life Test," IOOC/OFC'99, Postdeadline paper PD39, pp.PD39-1-PD39-3, 1999.
28. Victor V. Rossin, Ross Parke, and Jo S.Major, Jr., "Reliability of 980nm-pump laser module for submarine erbium-dope fiber amplifiers," OAA'99, ThD10, pp.175-178, 1999.
29. T.Higashi, S.Ogita, H.Soda, H.Kobayashi, H.Kurakake, O.Aoki and N.Okazaki, "Optimum Asymmetric Mirror Facet Structure for High-Efficiency Semiconductor Lasers," IEEE J.QE Vol.29 No.6, pp.1918-1923, June 1993.
30. Yoshio Tashiro, Satoshi Koyanagi, Kei-ichi Aiso and Shu Namiki, "1.5W Erbium Doped Fiber Amplifier Pumped by the Wavelength Division-Multiplexed 1480nm Laser Diodes with Fiber Bragg Grating," OAA'98, WC2, pp.213-215, 1998.
31. Yoshihiro Emori and Shu Namiki, "100nm bandwidth flat gain Raman amplifiers pumped and gain-equalized by 12-wavelength-channel WDM high power laser diodes", OFC/IOOC 99 Postdeadline paper, PD19, pp.PD19-1-PD19-3, Feb., 1999.
32. Kanji Tanaka, Katsuhiko Iwashita, Yoshio Tashiro, Shu Namiki and Shoichi Ozawa, "Low loss integrated Mach-Zehnder-interferometer-type eight-wavelength multiplexer for 1480nm band pumping," IOOC/OFC'99, TuH5, pp.88-90, 1999.
33. A.Kasukawa, "Recent Progress of High Power Laser Diodes for EDFA Pumping," IOOC/OFC'99, TuC1, pp.20-22, 1999.
34. M.Ohkubo et al., IEEE J. Quantum Electron., 30, pp.1936, 1994.

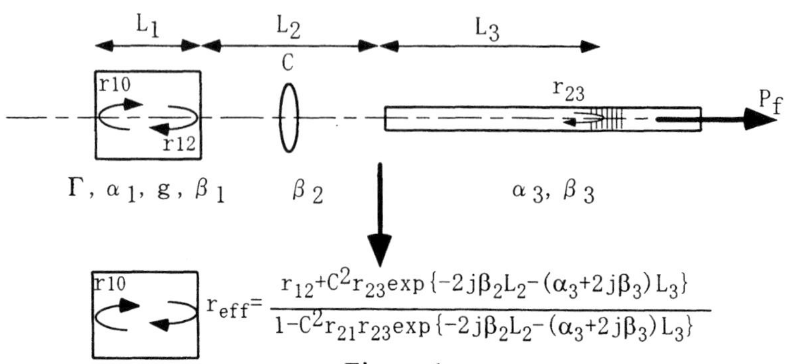

Figure 1

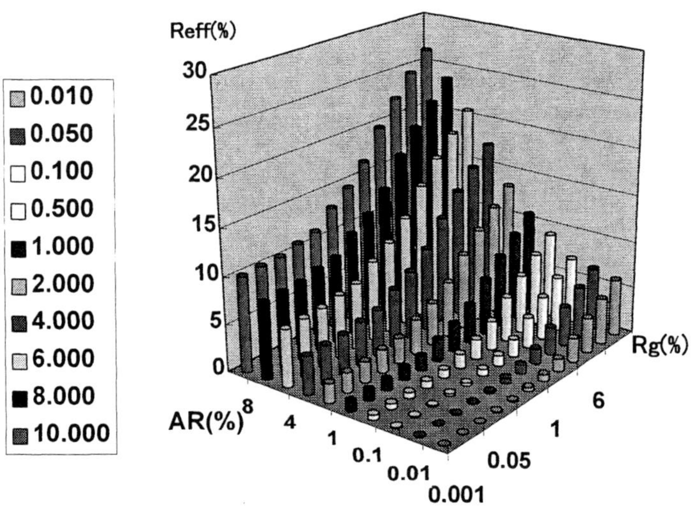

Figure 2

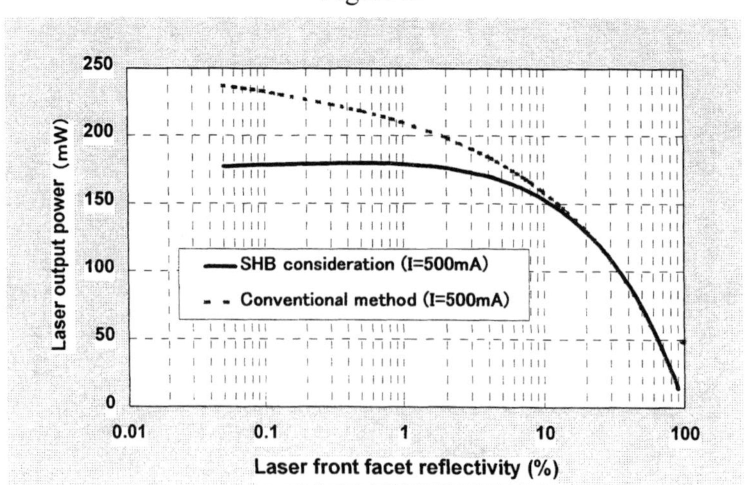

Figure 3

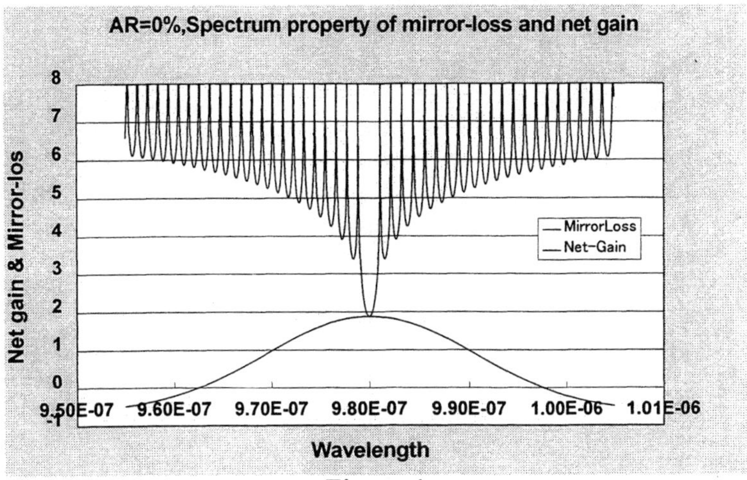

Figure 4

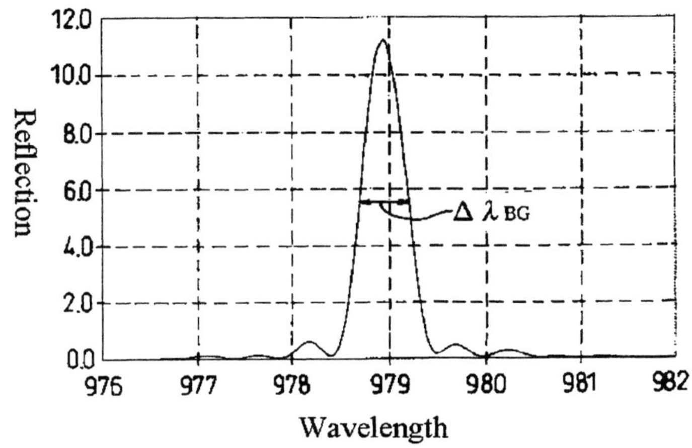

Figure 5

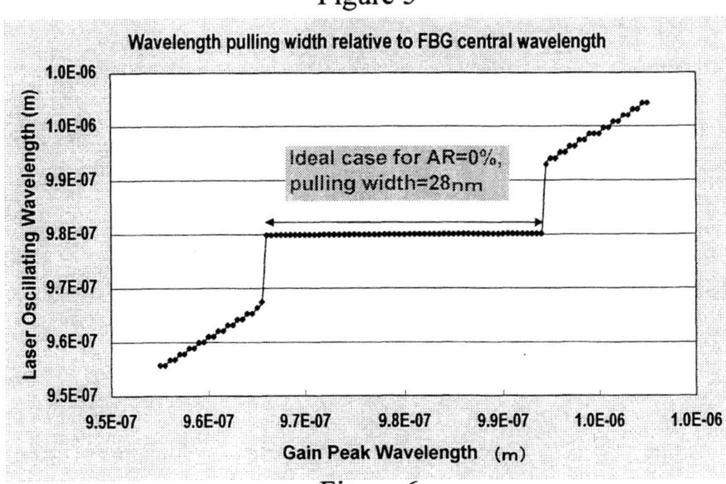

Figure 6

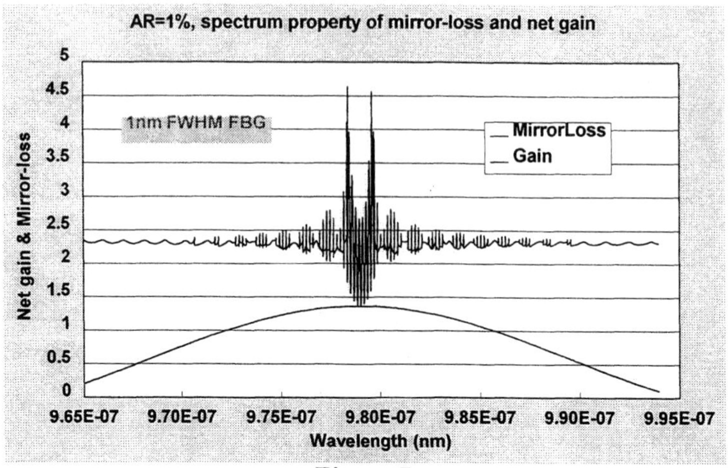

Figure 7

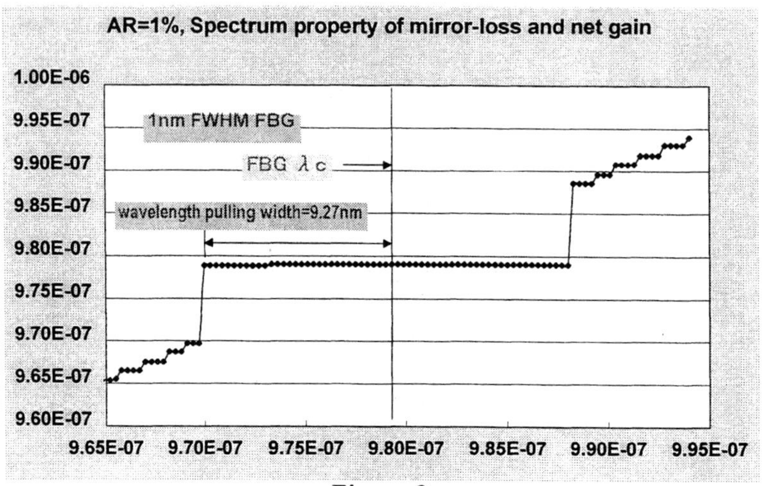

Figure 8

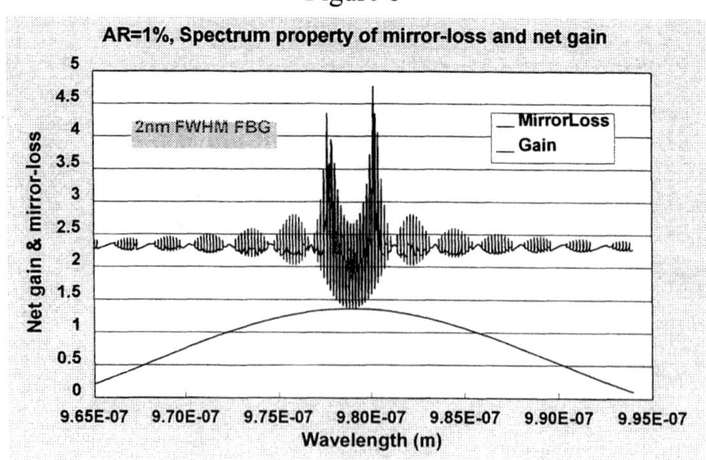

Figure 9

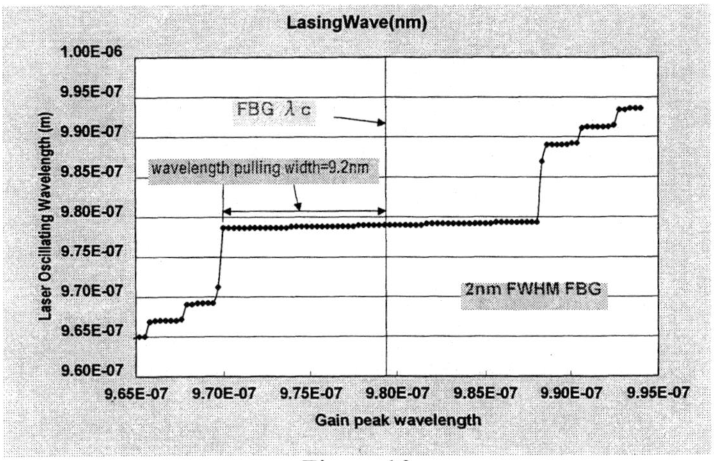

Figure 10

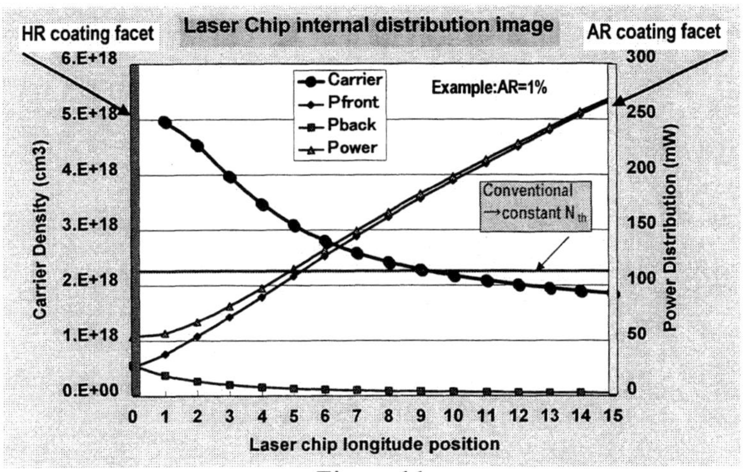

Figure 11

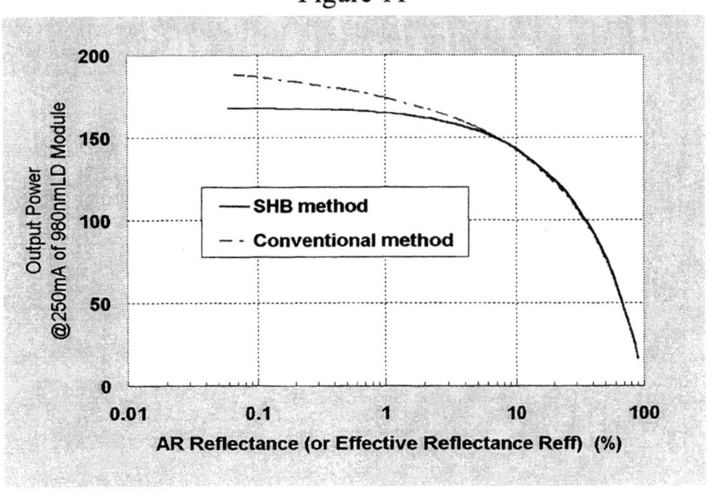

Figure 12

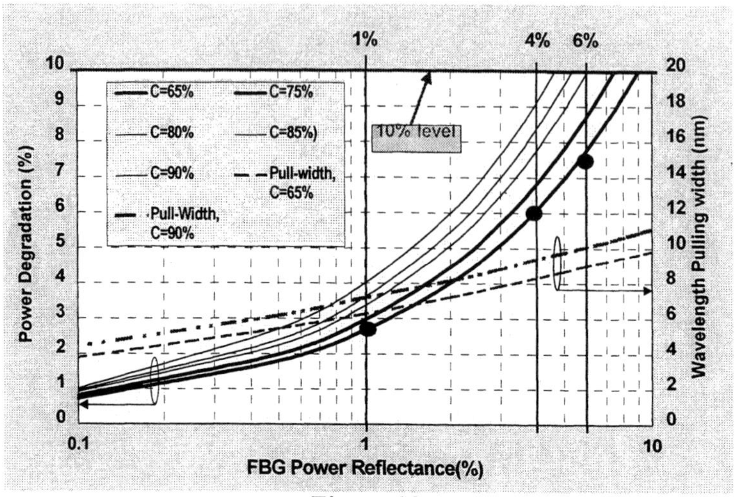

Figure 13

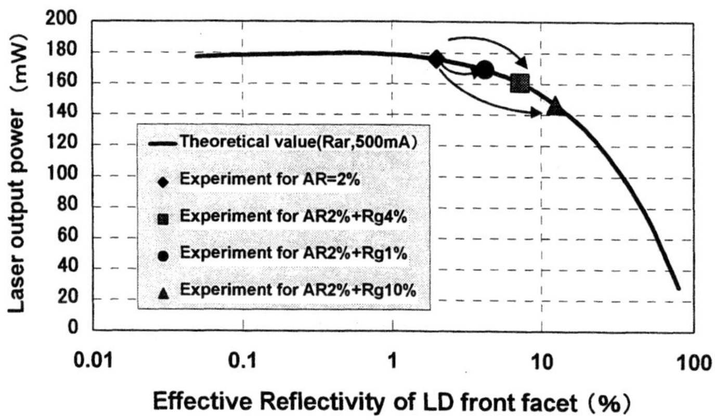

Figure 14

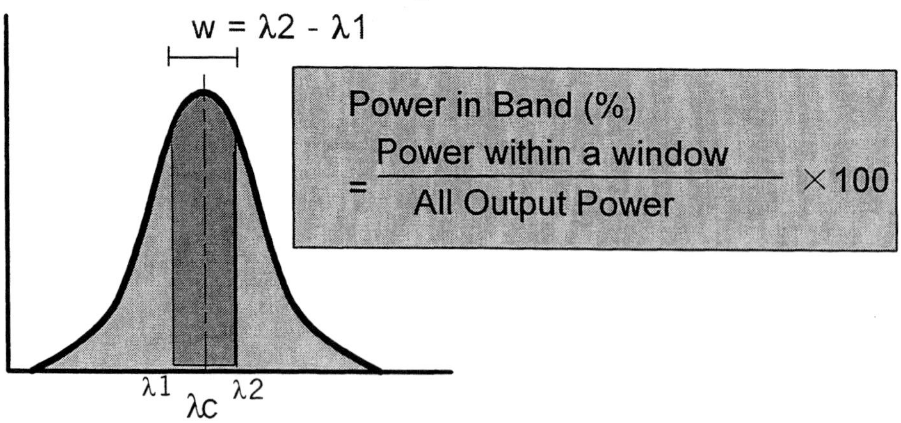

Figure 15

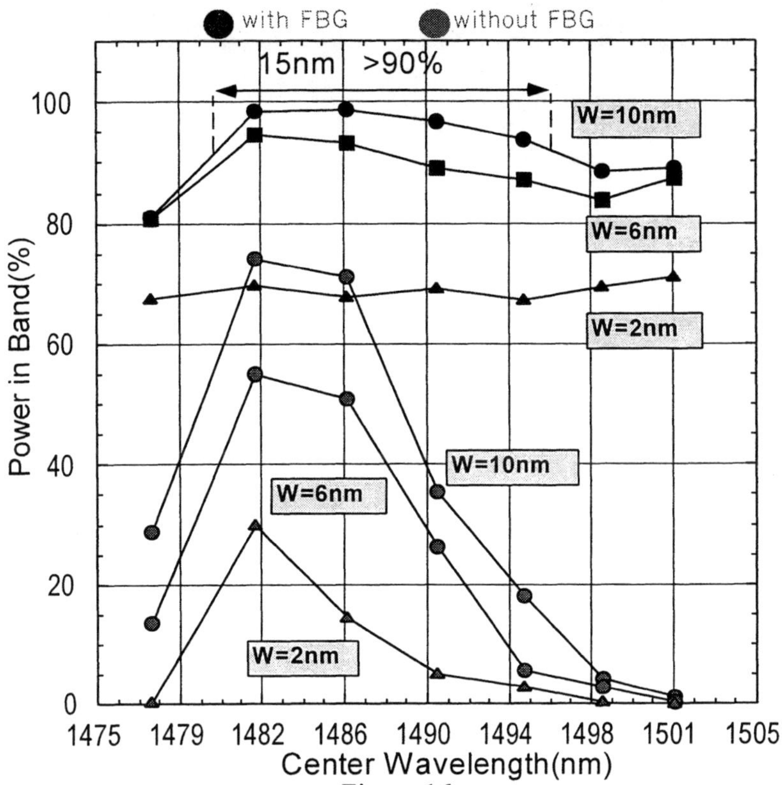

Figure 16

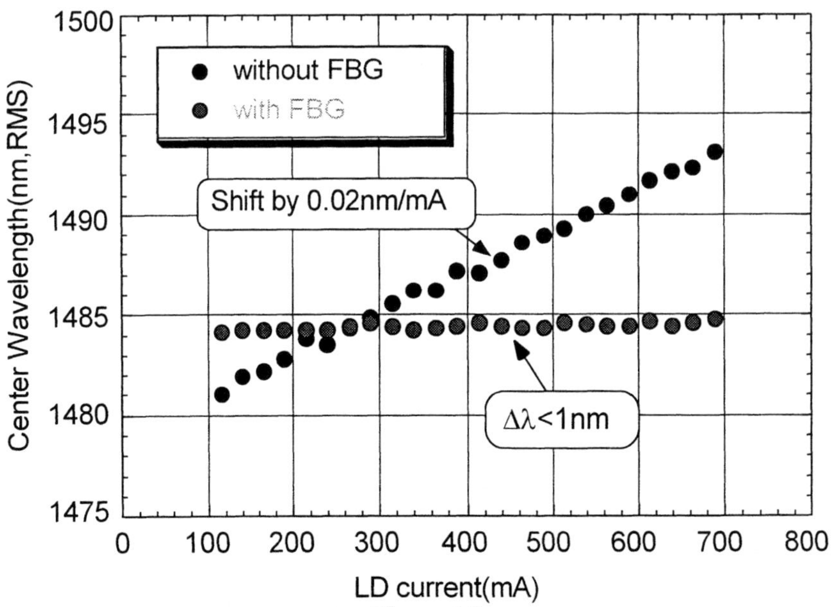

Figure 17

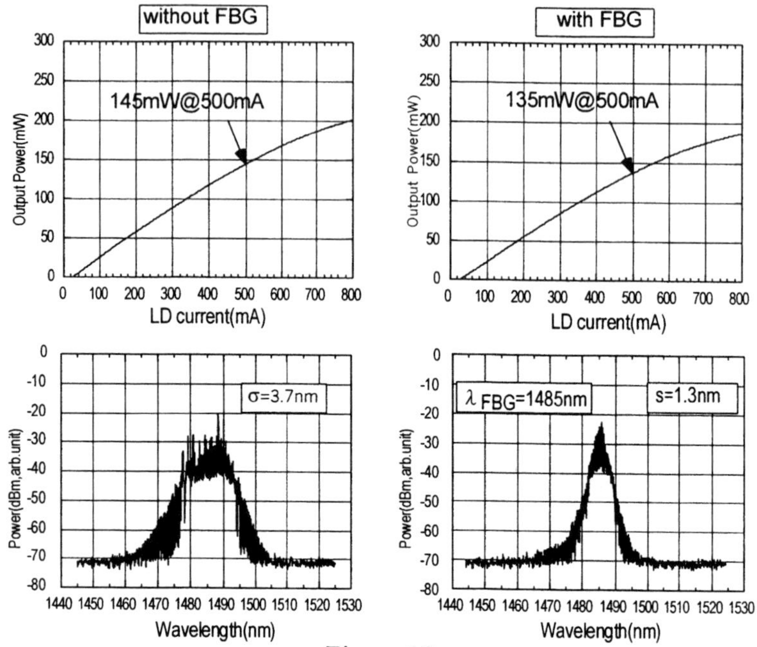

Figure 18

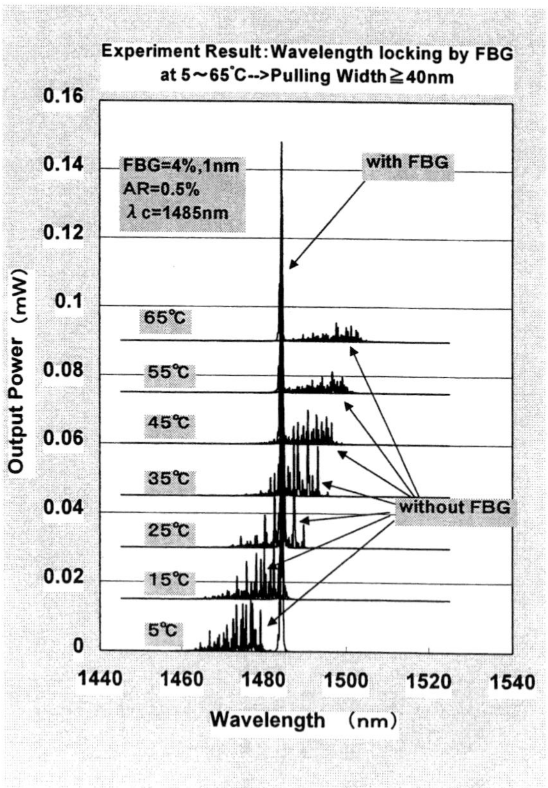

Figure 19

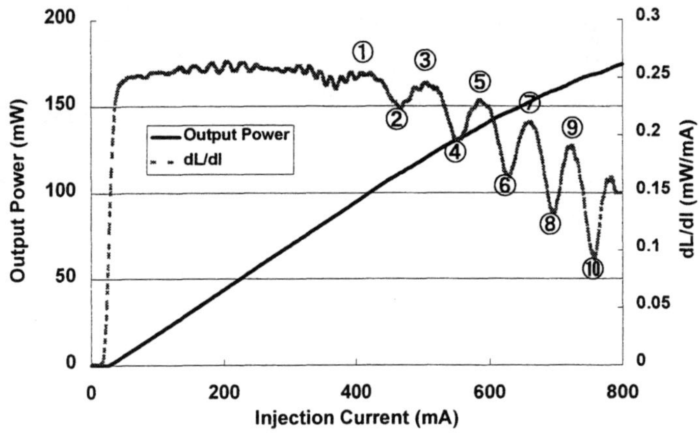

Figure 20

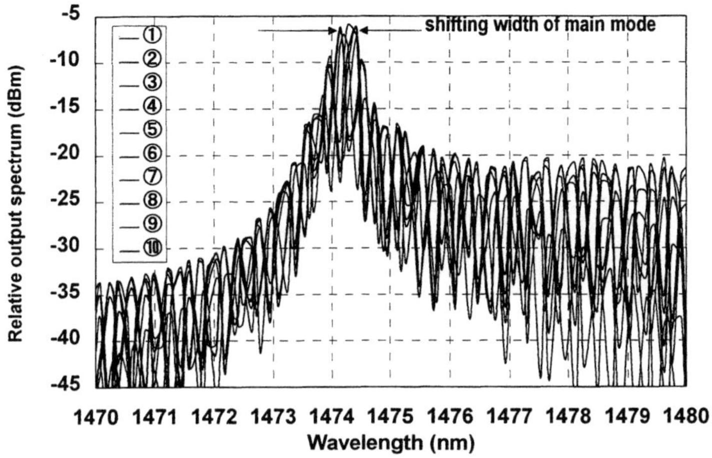

Figure 21

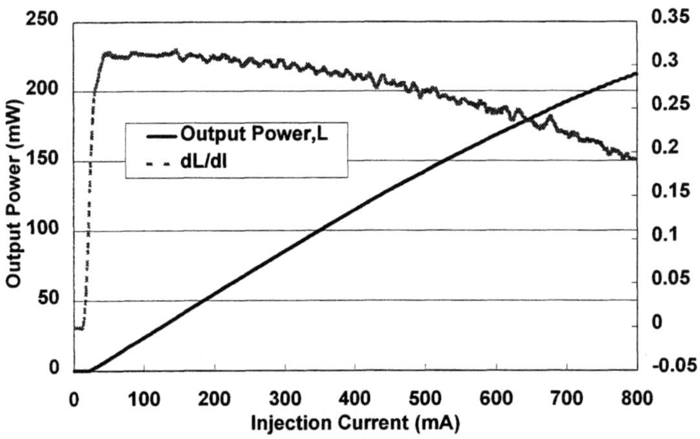

Figure 22

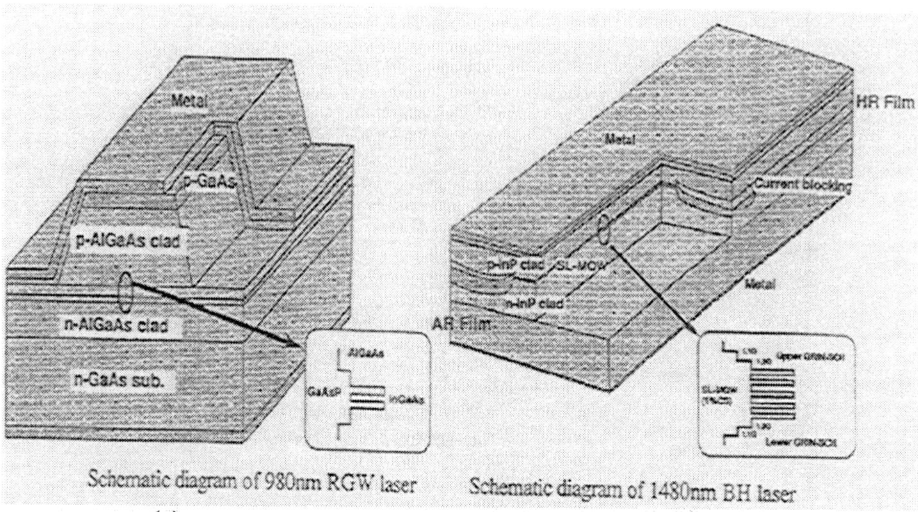

(a) (b)

Figure 23

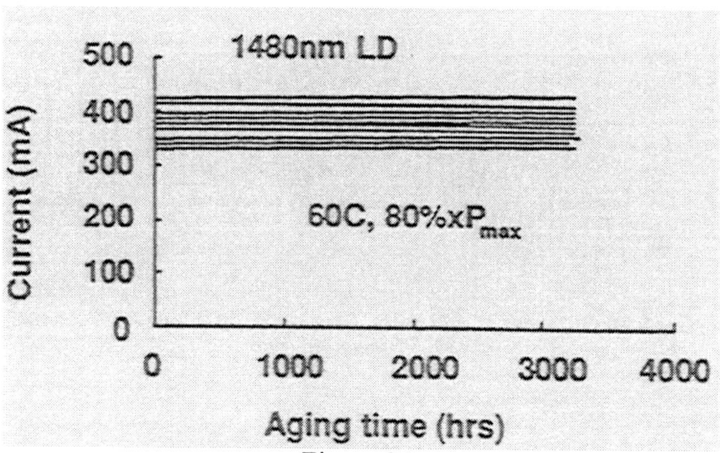

Figure 24

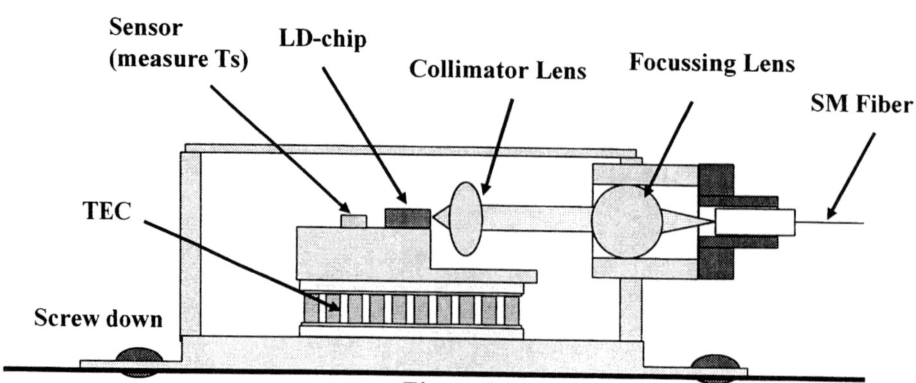

Figure 25

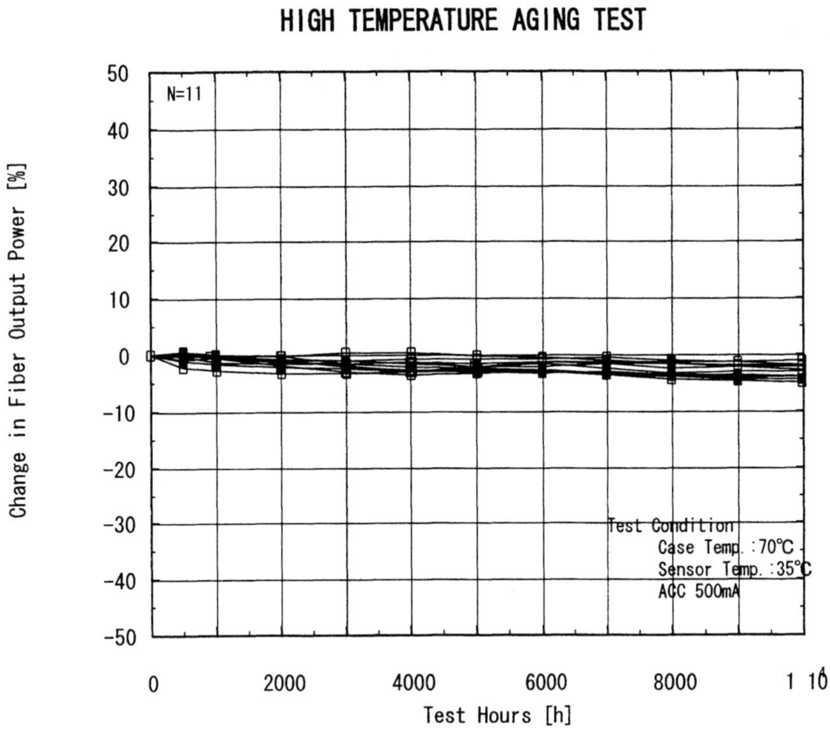

Figure 26

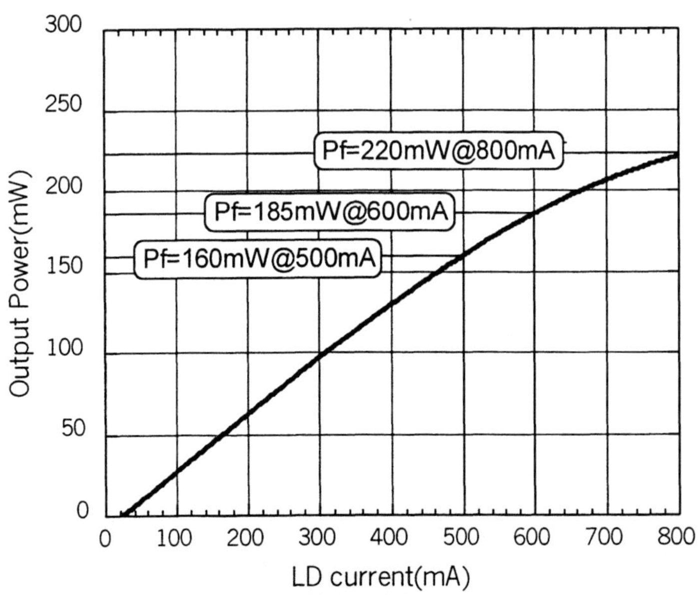

Figure 27

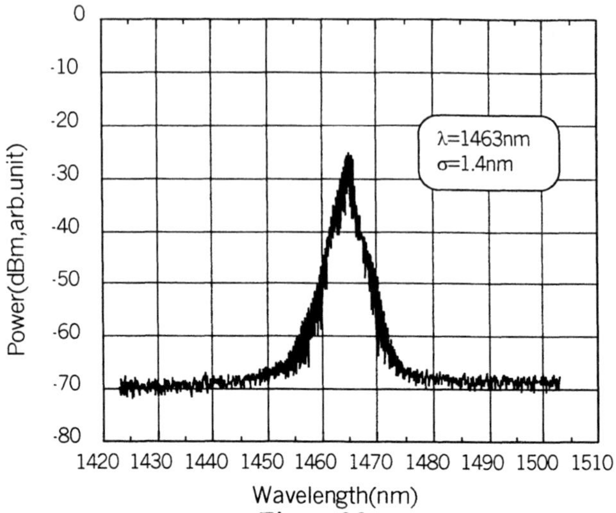

Figure 28

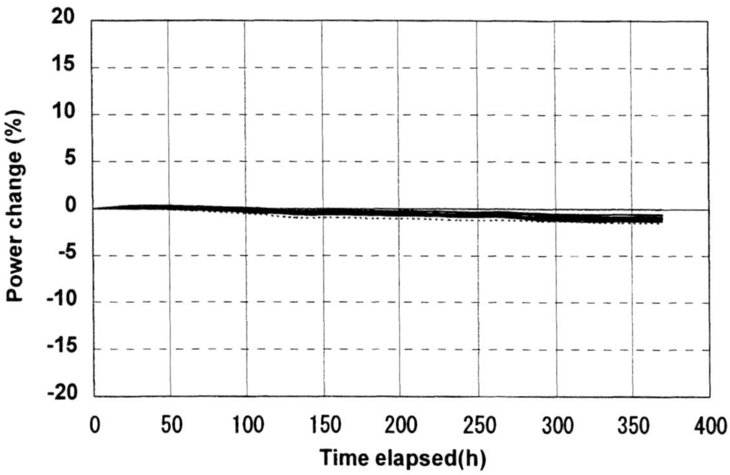

Figure 29

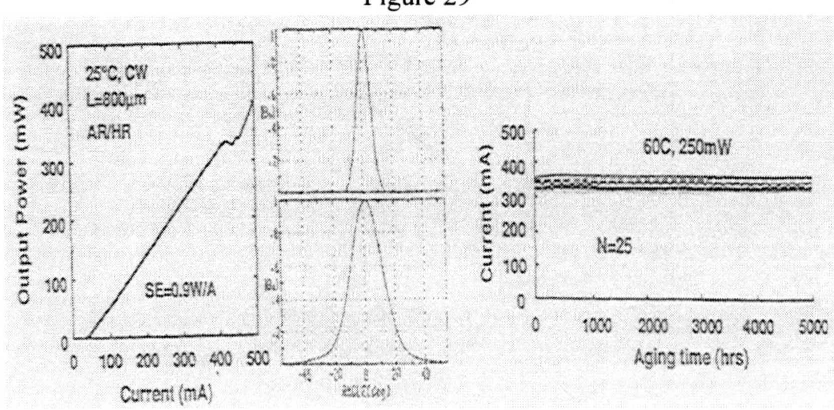

Figure 30

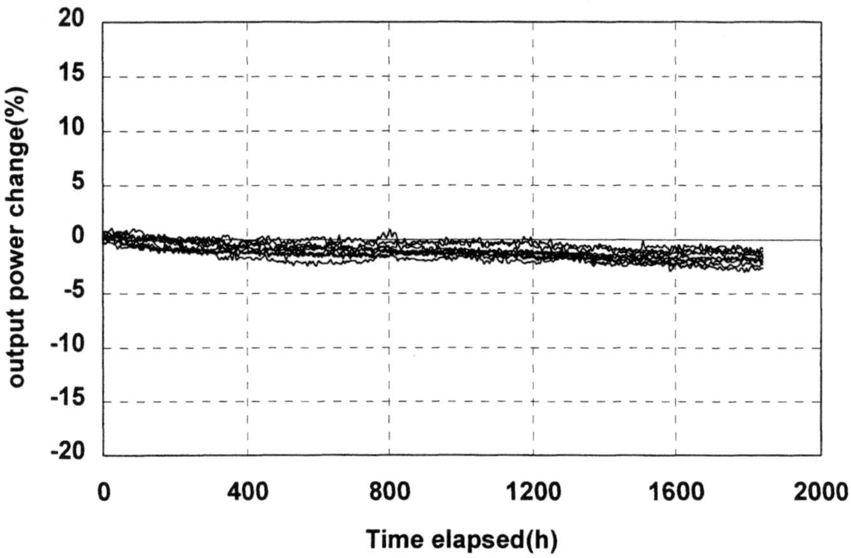

Figure 31

Figure 32

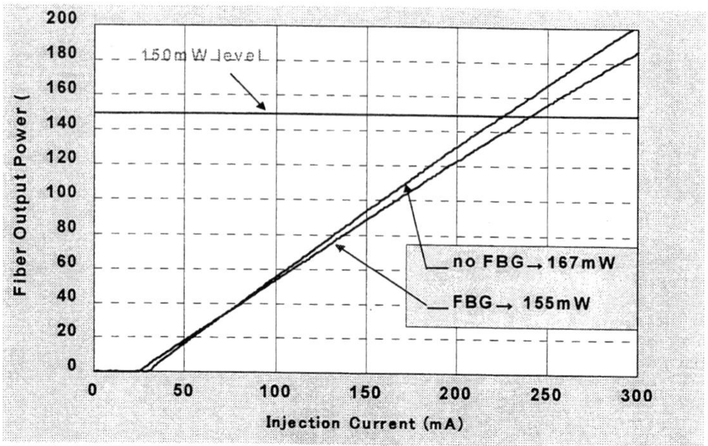

Figure 33

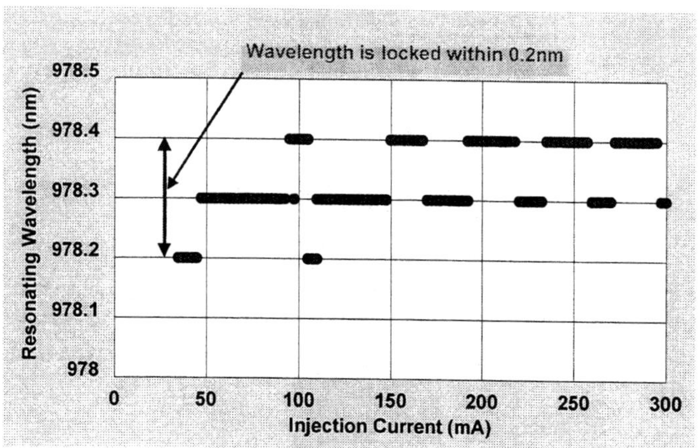

Figure 34

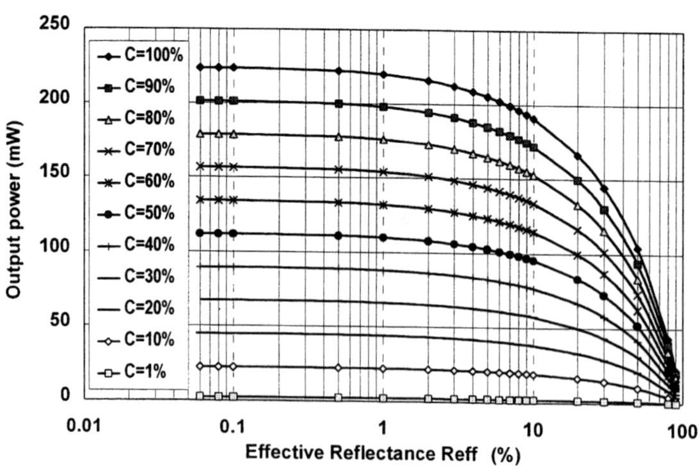

Figure 35

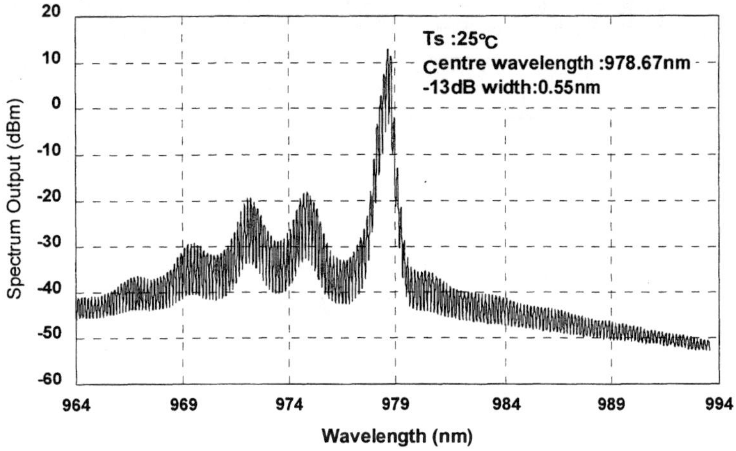

Figure 36

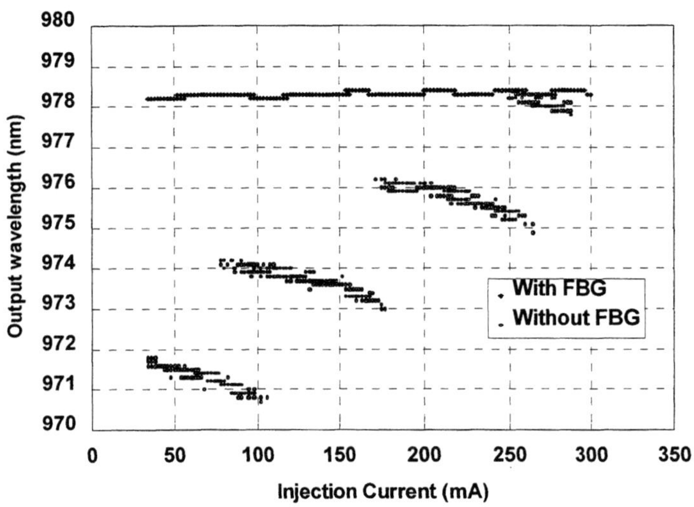

Figure 37

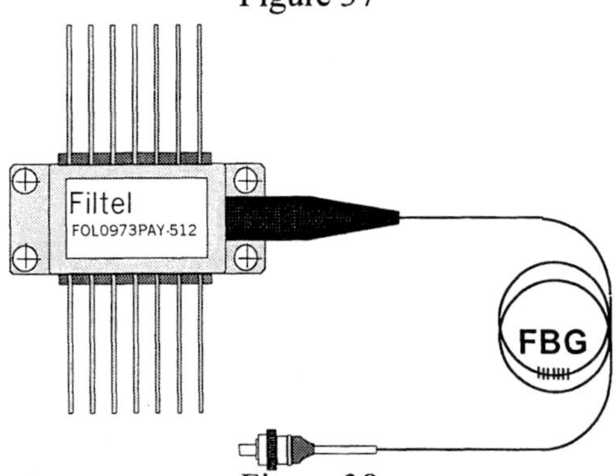

Figure 38

parameters	symbols	wavelength of laser chip	
		980nm	1480nm
Spontaneous_Recombination_Factor_(cm3/s)	(Beff)	1.50E-10	1.50E-10
Auger_Recombination_Factor_(cm6/s)	A1	7.50E-29	0
Spontaneous_Carrier_Lifetime_(ns)	τc	1.3	2
Differential_Gain_(cm^2)	G'	3.50E-16	1.00E-15
Carrier_Density_at_Transparent_(cm^-3)	N0	1.00E+18	1.00E+18
Gain_bandwidth_(nm)	$\Delta\lambda$gld	30	25
Gain_Center_Wavelength_(nm)	λc	1465 to 1490	980
Internal_Quantum_Efficiency	ηi	0.95	0.98
refractive index of LD	neff	3.4	3.4
absorption coefficient of LD (cm^{-1})	αi	15	8
confinement factor of LD active layer	Γac	2.50E-02	2.10E-02
Cavity length of LD (um)	L1	800	800
active layer width of LD (um)	W	2.5	4.3
thickness of active layer (nm)	d	5well × 4nm	2wellx7nm
Rear facet reflectance of LD (HR)	Rrear	0.98	0.92
front facet reflectance of LD (AR)	Rfront	0.02 to 0.06	0.01 to 0.06
Current Bias for fiber output Pf (mA)	I	500	250
Coupling power coefficient of LD and fiber	C	typically0.75	0.6 to 0.9
distance of LD to fiber coupling end (mm)	L2	12	0.5
refractive index of fiber	nfiber	1.54	1.54
propagation constant of fiber (dB/km)	lossf	0.2	0.2
distance of fiber endface to grating center (cm)	L3	100	100
Reflection center wavelength of FBG (nm)	λg	1485	980
reflectance of FBG	Rg	0.01 to 0.1	0.01 to 0.1
FWHM of FBG (nm)	$\Delta\lambda$g	1	1

Table 1

	Coupling efficiency C=75%											
	Rg (%)											
AR(%)	0.001	0.005	0.01	0.05	0.1	0.5	1	2	4	6	8	10
0.010	0.02	0.02	0.03	0.07	0.11	0.40	0.72	1.34	2.54	3.73	4.90	6.06
0.050	0.06	0.08	0.09	0.15	0.21	0.57	0.94	1.64	2.94	4.20	5.43	6.65
0.100	0.12	0.14	0.15	0.23	0.31	0.71	1.13	1.88	3.26	4.57	5.85	7.10
0.500	0.53	0.58	0.61	0.76	0.89	1.52	2.10	3.07	4.76	6.29	7.75	9.15
1.000	1.05	1.11	1.15	1.36	1.52	2.31	3.01	4.15	6.05	7.75	9.32	10.82
2.000	2.07	2.15	2.21	2.49	2.71	3.72	4.58	5.94	8.13	10.02	11.76	13.39
4.000	4.09	4.21	4.29	4.67	4.96	6.26	7.33	8.97	11.53	13.68	15.60	17.39
6.000	6.11	6.25	6.35	6.79	7.13	8.65	9.86	11.69	14.49	16.80	18.85	20.73
8.000	8.12	8.28	8.39	8.89	9.27	10.94	12.27	14.24	17.22	19.64	21.77	23.71
10.000	10.13	10.30	10.43	10.97	11.38	13.18	14.59	16.67	19.78	22.29	24.48	26.46

Table 2

DC to 50 GHz Wide Bandwidth InGaAs Photodiodes and Photoreceivers

Abhay M. Joshi and Xinde Wang
Discovery Semiconductors, Inc., Princeton Jct, NJ 08550

ABSTRACT

Discovery Semiconductors has developed 50 GHz "Dual-depletion InGaAs/InP Photodiodes". The PIN operates at -3V reverse bias and has minimum responsivity of 0.7 A/W at 1.3 and 1.55 um wavelength. The Ripple Factor is less than ±1 dB for a wide band of frequencies, DC to 50 GHz. The salient feature of the PIN is an on-chip co-planar waveguide output for proper impedance matching. The PIN exhibits group delay of less than ± 20 psec across the entire bandwidth. Discovery Semiconductors has also designed 50 GHz InGaAs PIN / p-HEMT Amplifier Photoreceiver Opto-electronic Integrated Circuit (OEIC) with a voltage conversion gain of 60 V/W at 1550 nm. The Photoreceiver OEIC exhibits an electric back reflection (S_{22}) of less than –10 dB across the entire 50 GHz bandwidth. The optical back reflection is better than –30 dB at 1300 and 1550 nm.

Keywords: InGaAs Photodiodes, Photoreceivers, Wide Bandwidth, OEIC, InP p-HEMT.

1. INTRODUCTION

Telecommunications has been one of the primary technology sectors of the 1980's and 1990's. Fiber Optic Communications allows information to be sent faster and cheaper than conventional electronic components and cables. Telephone, CATV and computer information can thereby be transmitted without the heavy cables and interference & signal attenuation problems, long associated with copper electrical wires, (twisted pairs). The annual growth rate has been 30 - 40% in the fiber optics industry becoming a multi-billion dollar business market in the 1990's and growth will continue to explode into the next decade.

SONET/STM standards have been developed over the last decade to allow for global compatibility of fiber transmission systems and the advantage of "drop and insert" flexibility. The ANSI SONET and CCITT SDH (Synchronous Digital Hierarchy) based transmission systems on optical fiber serve as the principal long haul and inter-office structures for virtually all new and upgraded telecommunications networks. SONET/SDH networks will allow capacity upgrading from the present OC-48 rates (2.5 Gb/s) to OC-192 rates (10 Gb/s) & OC-768 (40 Gb/s) by changing either the transmission terminal equipment or, in many cases only selected subsystems **(High Speed Laser and Photodiode/Photoreceiver Components).** Both of these standards and others specify the particular requirements of the optoelectronic components functioning at 1300 and 1550 nm wavelength and speed from 2.5 Gb/s to 10 Gb/s and even 40 Gb/s including reliability standards for the first generation systems entering into the "Information Super Highway". This SONET structure allows for future growth where high speed and more reliable components can be easily inserted to upgrade the present slower networks.

Demands imposed on SONET/SDH transmission systems have been for higher speed and performance for long haul systems and has grown steadily for more than a decade. The trend is expected to continue into the year 2000 and beyond. A majority of the present systems will continue to operate utilizing the presently supplied components modules (622 Mb/s to 2.5 Gb/s). However, the demands for wider bandwidth (10 Gb/s to 40 Gb/s) will continue to drive the market place for higher speed and reliable component modules. The main driving force is the "Information Super Highway", where many application goals have yet to be achieved due to the lack of speed and reliability. The need for speed is based on the emerging applications (such as E-Commerce, Video Conferencing, Web TV) that require larger bandwidths, greater speeds and high reliability. Multimedia requirements at video rates will force the network companies to expand their capacity using both higher bit rates and multiple wavelengths on each fiber. Optoelectronic integration using Opto-electronic Integrated Circuits (OEICs) will enable the rapid upgrade of the SONET/SDH networks with reliable, cost effective front end modules.

The combination of optical, microwave, and digital functions on the same chip is a technology that has significant potential for commercial applications such as ethernet fiber local area networks and optical communications systems (Synchronized Optical Network SONET, ISDN, telephony and digital CATV). Important inter-service military applications are optically fed phased array systems and optically controlled microwave networks for airborne and spaceborne systems.

These applications require components with wide bandwidth, reproducibility, and low cost. While the maturity of MMIC technology makes it feasible to combine optical components such as photodetectors and front end receivers [1-3], the demonstrations of integrated optical receivers reported so far have mainly been at frequencies lower than 10 GHz [4-6]. Although high speed p-i-n photodetectors and narrow band MMIC amplifiers with frequency response in excess of 20 GHz are commercially available, wide band amplifiers with frequency & gain performance necessary for photoreceivers above 2.5 Gb/s are currently difficult and expensive to manufacture. Integrated photoreceivers with frequency performance in the 40 Gb/s range have only recently been demonstrated and truly monolithic photoreceiver OEIC chips operating at any frequency are not commercially available.

2. ULTRAFAST, HIGH POWER InGaAs PHOTODIODE DESIGN & MODEL DEVELOPMENT

The basic issues for high bandwidth detectors are the electrical capacitance of the detector and the transit time for carriers in the detector. For high optical input power, we must also consider the non-linear effects of space-charge effects, barrier trapping, and saturation. In the design process, one must balance all of these factors to achieve the optimum detector performance. This balance can be found by analyzing the underlying physics of carrier motion through a p-i-n structure.

When the semiconductor is operating in the linear regime, the two major limits on device speed are the capacitive time-constant and the carrier transit-time. As analyzed in our previous work[7], there are two major capacitances to consider in optical detectors: the intrinsic capacitance of the diode itself, and the parasitic capacitance of the preamplifier circuitry. The intrinsic capacitance, $C_{Int.}$, is given by the simple parallel plate capacitance of the p-i-n structure, and is proportional to the area and permittivity, A and ε, and to the inverse of the depletion region thickness, d.

$$C_{Int} = \frac{\varepsilon A}{d} . \tag{1}$$

This capacitance, in association with the load resistance of the amplifier, R, determines the electrical time constant of detection, τ_{RC}.

$$\tau_{RC} = RC_{Int.} . \tag{2}$$

Hence, the capacitance needs to be minimized and requires the depletion region (d) to be thick. The transit time, τ_{TR}, is the time required for the photogenerated carriers in the i region of the detector to move to the electrode surfaces. Given that the carriers are moving at their saturation velocities, v, the transit time is proportional to the intrinsic thickness, as given by Eq. (3),

$$\tau_{TR} = \frac{d}{v} . \tag{3}$$

This time gives the limit of the duration of the impulse response of the detector. Thus, a thick detector will have a long transit time, and thus a slow response.

Fortunately, the properties of InGaAs and InP are such that the electrons can travel as much as ten times faster than the holes at certain fields[7]. By fabricating the detector such that the carriers are generated only near the p contact, as shown in Fig. 1(a), a large speed increase can be realized. When the carriers are generated near the p contact, the slower holes travel a shorter distance than the faster electrons. This allows a thicker intrinsic region without the transit time penalty.

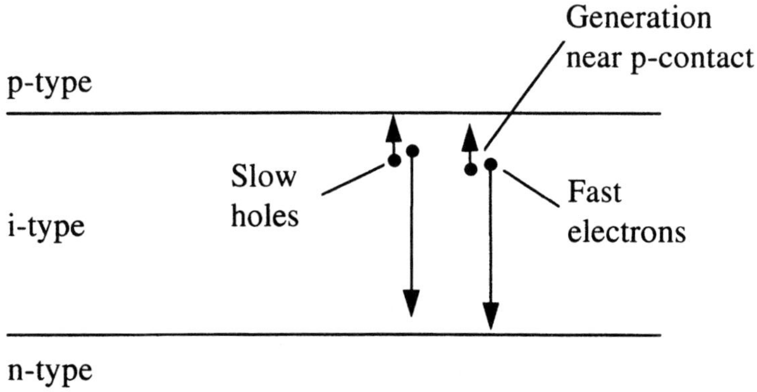

Fig. 1(a). A generic p-i-n detector structure, illustrating how selective carrier generation can accelerate the transit time effect because of the difference in hole and electron velocities.

Our previous work involved making detectors of 25 μm diameter and 1 μm thickness, with response speeds up to 40 GHz[7]. The selective carrier generation was accomplished by fabricating an InGaAs / InP dual-depletion detector as shown in Fig. 1(b). This works by using two different compositions of semiconductor in the depletion region: one absorbing (InGaAs) and one transparent (InP). The carriers are generated only in the absorbing material, and by placing the absorbing material adjacent to the p contact, the transit time is improved.

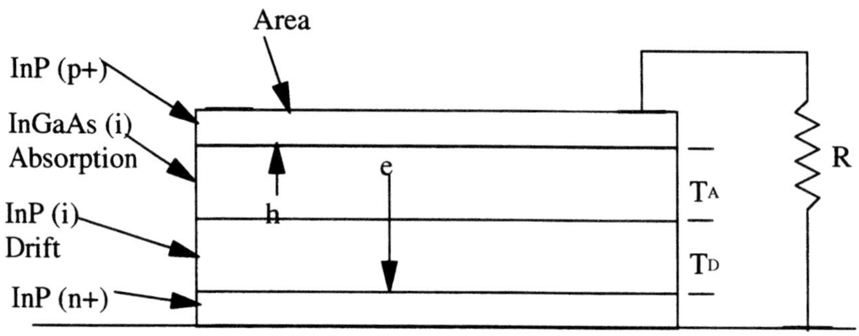

Fig. 1(b). Discovery Semiconductors' dual-depletion layer InGaAs/InP high speed p-i-n detector.

Using the carrier drift models developed previously, we can re-optimize the design of the detector to achieve a 50 GHz response. In past one year, our Ultra-fast InGaAs/InP dual-depletion photodiode[7] impulse response modeling has been improved in several ways:

1. First, additional physical effects such as *parasitic inductance of the wire-bonds and potential barriers at the contacts* have been added to the model to make the resulting data more representative of the actual measurements.
2. Second, the model has been migrated to a user-friendly "Visual-Basic" software platform to make the model more accessible to the device designer.
3. Third, the theoretical basis for extending the model further, so that it includes space charge and bandgap barrier effects, has been developed. This theory has helped to guide the development of faster and more efficient detector elements, and will continue to do so into the future.

The first improvement to the model was to add extra details regarding the device that were not included in the original model, such as parasitic inductance and contact potential barriers. These additions were motivated by the experimental observations of fabricated devices. Many diodes had responses that went to frequencies higher than predicted by the model, and the form of the response indicated pronounced 'peaking'. This suggested that the devices had a significant series inductance. In addition, it was found that all of the devices required a bias (9V) that was considerably higher than that predicted as optimal (3 to 5V) by the previous model. This suggested that the devices had large potential barriers at the contact.

To better fit the experimental data, a parasitic inductance was added to the equivalent circuit model for the diode and its associated load. This additional parasitic reactance causes the response to peak much like the experimentally observed outputs. In the current experimental situation, the value of the inductance depends on the number and arrangement of the wire bonds used to connect the detector chip to the RF output of the package. Judging from the experimental results to date, the inductance is about 0.5 nH. For scale, 1 nH is the inductance of 1 cm of straight conductor in free space. In the integrated case, the inductance will depend on the layout of the anode metal on the chip.

The second improvement to the model concerned addressing the usability problem of the previous model. Our old model was written in FORTRAN and had a file based input-output system. This prevented the user from asking what-if questions and thereby moving towards a solution quickly. This difficulty was solved by re-casting the model in the "Visual Basic" language, and allowing the user direct interactive control of all the model parameters and immediate observation of the frequency response curve that results from these parameters. A screen capture of a typical session is shown above in Fig. 2.

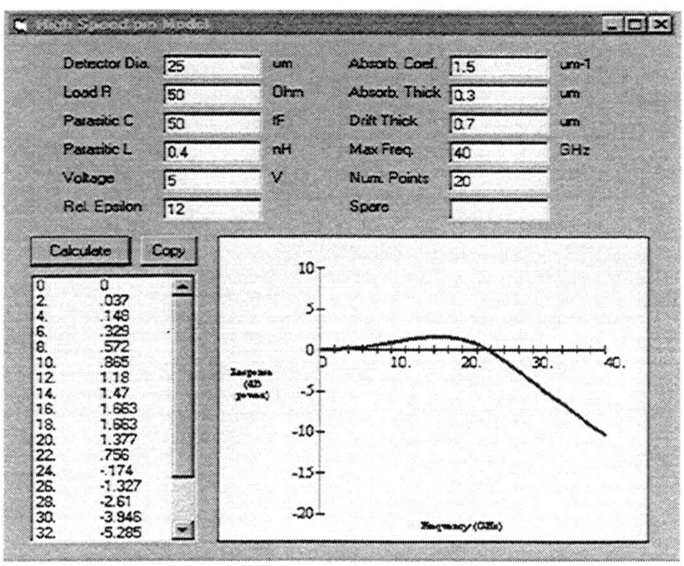

Fig. 2. An example of a typical session of Discovery Semiconductors' high speed InGaAs diode computer simulation model.

The introduction of the graphical user interface has improved the utility of the model, and has led to several realizations about the underlying mechanisms of device speed limitations. The most important of these discoveries is that the response is not as sensitive to InGaAs absorbing layer thickness as previously thought. This implies that the InGaAs absorbing layer may be thickened considerably (perhaps by a factor of 2) without serious speed penalty. This will increase the quantum efficiency of the detector at little or no cost.

The third improvement to the model is the incorporation of space charge and barrier effects. The current model uses several assumptions (e.g. constant carrier velocity, constant electric field) that prevent the inclusion of these effects. Hence, a completely new analysis method must be used. We have considered several methods, and have determined that a modified drift-diffusion conduction model solved via finite difference methods is most likely to work for our ultrafast InGaAs PINs.

The model described so far can capture most of the effects that we want to look at, such as space charge saturation and non-uniform bias field. However, one important thing not included is the effect of potential barriers. These barriers occur wherever two dissimilar semiconductor materials meet, such as where the absorbing InGaAs layer joins with the transparent (and higher bandgap) InP drift layer. At such heterojunctions, the carriers encounter a potential barrier, as shown in Fig. 3. The important factors at play at these junctions are the energy discontinuity (the height of the barrier), the width of the barrier (dependent on the electric field and/or the structure parameters), and the energy of the carriers. These parameters determine the probabilities of thermionic emission over the barrier and quantum tunneling through the barrier. These probabilities determine the carrier current through the barrier given the concentration of carriers on the low side.

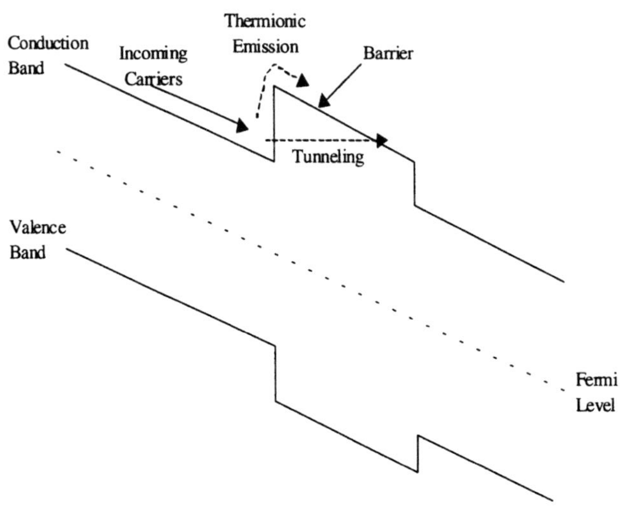

Fig. 3. A diagram of an energy barrier found in heterojunction semiconductor diodes.

These barriers can be incorporated into the finite difference model in the following way:

- First, the device is partitioned into segments in the normal way, and all the segment junctions that have energy barriers are found, and these heterojunctions are tagged as 'special'.
- The model then begins to model the dynamics of the carrier motion. At each step in this process, the current and carrier densities are determined for every segment junction.
- At the special junctions, the current density predicted by the drift conduction equation is compared with that predicted by barrier emission and tunneling equations. If the barrier current is less than the drift current, then the actual current (used in the conservation relationships) is reduced to the barrier current. This will cause carrier pile-up on the low energy side of the barrier, just as it happens in the real device. Eventually, the carrier concentration will increase to the point where the barrier current is large enough to support the drift current, and the barrier will reach a

steady state. This chain of events will produce distortions in the signal, and limit the response speed of the devices, which are the important device characteristics that we want to study and control.

Currently, the numerical 'core' of this model is being coded and tested. Once this basic module is complete, it will be incorporated into a graphical user interface much like the preceeding model. The eventual product will allow the designer to create arbitrary structures of varying semiconductor composition and doping, and then model the impulse response of the resulting diode. This will be a significantly higher level of flexibility than any other model of its kind.

3. InGaAs PHOTODIODE FABRICATION AND CHARACTERIZATION

Discovery Semiconductors' InGaAs PINs have several advantages as listed below for the DC to 50 GHz wide bandwidth applications:

- Use of semi-insulating InP substrate for reducing parasitic effects.
- Electronic co-planar waveguide output with 50 ohm output impedance matching on the chip.
- Microwave foot print with Signal - Ground - Signal (100 microns spacing) for easy on-wafer RF testing using standard microwave probes.
- Front illuminated, mesa diode which eliminates any "tail effects" as seen in the planar diode technology.

The use of "Air Bridge" technology allows the InGaAs PIN detectors to be front illuminated for maximum coupling efficiency and photo responsivity without increasing the parasitics. Please see Fig. 4 below. The above innovations are helping the company vigorously pursue the design and manufacturing of wide bandwidth InGaAs PIN / InP p-HEMT Amplifier, Opto-electronic Integrated Circuit (OEIC) Photoreceiver technology for the 21st century demands of optical & microwave communications. Discovery Semiconductors has commercially introduced these OEICs from January 1999.

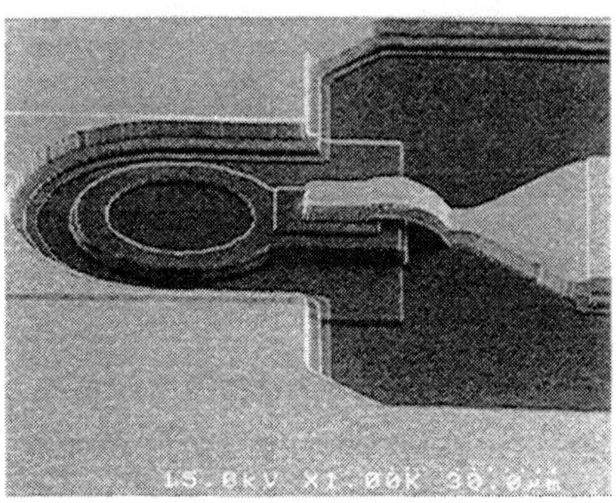

Fig. 4. A Scanning Electron Microscope (SEM) photograph of the co-planar waveguide output (CPW), front illuminated InGaAs PIN detector.

There is presently a strong and growing need for semiconductor opto-electronic components (Lasers, Photodiodes, Amplifiers) which are the main sources of success for the present day telecommunication systems. The future success of the fiber optic market will be, on developing new lasers, photodiodes, and amplifier components to operate at high speeds (Wide bandwidth of 10 Gb/s to 40 Gb/s), thereby expanding the growth of the telecommunications market place. Discovery Semiconductors, Inc is addressing the present and future needs of this market by manufacturing InGaAs Photodiodes with bandwidths of DC to 50 GHz and the first of a new family of InGaAs Photoreceivers, with a bandwidth of 100 KHz to 50 GHz. The eye diagram of a DC to 50 GHz wide bandwidth photodiode is shown below in Fig. 5.

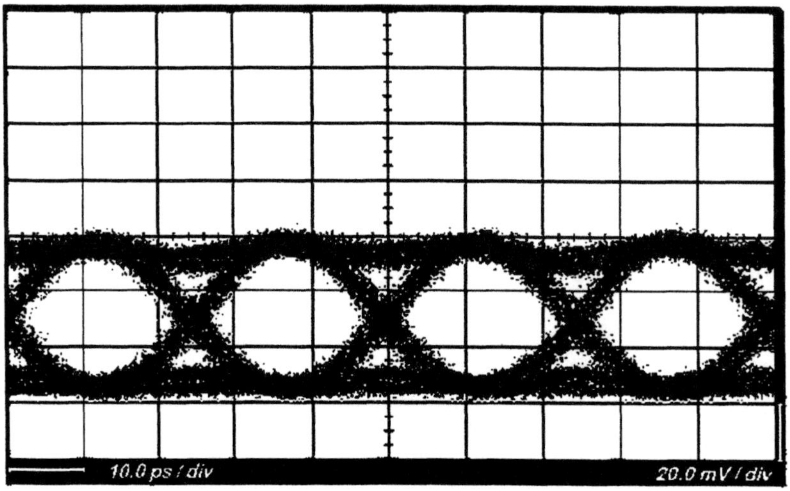

Fig. 5. 40 Gb/s NRZ eye diagram of Discovery Semiconductors' 50 GHz InGaAs Photodiode. The clear eye diagram illustrates the technical advancement and maturity of these InGaAs photodiodes.

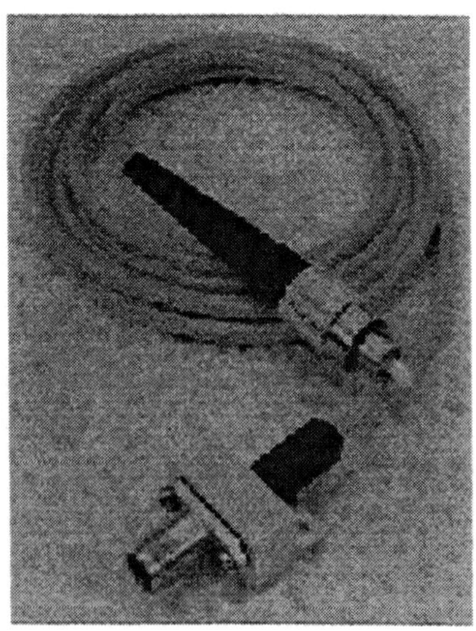

Fig. 6. Fiber Pigtailed Microwave Package with a K or V RF connector for Wide Bandwidth Photodiodes and Photoreceivers.

The AC characteristics expected from a photodiode for 40 Gb/s SONET are wide bandwidth with a very low ripple factor & excellent phase stability across the entire pass band. This requires skillful microwave packaging of the photodiodes and photoreceivers. Fig. 6. shows the fiber pigtailed microwave package with a RF connector, while Fig. 7. exhibits frequency response plot of a 50 GHz microwave packaged photodiode. The flat response with less than ±1 dB variation plus a smooth roll-off are both necessary to produce a clear eye diagram as shown before in Fig. 5. The phase stability is illustrated in Fig. 8.

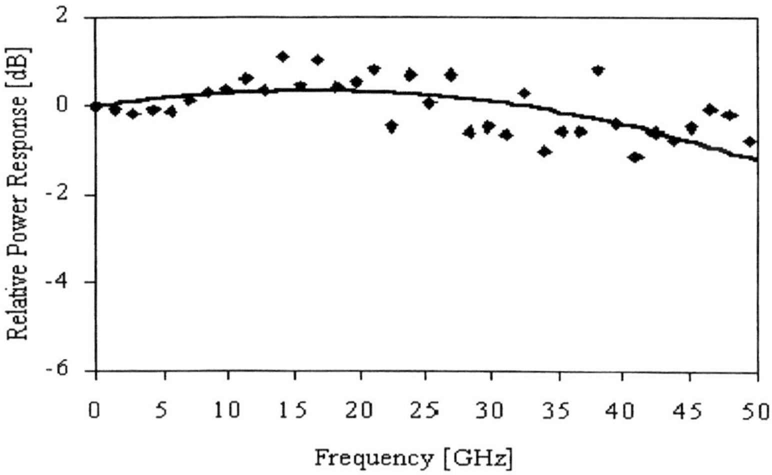

Fig. 7. Frequency response of InGaAs co-planar waveguide (CPW) output 50 GHz microwave photodiode.

From the above frequency response plot it is clear that the ripple factor is less than ± 1 dB for a wide band of frequencies, DC to 50 GHz. The accurate High Frequency InGaAs PIN computer model described in the previous section has vastly improved the performance to the best ripple factor of any commercially available microwave InGaAs PIN diodes. This results in typical group delays of less than ± 20 psec as shown in Fig. 9.

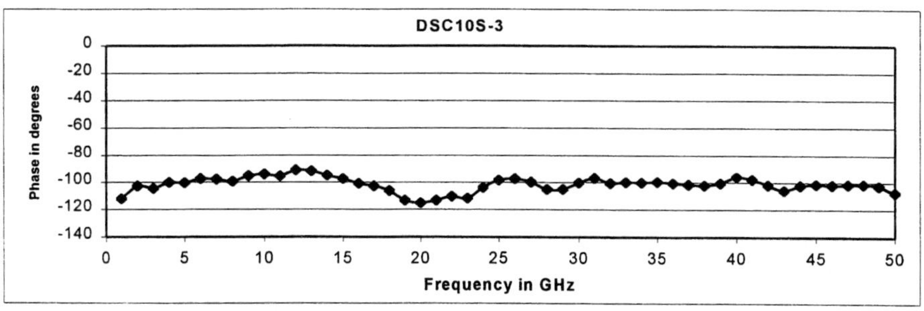

Fig. 8. DC to 50 GHz Phase Stability for InGaAs co-planar waveguide (CPW) Photodiode.

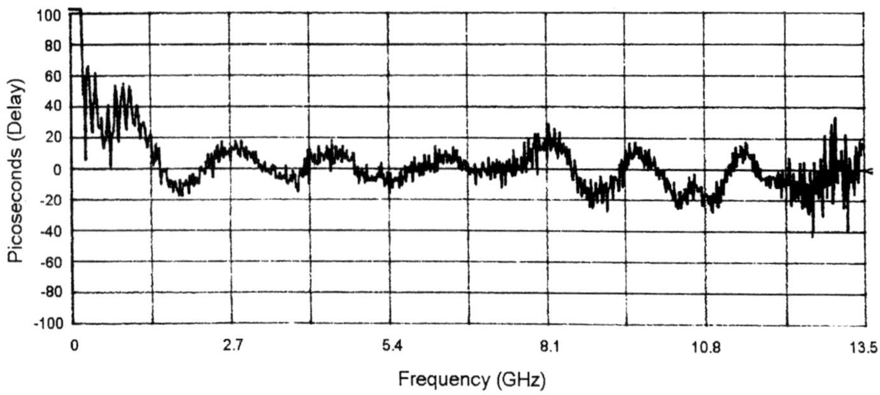

Fig. 9. Typical Group Delay for Discovery's PIN Diodes of less than ± 20 picoseconds within the valid system window. Note: The group delay between 0 to 1.35 GHz and 12.15 to 13.5 GHz is due to the measurement system and not a result of the PIN diode.

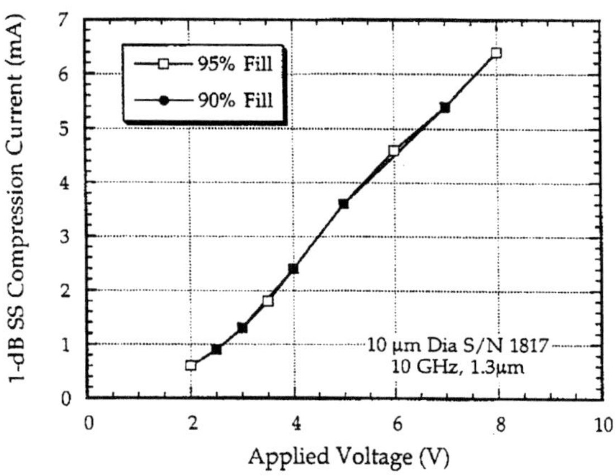

Fig. 10. The maximum average current handling capacity of 50 GHz InGaAs photodiode. The diameter of the device is 10 µm.

At 50 GHz bandwidth, the lack of good front end amplifiers has created a need for high optical power photodiodes. These photodiodes are expected to handle large amounts of optical input power (10 dBm or higher) and generate enough photocurrent to directly drive into a 50Ω load. Fig. 10 exhibits the average 1-dB compression current handling capacity of these InGaAs PIN diodes. At a reverse bias of 6V or more, these photodiodes can handle 5 mA or higher average photocurrent without exhibiting significant non-linearties. At lower duty cycles, these PIN diodes can handle even larger amounts of maximum optical input power without any degradation of the impulse response as shown in Fig. 11. This is due to the excellent microwave device design of these PIN diodes. Furthermore, as these devices are front illuminated, the space charge effects are far less severe than the back illuminated photodiodes[8,9] and hence, they handle larger optical input power.

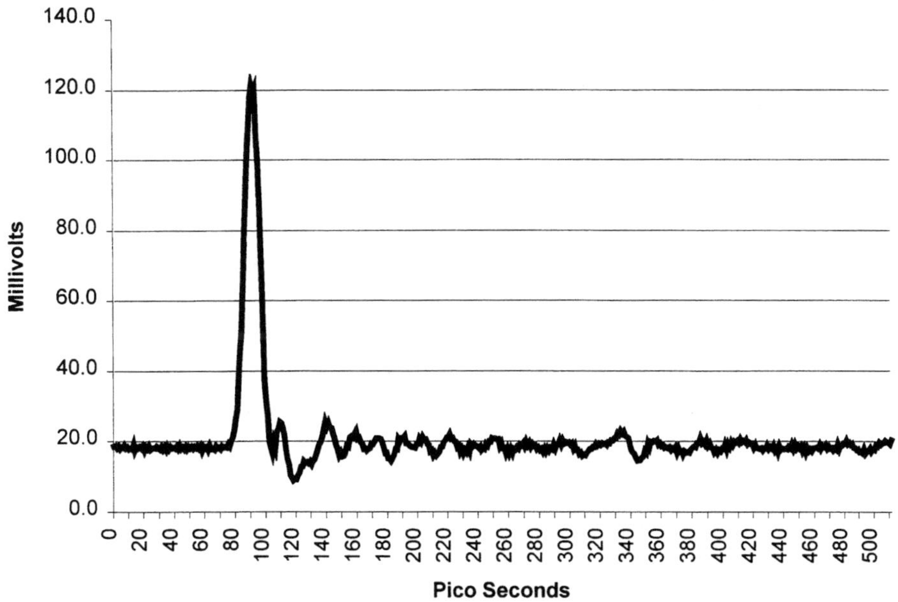

Fig. 11. The impulse response of a 50 GHz InGaAs photodiode for 2 psec wide pulses having a peak optical power of 95 mW and an average power of –11.3 dBm. The operating reverse bias is 3V and the observed rise time is 9 psec. Note: Very little ringing for the impulse response.

4. WIDE BANDWIDTH PHOTORECEIVERS

In addition to the InGaAs photodiodes, we have built and tested a number of wide bandwidth photoreceivers, for both the OC-192 (10 Gb/s) and OC-768 (40 Gb/s) systems as shown in Table I. Applications could include wide band communication links for digital and analog applications.

Table I: Discovery Semiconductors' InGaAs Wide Bandwidth Photoreceivers

MODEL	- 3 dB Bandwidth		Voltage Gain Typical @ 1.55 μm	Power Gain		Bias Voltage
	Min	Typical		Min	Typical	
DSC R401	12	14 GHz	80 V/W	7	10 dB	+5V-0.5V
DSC R402	10	11 GHz	350 V/W	15	20 dB	+8V
DSC R403	9	10 GHz	1000 V/W	25	30 dB	+8V/-0.5V
DSC R404	45	50 GHz	60 V/W	7	10 dB	+5V/-0.5V

The schematic of our 50 GHz Photoreceiver is shown in following Fig. 12. The InGaAs PIN photodetector is DC coupled to the amplifier with a 50Ω input and output impedance, and about 10 dB power gain. The voltage conversion gain of the amplifier is about 60 V/W at 1550 nm. Thus, 10 dBm optical input power will result in a signal ouput of 600 mV sufficient to drive the following decision circuits.

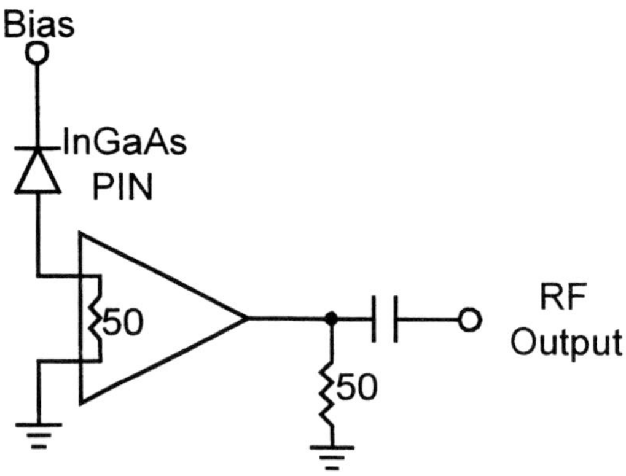

Fig. 12. 50 GHz wide bandwidth photoreceiver consisting of an InGaAs PIN diode and GaAs MMIC amplifier.

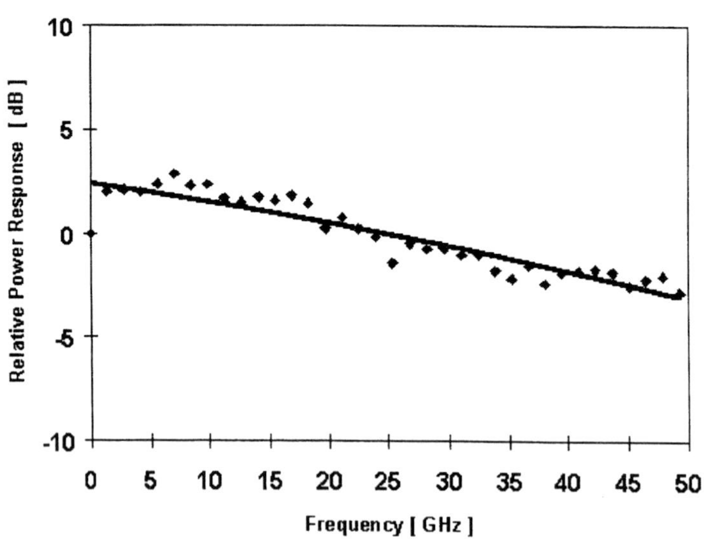

Fig. 13. Response of the 50 GHz wideband photoreceiver as a function of frequency. The measurement was made at discrete frequency points as set by the optical heterodyne system. The response at the low frequency is limited by the input capacitor and at high frequency by the amplifer.

Although the 50 GHz InGaAs PIN diodes can handle large optical input power as described earlier, the development of the Model R404, 50 GHz InGaAs Photoreceiver is essential to prevent the problems of "Thermal Runaway" and better reliability[10]. The active diameter of a 50 GHz PIN diode is only 10 μm and exhibits far higher photocurrent and electric power density. This may lead to "Thermal Runaway" and cause catastrophic failure of the PIN diode.

The wideband photoreceiver was tested for its frequency response from DC to 50 GHz as shown in Fig. 13. The photoreciever package was placed in a fixture so that the fiber could be aligned to the detector. The RF set up consisted of a 50 GHz spectrum analyzer and optical heterodyne source for frequency generation. The frequency was generated by either current control mode or temperature control mode. Since the analyzer could not have a DC input, a chip capacitor was placed in series with the output, thus limiting the lower frequency response to 100 KHz. The photoreceiver performed as expected showing an approximate 10 dB gain over the entire bandwidth compared to a photodetector alone. This result demonstrates that 50 GHz wideband photoreceiver can be manufactured using current technology and can soon go from 50 to 100 GHz with the improvement in the speed (f_T) of the p-HEMTs.

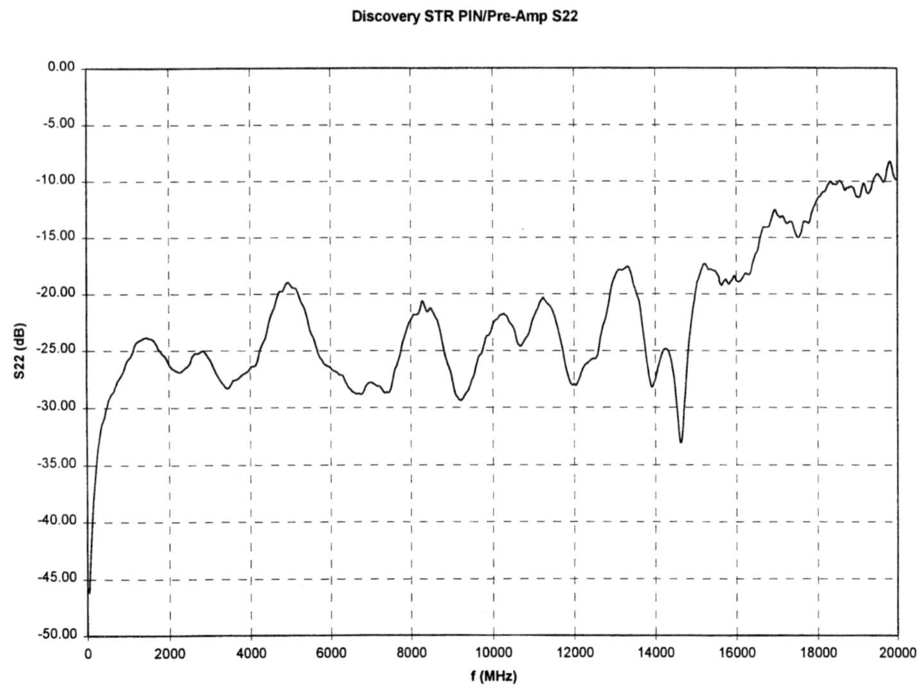

Fig. 14. Electric back reflection (S_{22}) data for the 14 GHz Model R401 wide bandwidth photoreceiver. The back reflection is less than –15 dB for the entire bandwidth.

The electric performance of the 10 Gb photoreceivers, Model R401 & R402 is exhibited in Figs. 14 and 15. Fig. 14 illustrates the superior electric back reflection (S_{22}) of less than -15 dB for the entire bandwidth. Fig. 15 shows the average input equivalent noise current density measured with hp spectrum analyzer 3582A. At the lower frequency cut-off of ~25 KHz, the receiver exhibits the noise current density of approximately 10 pA/rt. Hz. The measured sensitivity for this receiver is –17 dBm for 10 Gb/s PRBS 2^{31}-1 modulation at BER of 10^{-12}.

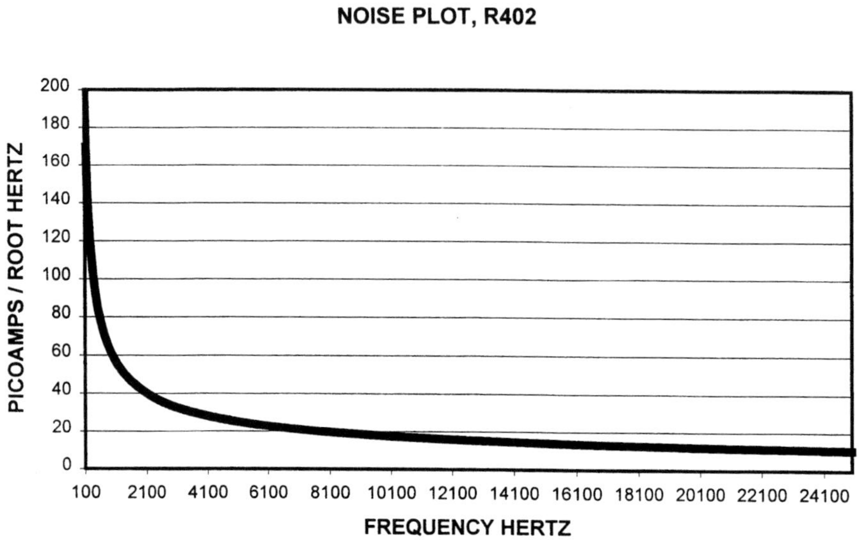

Fig. 15. Average input equivalent noise current density for 10 Gb/s InGaAs photoreceiver.

5. FUTURE MONOLITHIC PHOTORECEIVERS

Classically, high speed InGaAs detectors and GaAs amplifiers have been manufactured as discrete components, and subsequently wirebonded together using hybrid technology. While this is sufficient for most applications up to 2.5 Gb/s (OC-48), the hybrid approach has several drawbacks:

- increased labor,
- low yield,
- low reliability, and
- increased parasitic losses.

These problems become worse at 10 Gb/s (OC-192) or higher frequencies (OC-768). Currently, Discovery Semiconductors is actively developing totally integrated InGaAs detector/ InP p-HEMT Amplifier OEICs for fiber-optic applications. A photograph of one of our prototype OEICs is illustrated in Fig. 16. This approach will not only eliminate the drawbacks of the hybrid method, but also produce higher speed transistors (f_T > 100 GHz) owing to the use of InP as the active material. Over the next few years, the company plans to commercially manufacture these monolithic OEICs for the 10 Gb/s (OC-192) and 40 Gb/s (OC-768) standard.

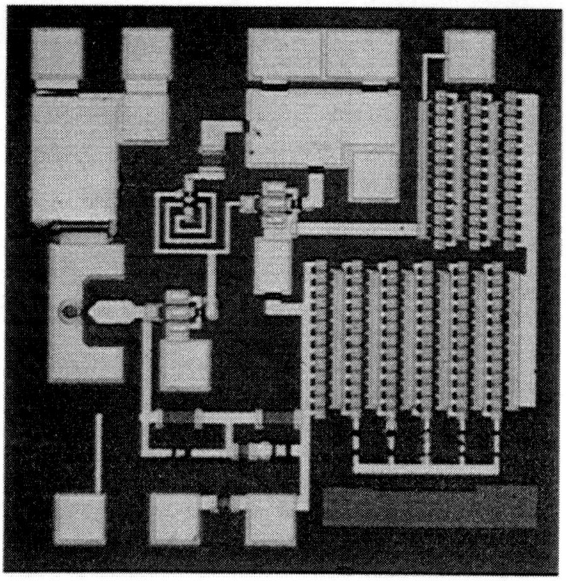

Fig. 16. Discovery Semiconductors' next generation monolithically integrated InGaAs Photodiode and Amplifier OEIC for wideband telecommunications, RF photonics, and CATV applications.

5.1. Wide Bandwidth OEICs Manufacturing Challenges

To develop photoreceivers with 10 Gb/s and 40 Gb/s bandwidth, a wide band amplifier technology must be in place. Technology improvents are needed in the following areas:

- The wide band amplifier for the proposed photoreceiver must be based on device technologies with high f_T (100 GHz or higher), and Field Effect Transistor (FET) capacitances (C_{FET}) lower than 50 fF. We have chosen the InP HEMT technology because it is the highest frequency technology available that is compatible with MMIC fabrication approaches. For 40 Gb/s amplifier, we will optimize InP HEMT technology by reducing C_{FET} below 50 fF.

- Gate leakage current can severely degrade the noise performance of a photoreceiver. This is particularly true of HEMTs whose Schottky contacts exhibit poor characteristics. Thus, reducing the gate leakage currents of HEMTs will result in low noise amplifiers and thus, improve the sensitivity of the photoreceivers.

- The design of high speed photoreceivers requires a compromise between the bandwidth and noise. In addition, the open loop gain of the amplifier must be maximized while the photodiode and p-HEMT capacitances have to be minimized. The open loop gain of the amplifier is maximized by using InP HEMT technology and a maximum single stage gain of over 10 can be easily achieved. Therefore, we should be able to achieve open loop gains well in excess of 100 using three stage amplifiers.

Using all the above three technology improvements, high sensitivity, high voltage gain (V/W), wide bandwidth photoreceivers will be manufactured.

6. SUMMARY & CONCLUSIONS

To summarize, wide bandwidth photodiodes and photoreceivers are key components of the upcoming SONET/STM standards of OC-192 and OC-768. For the new and better OC-192 and OC-768 systems to be a reality, new technology in InGaAs PIN and OEIC/MMIC chip fabrication as well as microwave packaging has to be developed and inserted. Through its innovative wide bandwidth products, Discovery Semiconductors is filling this gap very effectively.

7. ACKNOWLEDGMENT

The authors would like to thank their colleagues, William F. Stacey, Yili Wang, Dan R. Mohr and Doug Malchow for helping prepare this manuscript.

8. REFERENCES

1. H.C. Huang, P. Laux, J.F. Bass, S.W. Chen, T.T. Lee, S. Tadayon, J.L. Singer, J. Kearney, and O.A. Aina, "A W-band Multifunction MMIC," *IEEE MTT Conf.*, San Diego, June 1994.
2. M.G. McDermott, C.N. Sweeney, J. Borelli, G. Dawe, and L. Raffaelli, "Integration of High-Q Varactor Diodes and 0.25 um GaAs MESFETs for Multifunction Millimeter Wave Monolithic Circuit Applications," *IEEE Trans. on Elect. Dev.*, Vol. 38, No. 9, pp. 1183-1190, 1990.
3. W.J. Ho, E.A. Sovero, D.S. Deakin, R.D. Stein, G.J. Sullivan, and J.A. Higgins, "Monolithic Integration of HEMTs and Schottky Diodes for Millimeter Wave Circuits," *IEEE GaAs Integrated Circuits Symp. Dig.*, pp. 301, 1988.
4. G. Sasaki, K.I. Koike, N. Kuwata, and K. Ono, "Optoelectronics Integrated Receivers on InP Substrates by OMVPE," *J. of Lightwave Tech.*, Vol. 7, No. 10, pp. 1510, 1989.
5. H. Yano, M. Murata, G. Sasaki, and H. Hayashi, "A High Speed Eight Channel Optoelectronic Integrated Receiver Array Comprising GaInAs p-i-n PD's and AlInAs/GaInAs HEMTs," *J. of Lightwave Tech.*, Vol. 10, No. 7, pp. 933, 1992.
6. S.D. Walker, L.C. Blank, R.A. Garnham, and J.M. Boggis, "High Electron Mobility Transistor Lightwave Receiver for Broad-Band Optical Transmission System Applications," *J. of Lightwave Tech.*, Vol. 7, No. 3, pp. 454, 1989.
7. F. J. Effenberger, A. M. Joshi, "Ultrafast, Dual-Depletion Region, InGaAs / InP p-i-n Detector," *J. of Lightwave Tech.*, Vol. 14, pp. 1859-1864, (1996).
8. K.J. Williams, R.D. Esman, and M. Dagenais, "Effects of High Space-Charge Fields on the Response of Microwave Photodetectors," *IEEE Photon. Tech. Lett.*, Vol. 6, No. 5, pp. 639-641, 1994.
9. K.J. Williams, R.D. Esman, R.B. Wilson, and J.D. Kulick, "Differences in p-side and n-side Illuminated p-i-n Photodiode Nonlinearities," *IEEE Photon Tech. Lett.*, Vol. 10, No. 1, pp. 132-135, 1998.
10. K.J. Williams, Naval Research Lab., *Private Communications*, July 1999.

Reliability of lithium niobate Mach Zehnder modulators for digital optical fiber telecommunication systems

David R. Maack

JDS Uniphase
Bloomfield, CT 06002

ABSTRACT

This paper is about the reliability of lithium niobate external modulators, one of the first major commercial successes for integrated optical circuits in fiber optics. Results from Bellcore type qualification programs along with long term aging tests at several temperatures for 2.5 Gb/s and 10 Gb/s modulator types will be presented., Additional published reliability data will be referenced and shown, however, limited information is available. Actual field data will also be presented. Where fitting or noteworthy, references will be made to other types of integrated optical circuits as this technology field is much greater than modulators alone. For the purposes of this paper, lasers, electro-absorption modulators, acousto-optic devices and optical detectors are not considered even though they too are highly commercially successful integrated photonic devices.

Keywords: integrated optical circuits, lithium niobate, external modulation, DWDM, fiber optics, Mach Zehnder interferometers, drift, reliability,

1. INTRODUCTION

The harnessing of the photon by tomorrow's integrated optical circuits will have as much impact on our world as does the harnessing of the electron by today's electronic integrated circuits. Integrated optical circuits and the whole emerging field of photonics are truly technologies in their infancy. The current state of the industry is analogous to the early days of the electronic integrated circuit and is projected to follow a similar growth curve. A wide array of device types have been demonstrated in the laboratory using a broad range materials, fabrication methods, design concepts and ideas. This industry is about to explode even though only a few products have had large commercial success.

The largest success story in integrated optical circuits for the fiber optics industry is lithium niobate modulators with tens of thousands being shipped per year into the emerging dense wavelength division multiplexing (DWDM) market. These components have finally reached sufficient market penetration such that adequate field data is available to confirm and corroborate engineering and laboratory reliability study predictions. Market size will pass $100,000,000 in the near future. Growth rate projections for the DWDM market are greater than 28% per year for the next five years.[1]

Performance improvement projections are even more aggressive. What started out as a low bandwidth component for Sagnac fiber optic gyroscopes will soon be a critical 40 Gb/s sub-component in commercial single fiber terabit transmission links. Stability, performance and reliability have reached levels only speculated of a few years ago.

The list of integrated optic circuit device types demonstrated is becoming quite large. It includes simple waveguides, MxN splitters, Mach-Zehnder interferometers, wavelength division multiplexers and demultiplexers, optical amplifiers, sensors of all types, analog and digital modulators, switches, and even logic gates. In addition, other components are being added to these optical platforms thereby significantly increasing functionality. These components include lasers, detectors, bulk optics, fibers and, of course, electronic circuitry. Most of the above devices have been demonstrated using lithium niobate although a number of other material systems have also successfully been utilized. These other systems include various glasses, silicon, semiconductor materials and even plastics. Lithium niobate, however, is by far the most successful to date.

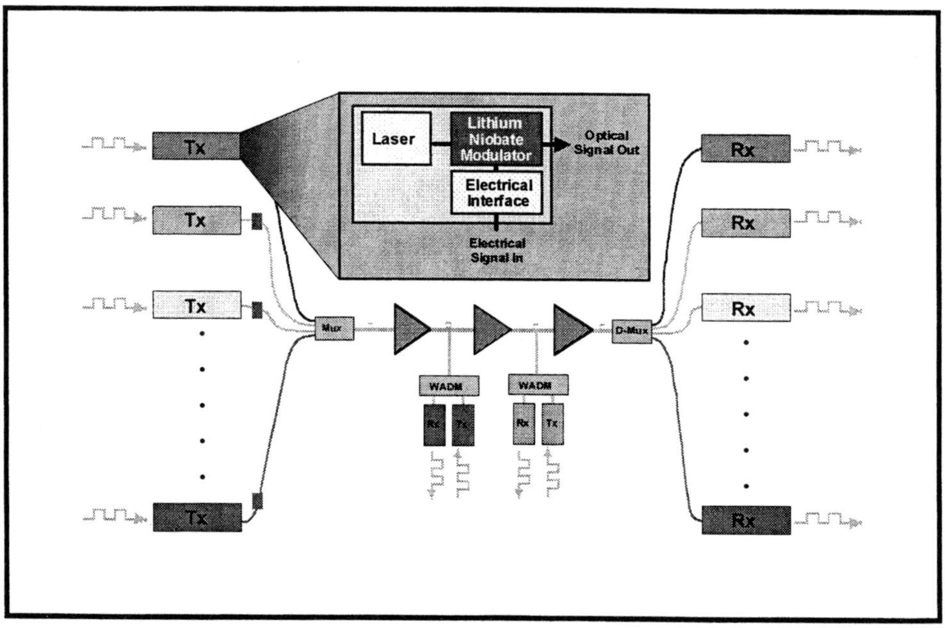

Figure 1. DWDM system using lithium niobate modulators

The major application for lithium niobate modulators is in the transmitters of DWDM telecommunications systems. A typical system design is shown in Figure 1. The DFB laser diode is operated in a steady state DC condition and coupled to the modulator through polarization maintaining fiber. A digital electrical signal is applied to the RF input of the modulator thereby modulating its optical output. Digital 1's are represented by full optical power and digital 0's by no optical power. (Analog electrical signals would produce an analog optical output.) The modulator's output fiber is then coupled to the transmission line. The major advantage of this configuration is the elimination of many of the non desirable effects and non-linearity's that occur when a laser diode is modulated directly. Longer distances between transmitters, amplifiers and receivers and closer wavelength spacing between the adjacent optical channels are just some of the advantages of external modulation. Reliability is another.

1.1 Lithium niobate

Lithium niobate is a highly developed optical material from a class of materials known as frozen ferroelectrics.[2, 3] It is used in the crystalline form and has what is most commonly referred to as a rhombohedral structure. The most important properties for integrated optical circuits are its relatively large electro-optic coefficients, low optical loss and chemical stability. Other characteristics include strong pyroelectric and piezoelectric properties which can play detrimental roles in achieving high reliability. All properties, including its mechanical parameters, are highly dependent upon the crystal orientation.

Lithium niobate has been available for many years in large uniform crystals and substrates primarily for the surface acoustic wave (SAW) device industry.[4] SAW devices have been used for many years as filters in television receivers and mobile telephones thereby creating a relatively large demand for lithium niobate. This established industrial base for high grade material is one of the key reasons that it has become the workhorse for integrated optical circuit substrates. Although the current demand for optical grade material is still relatively small, less than 10% of the total usage [5], the recent growth rate has been significant. This 'new' market has helped to drive material processing improvements resulting in lower cost, higher performing substrates.

Three inch and four inch lithium niobate substrates or wafers are available in all three of the crystal orientations. These are X-cut, Y-cut and Z-cut with the axis indicating the direction normal to the substrate surface. The most commonly used for modulators are X-cut and Z-cut. The substrates are about 1 to 2 mm in thickness with highly polished surfaces. Great care is taken to eliminate or reduce bulk and surface defects.

1.2 Writing waveguides and optical circuits

In order to create an optical integrated circuit, optical waveguides must be 'written' into the substrate. This is done using a number of techniques[5] with most based on photo-lithography like that used in the manufacture of electronic integrated circuits. The waveguides are created by diffusing various ions into the waveguide region replacing some of the lithium ions. An area of reduced index of refraction is created forming a light guiding medium. Most commercially available modulators use either titanium ions or protons (hydrogen ions), however, other elements have been successfully demonstrated.[6] Suffice to say that the details of these processes are highly proprietary.

The bulk of the commercially deployed modulators are confined to three material systems. These are titanium ion waveguides in X-Cut substrates, proton waveguides in X-Cut substrates and titanium ion waveguides in Z-Cut substrates. This is very important as each of these systems has different reliability and test considerations.

The waveguides are written to form a Mach Zehnder interferometer as shown in Figure 2 for X-Cut devices and Figure 3 for Z-Cut devices. Incoming light is split into two paths and recombined some distance later. If the light is in phase at the combiner, constructive interference occurs and all the light, less waveguide losses, appears at the output. If the light is 180 degrees out of phase, then destructive interference occurs and the optical power is launched into higher order modes. These higher order modes are not supported in the output waveguide. The optical power radiates into the substrate and no light appears at the output. By placing control electrodes on the waveguides, the phase differential between the two paths can be changed with an electrical signal. This phase difference is produced by the electric field changing the index of refraction in the waveguide due to the to the high electro-optic coefficient of lithium niobate. The light is, therefore, modulated in direct proportion to the electrical signal, either analog or digital, although not necessarily linear. Analog modulators are used primarily in the CATV market while digital modulators are used in 2.5 Gb/s and 10 Gb/s digital systems with 40 Gb/s close to commercial deployment. In these digital systems, the modulator acts as a simple, high speed shutter.

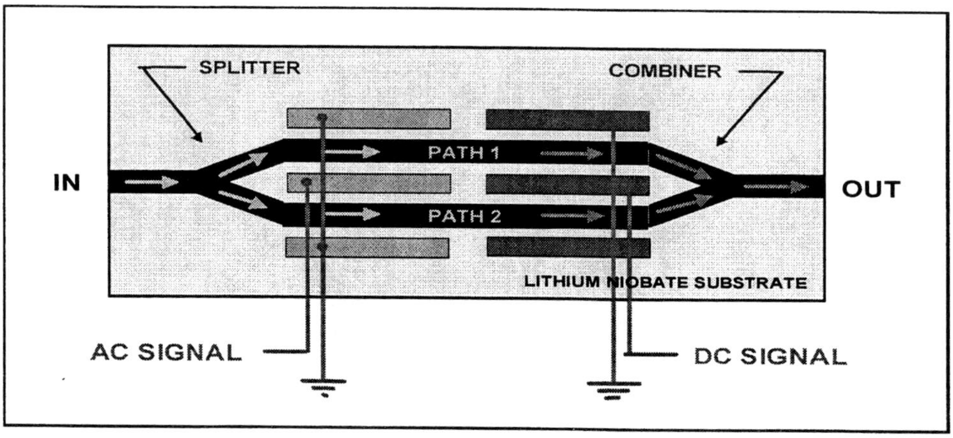

Figure 2. Simple X-Cut Mach Zehnder interferometer

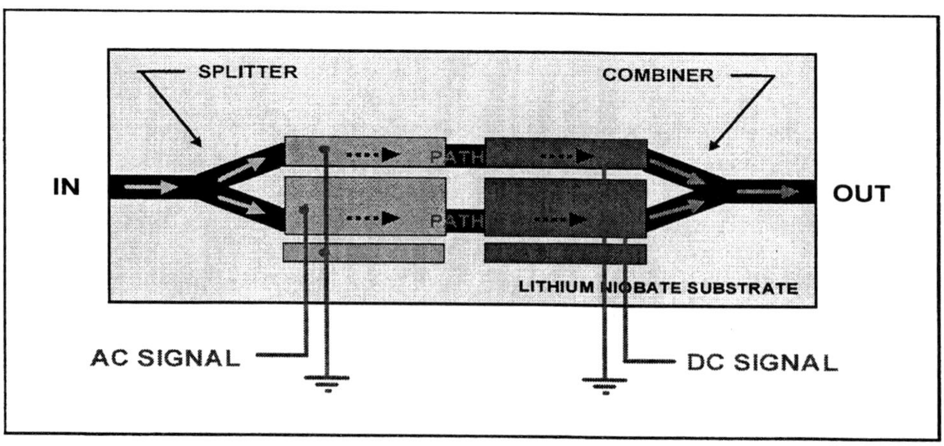

Figure 3. Simple Z-Cut Mach Zehnder interferometer

As stated earlier, the most commonly used substrates for modulators are X-cut and Z-cut. In both of these orientations, the electric field across the wave guide is parallel to the Z-axis because of its higher electro-optic coefficient. In the X-Cut orientation, the X-axis is perpendicular to the substrate surface, light propagation is in the Y-axis and the Z-axis is on the surface plane perpendicular to the waveguide. The electrodes are placed on either side of the wave guide and produce an E-field perpendicular to the waveguide and parallel to the surface of the substrate.

In the Z-Cut orientation, the Z-axis is perpendicular to the substrate surface and light propagation is in either the X or Y-axis. The electrodes now must be placed over the waveguides to produce the E-field parallel to the Z-axis and perpendicular to the surface of the substrate. This has a major impact in that a buffer layer with the proper optical and electrical properties must be placed between the electrode and the waveguide to maintain total internal reflection.

The state of optical polarization also plays a critical since the electro-optic coefficients are polarization dependent. It is desirable to modulate the most sensitive orientation. Further, light in the waveguide can be polarized in two directions, but the E-field is primarily only in one of these axis. Consequently, one of the polarization states is not well modulated increasing the residual or background light. In order to achieve maximum efficiency, polarization maintaining fiber and properly orientated light sources are used on the input. The output fiber is generally standard singlemode, however, certain applications can require polarization maintaining fiber. In X-cut, the TE optical mode is launched into the waveguide while in Z-Cut the TM mode is launched.

1.3 Quadrature and Drift

A very desirable condition is to have the modulator produce full optical power for a digital '1' and no optical power for a '0'. This requires a knowledge of the electronic RF driver which provides the signal to the electrodes of the modulator. For standard X-cut, this driver supplies a positive and negative going square wave signal where ones are represented by the positive swing and zeros are the negative swing. For Z-cut and advanced X-Cut devices, more complex dual drive schemes are used, however, the concepts of a transfer function and operation at quadrature are the same.

Optical output power from a modulator changes from a maximum to a minimum over 180 degrees or π radians of optical phase shift between the two paths in the interferometer. The voltage required to change the phase by π radians is called $V\pi$. In order to optimize the driver/modulator efficiency, the modulator needs to be at quadrature ($\pi/2$ absolute phase difference) at zero volts. Figure 4 shows this in greater detail. Starting at t=0, the electrical signal is driven to its maximum negative value and the optical phase difference goes to an absolute phase difference of π radians or $-\pi/2$ radians from quadrature. Destructive interference occurs and no light exits the modulator. At some time later, the electrical signal is driven to its maximum positive value and the optical phase difference goes to an absolute phase difference of 0 radians or $+\pi/2$ radians from quadrature. Now constructive interference occurs and all the light exits the modulator.

The modulator in its quiescent or zero input voltage state needs to be at quadrature. Ideally, this phase difference would be generated in manufacturing. Unfortunately, two issues prevent this. The first is that current manufacturing processes cannot yet hold the physical tolerances required. Second, the quadrature point can drift with time. To correct for both of these, various schemes have been developed. For modulators using titanium waveguides, a bias voltage is applied either to a second set of DC electrodes or in a bias-T circuit to the RF electrodes. A control loop maintains the modulator at quadrature by constantly varying the DC bias voltage. For modulators using proton waveguides, passive biasing techniques are used to achieve quadrature and no DC voltage is applied. These bias free proton waveguide devices inherently exhibit much lower drift. The absence of DC voltage on the electrodes also helps reduce drift.

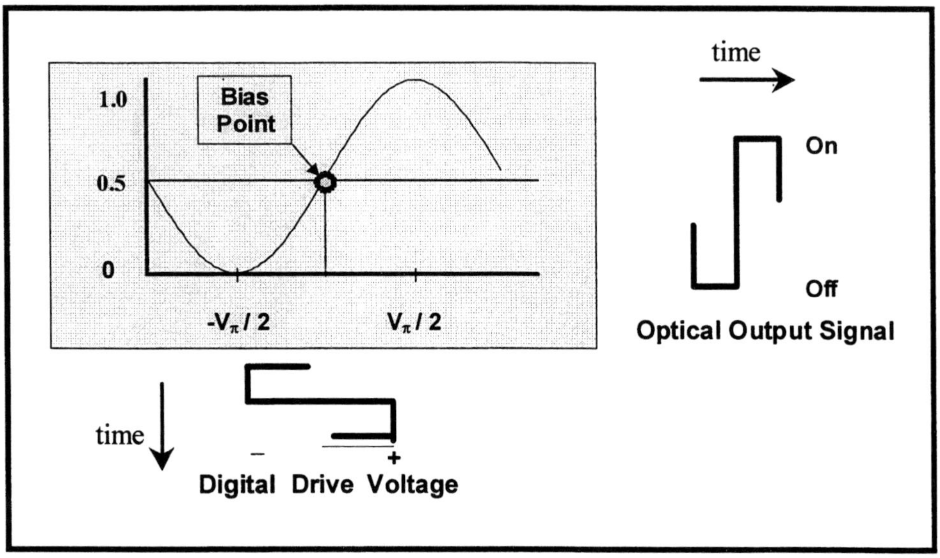

Figure 4. Transfer function

Drift in lithium niobate modulators is an issue and represents a major reliability risk. As long as the bias voltage necessary to maintain quadrature is within the rail limits of the power supply, typically between 10 and 15 volts, quadrature can be maintained. If, however, this rail voltage is exceeded, the modulator will drop off of quadrature causing the optical power for a 1 to be less than maximum and that for a 0 to be some finite value. At the optical receiver, the difference between the 1's and the 0's diminishes and the bit error rate (BER) of the system increases. At this time the system can be "reset" by reducing the bias voltage by a factor of 2 times $V\pi$, or 360 degrees of phase shift, however, this will cause a loss of data. In addition, this voltage upset to the modulator can cause a number of detrimental effects, most particularly faster bias drift.

Bias drift has been an issue for lithium niobate Mach Zehnder modulators from the beginning. Much empirical improvement was made to reduce drift and a number of theories were postulated, however, it was the work of Korotky and Veselka[7] of Lucent that finally solidified a general mathematical model. This model reduces the equivalent electrical circuit of the modulator to a ladder network of RC circuits, see Figure 5.

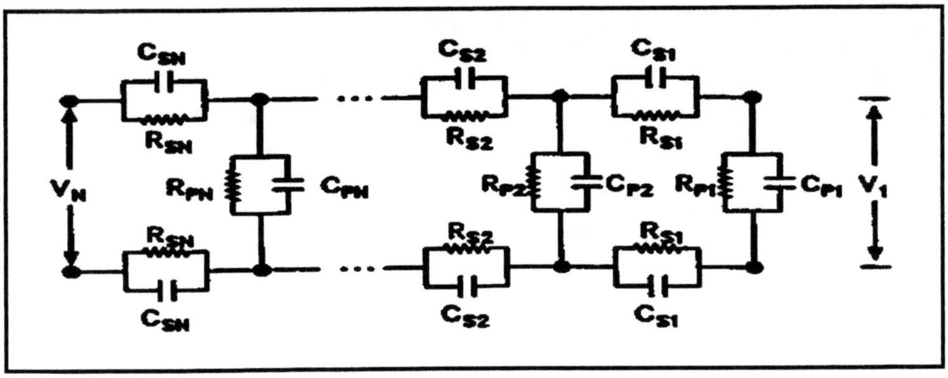

Figure 5. RC ladder model for lithium niobate modulators, from Korotky and Veleska [7]

In this model, "V_N is the externally voltage applied to the modulator while V_1 is the effective voltage drop that directly determines the electro-optically induced phase shift"[7]. In order to keep phase shift constant, i.e., no bias drift, then V_1 needs to be kept constant. Unfortunately, all of the RC circuits continue to charge and discharge at different rates with time constants from seconds to years and affect V_1. This causes V_1 to drift. We can correct for this drift by changing V_N as long as we can maintain V_1 constant, however, that depends greatly on the values of the individual resistors / capacitor pairs. It is easy to visualize a situation where the value of V_N exceeds the power supply rail limits.

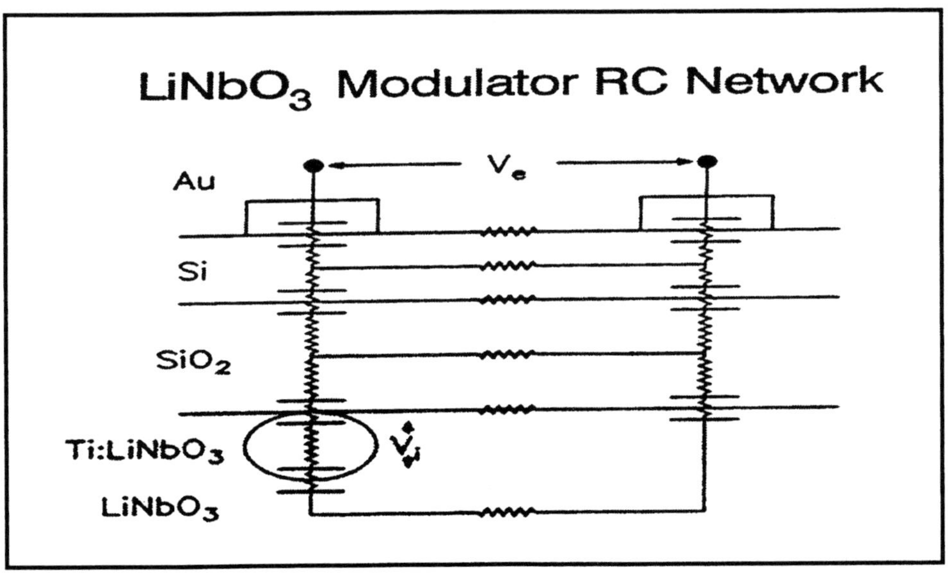

Figure 6. Cross section of a typical Ti:$LiNbO_3$ electro-optic device, from Korotky and Veleska [7]

The physical interpretation of these resistor/capacitor pairs are the various layers, surfaces, interfaces and impurities in the modulator. Figure 6 shows the equivalent circuit for a simplified Z-Cut modulator. Lithium niobate is an excellent dielectric and can hold very high surface and internal charges. This charge can also migrate and dissipate from one place to another. In addition, it is piezo-, pyro- and opto- electric meaning that any outside stress whether mechanical, thermal gradient or optical can generate electrical charge. All of these can directly affect and change the value of V_1 forcing the bias control loop to change the value of V_N to compensate. There is also empirical evidence that as V_N increases, the drift rate increases. In work done by Nagata, et. al. of Sumitomo, see Figure 7, Z-Cut modulators were subjected to different initial fixed DC voltages at 130°C [8]. The drift rate was recorded and plotted. The drift rate is greater for the sample at 8 volts than the one at 4 volts strongly suggesting that drift rate is proportional to bias voltage. Bias voltage needs to be kept to a minimum. More work is needed on this subject.

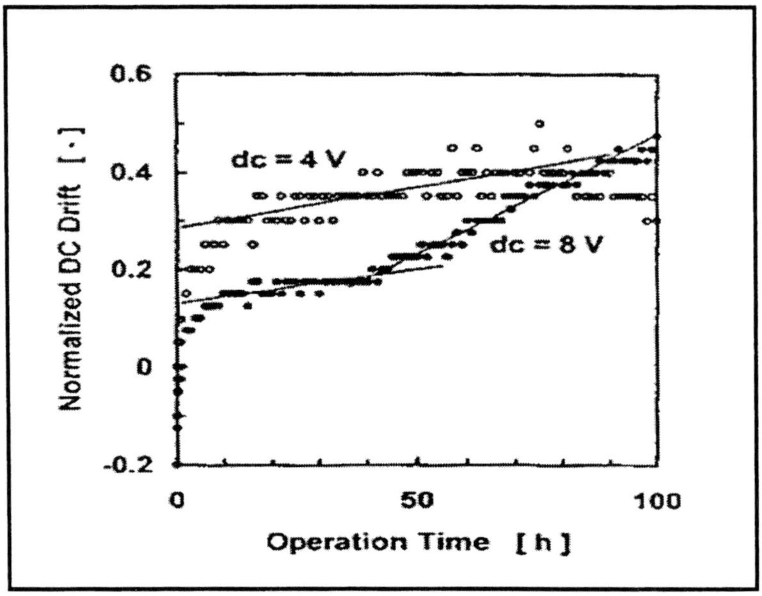

Figure 7. Normalized dc drift measured by a fixed dc bias method at 130°C, from Nagata, et. al.[8]

There are other reliability considerations with lithium niobate other than drift. These would include mechanical breakage, changes in loss, changes in Vπ, and the stability of all of the electrical parameters associated with the electrodes. The electrical parameters include resistance, impedance, frequency response and electrical back reflection. By using well established electronic integrated circuit processing technology along with carefully developed lithium niobate processing technology, these have not been a serious reliability issue.

1.4 Packaging

The final area to review that can affect the reliability is the packaging technology. Optical fibers, DC bias control connections and high speed RF connections must be made to the outside world. The lithium niobate substrate must be protected from the environment especially from those stresses, like mechanical, thermal change or optical, that can promote drift. In addition, the entire modulator needs to be in a package that protects it from dirt and other contaminants. And, if moisture is a failure mechanism for a particular design, the package must be hermetic.

In general, there are two points of interconnection. The first is at the integrated optical circuit itself and the second is through the wall of the package. The optical connections are generally achieved by attaching a carrier block on the end of the fiber not unlike that done in attaching a connector ferrule to a fiber. The end is optically finished. This assembly is then aligned to the substrate and epoxied in place. This attachment must be capable of maintaining sub-micron alignment and stability over the entire life of the modulator. Further complicating the alignment is the polarization maintaining fiber on the input. Its axis must also be aligned properly.

Now that the fiber has been attached to the optical waveguide, it still has to penetrate the package wall. In non-hermetic devices, this can be as simple as a hole in the wall, a sealing method and a strain relief boot on the outside. In hermetic devices, however, a glass to metal seal must be made to the surface of the fiber. This can significantly increases the potential modes of failure as the fiber must be stripped of its buffer exposing bare glass. This process usually involves coating the bare fiber surface with a metal followed by subsequent soldering into a ferrule.

The electrical connections to the electrodes are also very important, however, this technology is borrowed from the electronic IC world. Wire bonds are used to connect the electrodes to a circuit board. This board can be quite specialized and sophisticated as it needs to accomplish a number of tasks. For digital applications, the frequencies of operation are 2.5 GHz and 10 GHz respectively. This requires complex wideband RF waveguide technology and good impedance matching. The DC leads are somewhat simpler form the electrical point of view, but still have the same mechanical reliability considerations.

The modulator is placed in a final package usually made of a metal such as aluminum or Kovar®. Care is taken to protect the substrate from environmental stresses by careful mounting techniques. Electrical feed throughs, whether they be wire or electrical connectors, are soldered to the circuit board. Non-hermetic packages are typically closed with screws or epoxies while hermetic packages are sealed with a seam welding technique. In general, X-Cut does not require hermeticity while Z-Cut does.

2. RELIABILITY STUDIES

A number of reliability studies have been reported, however, most of the data is for long term aging of bias drift at a limited number of temperatures. There are other tests and failure modes that are also important. Most of the data presented in this paper has been generated by internal qualification and reliability programs. These programs are in accordance with the appropriate Telecordia specifications. Specifically, the devices tested were:

1. 2.5 Gb/s (OC-48), X-Cut lithium niobate substrates with titanium diffused waveguides in butterfly packages with solder pin electrical terminals in accordance with Telecordia TR-NWT-000468, Issue 1, December 1991.[9]

2. 2.5 Gb/s (OC_48), X-Cut lithium niobate substrates with proton diffused waveguides in SMA packages with coaxial connector electrical terminals in accordance with TR-NWT-000468, Issue 1, December 1991. [9]
3. 2.5 Gb/s (OC_48), X-Cut lithium niobate substrates with proton diffused waveguides in butterfly packages with solder pin electrical terminals in accordance with TR-NWT-000468, Issue 1, December 1991. [9]
4. 10 Gb/s (OC-192), X-Cut lithium niobate substrates with titanium diffused waveguides in SMA packages with coaxial connector electrical terminals in accordance with GR- 468-CORE, Issue 1, December 1998. [10]

Item 4 is in progress at this time and not all tests are complete. Information on 10 Gb/s (OC-192), Z-Cut lithium niobate substrates with titanium diffused waveguides was obtained from published data. Unfortunately, not a lot of data is available as few manufacturers publish qualification and reliability information. Much of the data on items 1, 2, 3 and 4 above is being published for the first time.

2.1 Measurements

The rapid entry of narrow band WDM technology into today's telecommunication systems has introduced a number of new components, such as lithium niobate modulators, into the hardware infrastructure raising issues of which measurements, measurement procedures and pass/fail criteria to use in reliability and qualification testing. With many of these components, industry standard measurement techniques have not yet been established. One of the most important issues for any reliability study are the performance parameters that will be measured and the pass/fail criteria for each.

Lithium niobate modulators are complex components that combine optical, electrical and RF functions in very interdependent ways Each of these functions has a number of measurable parameters which when combined, fully characterize the modulator's overall performance. These are known as the modulators intrinsic parameters. Unfortunately, the pass/fail criteria and relative importance of each individual parameter can be very system dependent. Consequently, there is a need to establish a set of measurements for pass/fail evaluation in reliability testing that properly predicts system performance. These are known as extrinsic measurements. A set of definitions for key performance parameters can be found in Table 1. In addition to the parameters in this table, hermeticity needs to be measured on hermetic devices.

Although the above parameters fully define the performance of a modulator, they do not always represent a complete set of measurements that fully predicts transmission system performance. The best measurements to predict how the modulator performs in a system would be bit error rate (BER) and insertion loss. Unfortunately, BER is a very difficult, expensive and time consuming measurement. Fortunately, standard 2.5 Gb/s pseudo random binary sequence (PRBS) extinction ratio "eye diagram" measurements with the proper modified procedures and setup can accurately predict how well this component will operate in a system[11]. One of the big advantages of the eye diagram is that it integrates most of the intrinsic parameters into a single extrinsic measurement. Many times, the PRBS extinction ratio along with insertion loss are adequate measurements to evaluate performance for a reliability test. In some studies, the intrinsic measurements may not even be performed.

S_{11}	The ratio of incident electrical power to power reflected from the modulator back into the source and is a function of frequency.
S_{12}	The frequency response of the modulator.
$V\pi$ RF	The RF voltage required to drive the modulator through π radians of phase shift. When the modulator is at quadrature, a voltage change of $\pm V\pi/2$ will drive the optical output state from maximum to minimum. This is generally specified at one frequency or over a range of frequencies.
$V\pi$ DC	The DC voltage required to drive the modulator through π radians of phase shift.
Phase	The absolute amount of phase shift away from quadrature (Measured only on non-electrically biased devices)
Optical Loss	The total optical power loss through the modulator
Return Loss	The ratio of the launched light to the reflected light
Extinction Ratio	The ratio of maximum optical intensity to the minimum optical intensity usually specified at a low frequency or DC.

Table 1. Intrinsic performance parameters

The accuracy of these measurements is an important issue. Some of these measurements are under as much development as the component. For the long term aging tests, the measurement equipment and procedures could easily undergo several generation of changes during the lifetime of the test. For example, the PRBS extinction ratio test still has a 2 dB uncertainty on measurements between 14 and 16 dB[12]. Most of these measurements reported in this paper are above 16 dB.

2.2 Tests

Classical reliability testing would suggest that a complete reliability study would subject multiple components to multiple high levels of all potentially life reducing stresses. The tests would be run until all samples failed or a time to failure could be extrapolated. From this data, lifetime prediction could be made knowing the stress levels of the environment of interest. This is not realistic for lithium niobate modulators. For one, the cost would be exorbitant. In addition, the relationships between the real world and many of the stresses is not well understood. The question is: What represents a reasonable set of tests to assure an adequate level of reliability?

Telecordia (Bellcore) has attempted to address this question with "GR-468-CORE, Issue 1, December 1998, Generic Reliability Assurance Requirements for Optoelectronic Devices Used in Telecommunications Equipment" [10] (GR-468), currently out for industry comment, and its predecessor, "TR-NWT-000468, Issue 1, December 1991, Reliability Assurance Practices for Optoelectronic Devices in Interoffice Applications" [9] (TR-468). The new GR-468-CORE adds, for the first time, a separate section addressing external modulation. See Table 2 for a description of the specific tests in GR-468 that are applicable to external modulators in a central office environment. The pass/fail level in these qualification programs is based on a 20% lot tolerance percent defective from MIL-M-36501F (October 1983)[13]. Minimum lot size is 11 samples with no failures, however, one failure is allowed with a lot size of 18. Most customers for modulators reference the Telecordia specifications even if they do not follow it explicitly.

Telecordia (Bellcore) GR-468-CORE, Issue 1

CATEGORY	TEST		REQUIREMENTS
Other	Characterization	Test Qty:	20
		Reference:	GR-468, Section 10.3.1
		Conditions:	min, room, max. op. temp
Mechanical	Drop Shock	Test Qty:	11
		Reference:	MIL-STD-883E Method 2002, Cond B
		Conditions:	500 G, 1.0 ms, 5 times/axis
	Vibration	Test Qty:	11
		Reference:	MIL-STD-883E Method 2007, Cond A
		Conditions:	20-2,000 Hz-min/cy 20G, 4 cy/ axis
	Thermal Shock	Test Qty:	11
		Reference:	MIL-STD-883E, Method 1010
		Conditions:	D100°C Liquid to Liquid, 10 cycles, 0°C to +100°C
		Note:	Hermetic only
	Solderability	Test Qty:	11
		Reference:	MIL-STD-883 Method 1011
		Conditions:	steam aging not required
		Note:	packages with solder pins only
	Fiber Pull	Test Qty:	11
		Reference:	GR-468 Table 24
		Conditions:	3 Pulls, 5 sec/pull PM: 0.5 Kg., SM: 1.0 Kg. If field accessible 2.0 Kg
Endurance	Accelerated Aging (high temp)	Test Qty:	11
		Reference:	GR-468, Section 5.18
		Conditions:	2000 hrs @ 70°C for pass/fail, 5000 hrs for info
	High-Temp. Storage	Test Qty:	11
		Reference:	GR-468, Table 24
		Conditions:	2000 hrs @ 85°C
	Low Temp. Storage	Test Qty:	11
		Reference:	GR-468 Table 24
		Conditions:	2000 hrs @ -40°C
		Note:	Optional

Table 2. Telecordia GR-468-CORE, *Generic Reliability Assurance Requirements for Optoelectronic Devices Used in Telecommunications Equipment*

Telecordia (Bellcore) GR-468-CORE, Issue 1

Endurance (cont)	Thermal Cycling	Test Qty:	11
		Reference:	GR-468, Section 5.20
		Conditions:	-40°C/+70°C, 100 cycles for pass/fail, 500 cycles for info
	Damp Heat	Test Qty:	11
		Reference:	MIL-STD-202F, Method 103B
		Conditions:	500 hrs @ 85°C/85%RH or 5000 hrs @ 50°C/85%RH
	Cyclic Moisture	Test Qty:	11
		Reference:	MIL-STD-883, Method 1004
		Conditions:	10 cycles with 5 sub zero cycles
		Note:	Hermetic products only
Misc	Internal	Test Qty:	11
		Reference:	MIL-STD-883, Method 1018
		Conditions:	max 5000 ppm water vapor
	ESD Threshold	Test Qty:	6
		Reference:	TR-870, Section 4.2.3
		Conditions:	> 500 volts

Table 2. Telecordia GR-468-CORE, *Generic Reliability Assurance Requirements for Optoelectronic Devices Used in Telecommunications Equipment*

GR-468 is a qualification program in that passing of all the requirements suggests a high degree of reliability. Although it represents the extensive experience of Telecordia, it does not quantify reliability. Still, these are very important tests to perform and pass, but they are not sufficient in themselves. Additional testing must be performed in accelerated aging at different temperatures and conditions. GR-468 requires only 2000 hours at ≥70°C. This test needs to be continued to failure. In addition, other temperatures such as 85°C or 100°C need to be run. Results for each of the GR-468 tests will be presented for a number of different types of modulators.

2.2 Characterization

Characterization is the first test that GR-468 requires. It is meant to show that the modulator meets all performance parameters throughout its temperature design space. The performance parameters are ultimately determined by the final customer's specifications, however, there is a suggested list in GR-468. A reasonable list of parameters are those found in Table 1 of this paper, however, individual applications can have a much larger list. All of these parameters are to be measured at the minimum operating temperature, typical room temperature and maximum operating temperature. Generally, this is -5°C, +20°C and +70°C. In addition, an additional 50 to 200 units from 3 different production lots must also be tested. The purpose of this test is to assure that the modulator meets its design performance parameters.

2.3 Mechanical integrity

The next section in GR-468 are the mechanical tests. These are extremely important because they test the mechanical integrity of the modulator. They are not reliability tests in the strictest sense of the word, but they do represent Telecordia's many years of experience and judgement as to what constitutes a reasonable test to verify a good mechanical design. The specific tests include drop shock, vibration, thermal shock, solderability and fiber pull.

2.3.1 Drop shock

The drop shock test is performed in accordance with MIL-STD 883[14], Method 2002, Condition B. The test level is 500 g's with a 1.0 ms pulse duration. Each sample is subjected to 5 shocks in both directions in all three orthogonal axis for a total of 30 'hits'. The minimum number of samples is eleven and all must pass. This test is a very good measure for the quality of the wirebonds, the fiber attachment method and the integrated optical circuit mounting technique. On hermetic units, it is a good measure for how well the fiber is stress relieved at the seal.

Table 3 outlines test results for a number of different modulator types. The 'before' column is the average initial value of all samples in the test. Delta (Δ) is the difference between the post test average value and the initial average value. Positive numbers mean that the final value denotes a deterioration of that parameter while negative numbers mean an improvement. This convention will be maintained for all results tables. As can be seen, the entire matrix of data is not available for various reasons. It is important to note that all four programs listed at the beginning of Section 2 passed the Telecordia requirements. Although little quantitative information is available for Z-Cut, those manufactures do state that their devices have passed the Telecordia requirements.

Parameter	2.5 Gb/s, X-Cut Ti Waveguide		2.5 Gb/s, X-Cut Proton Waveguide		10 Gb/s, X-Cut Ti Waveguide	
Test Level	1500 g		1500 g		500 g	
	Before	Δ	Before	Δ	Before	Δ
S_{11} (dB)	Pass		−9.9	0.2	−11.8	−.04
S_{21} (GHz)	Pass		Pass		Pass	
Vπ RF (V)	No data		No data		6.2	−.02
Vπ DC (V)	7.7	−.01	6.6	0.0	5.9	.01
Phase	n.a.		0	0.3	n.a.	
Loss (dB)	4.2	0.1	3.1	0.0	3.0	.05
ER DC (dB)	33	1.8	36	0.0	26	−0.8
ER PRBS (dB)	19	1.4	No data		16	−0.5

n.a. = not applicable

Table 3. Comparison of drop shock test results

The test level differences listed in Table 3 are because Telecordia has reduced the stress level requirement from 1500 g's to 500 g's in the recent release of GR-468. The original 1500 g level was designed to test much smaller components like laser diodes and detectors for a wide variety of environments. Modulators, on the other hand, are normally deployed in central offices. In addition, the cost, handling, packing, shipping and general care taken with modulators is much higher than other components. They are mounted on complex circuit boards that are handled very carefully thereby justifying the lowering of the requirement. This change was a good decision by Telecordia as they recognized that continued adherence an improper test level could impede the development of this entire industry.

2.3.2 Vibration

The vibration test is performed in accordance with MIL-STD 883, Method 2007, Condition A. The test level is 20-2,000 Hz-minute/cycle and peak forces reach 20 g's. Each sample is subjected to 4 cycles per axis. The minimum number of samples is eleven and all must pass. This test gives a good indication for a number of failure mechanisms. In particular, it will identify any mechanical resonance in the frequency range of the test. A typical lithium niobate substrate has a very large aspect ratio, its length is much larger than its width or thickness, requiring well though out mechanical designs. This test will quickly identify flaws in all the mechanical sub-components. Another weakness that the vibration test can highlight is fiber surface damage in hermetic seals. The vibrations can promote stress and crack growth.

Parameter	2.5 Gb/s, X-Cut Ti Waveguide		2.5 Gb/s, X-Cut Proton Waveguide		10 Gb/s, X-Cut Ti Waveguide	
	Before	Δ	Before	Δ	Before	Δ
S_{11} (dB)	Pass		−10.0	0.0	−11.9	.04
S_{21} (GHz)	Pass		Pass		Pass	
Vπ RF (V)	No data		No data		6.1	.03
Vπ DC (V)	7.5	0.0	6.7	0.0	5.9	−.02
Phase	n.a.		−5.1	0.8	n.a.	
Loss (dB)	4.0	0.3	3.1	0.0	3.1	.01
ER DC (dB)	36	−0.2	38	−0.2	26	0.3
ER PRBS (dB)	20	0.7	No data		16	0.2

n.a. = not applicable

Table 4. Comparison of vibration test results

The results from a number of thermal shock tests on different modulator types is shown in Table 4. Examination of the results shows that little change occurred and all samples meet the requirements. Again, no quantitative data is available for Z-Cut, however, it is reasonable to assume that Z-Cut will also pass the test.

2.3.3 Thermal shock

The third mechanical integrity test in GR-468 is thermal shock. It is required for hermetic devices only and is performed in accordance with MIL-STD 883, Method 1011. The procedure involves transferring the sample ten times between two liquids at temperatures of 0°C and 100°C respectively. The minimum number of samples is eleven and all must pass. Its purpose is to test the integrity of the hermetic seal as Telecordia does not require either electrical or optical measurements. The mode of failure is loss of hermeticity usually between the fiber surface and its initial metallic coating. This test can also promote fiber breakage.

Although this test is called out only for hermetic modulators, many customers want a substitute test for non-hermetic devices. The test usually chosen is a modified MIL-STD-883, Method 1010 with a transfer time of ≤ 1 minute. The modified temperature extremes are typically – 30°C and + 70°C for a total excursion of 100°C. Although the this air to air test is not as severe as the liquid to liquid test, it can reveal design weaknesses.

Parameter	2.5 Gb/s, X-Cut Ti Waveguide		2.5 Gb/s, X-Cut Proton Waveguide		10 Gb/s, X-Cut Ti Waveguide	
Type	Non-hermetic		Non-hermetic		Non-hermetic	
	Before	Δ	Before	Δ	Before	Δ
S_{11} (dB)	Pass		–10.0	0.0	Planned	
S_{21} (GHz)	Pass		Pass			
Vπ RF (V)	No Data		3.2	0.0		
Vπ DC (V)	7.5	0.0	6.6	0.0		
Phase	n.a.		0	–1.3		
Loss (dB)	4.3	–0.1	3.0	0.0		
ER DC (dB)	36	–0.4	37	0.1		
ER PRBS (dB)	20	–0.1	No data			

n.a. = not applicable

Table 5. Comparison of thermal shock test results

The results from a number of vibration tests on different modulator types is shown in Table 5. Examination of the results shows that little change occurred and all samples met the requirements. Again, no quantitative data is available for Z-Cut, however, it is reasonable to assume that Z-Cut passes this test.

2.3.4 Solderability

The solderability test is performed in accordance with MIL-STD 883, Method 2003 without steam aging. The original purpose of this test is to make sure that one can solder to electrical leads. It involves specific solders, temperatures, preparations, and times. Pass/fail is determined by microscope inspection. In addition, GR-468 adds that the solderability test is to check for mechanical integrity due to the large thermal shock.

It is important to note that electrical package feed throughs used here are a mature technology taken from the electronics industry. Properly made, they will pass this test without a problem. Passing this test requires that both the wetability of solder on the pin and the mechanical integrity of the glass to metal seal are good. Hermeticity can be affected if the thermal shock and high temperature break the seal. No failures have been observed in any of the device types with electrical feed throughs.

2.3.5 Fiber pull

Fiber pull is performed in accordance with the conditions stated in Table 24 of GR-468. No specific test procedure is given, however, test levels are dependent upon fiber type and application. Generally, this test is performed by securing the modulator in a vertical orientation and hanging the proper weight from the fiber. The weight is slowly released three times and left hanging for 5 seconds.

On applications where the end user, meaning the field personnel of the installer, must access the fiber to install it, then the fiber is to be subjected to a 2 kg. load. If the fibers are not end user accessible and are only handled by the equipment manufacturer, then standard singlemode fiber is subjected to a 1 kg. load and polarization maintaining fiber is subjected to a 0.5 kg. load. Modulators fall primarily into the latter category. The reason for the difference between standard singlemode fiber and polarization maintaining fiber is the difference in the coatings. The polarization maintaining fiber has a much softer and thicker primary buffer and, therefore, is not as rugged. Care must be taken in handling these fibers.

Parameter	2.5 Gb/s, X-Cut Ti Waveguide		2.5 Gb/s, X-Cut Proton Waveguide		10 Gb/s, X-Cut Ti Waveguide	
Level	½ Kg.		1 Kg.		½ Kg. & 1 Kg.	
	Before	Δ	Before	Δ	Before	Δ
Loss (dB)	5.0	.03	3.0	0.1	Planned	

Table 6. Comparison of fiber pull test results

The important measurements to perform are optical loss and hermeticity (on hermetic devices only) as no other properties of the modulator can be affected. Table 6 shows the results on a number of pull tests performed on different modulators. As can be seen, there are virtually no changes. The major modes of failure are fiber breakage and loss of hermeticity.

2.4 Endurance tests

The next section in GR-468 are the endurance tests. These include long term aging, high temperature storage, low temperature storage (an optional test), thermal cycling and damp heat. It also calls for a cyclic moisture test for products designed for uncontrolled environments, however, modulators are currently used only in central office applications and this test is not required.

2.4.1 Accelerated aging at high temperature

The accelerated aging test is one of the most important tests to do for lithium niobate modulators. Although GR-468 only calls for a single 2000 hour test at ≥ 70°C, longer times and multiple temperatures are necessary to quantify reliability. In addition, it turns out that two tests need to be done on actively biased devices (titanium waveguides). One of these tests needs to be at constant phase or maintained at quadrature and the other needs to be run with a constant voltage on the bias electrodes. The reason will be explained shortly. On passively biased devices (proton waveguides), only a constant phase test is needed. By definition, this is simultaneously a test at constant voltage, namely, zero volts.

GR-468 requires that the accelerated aging test shall be preformed with normal bias and maximum modulation rate. How this powering is done is a very important issue. On actively biased modulators, normal bias means that the modulators under test must be at quadrature. This requires a set up consisting of lasers, detectors and bias control electronics. On passively biased modulators, the electrodes are shorted to ground.

The importance of the proper bias control on actively biased modulators cannot be over emphasized. According to Korotky and Veselka[7], see Figure 5, V_1 must be kept constant or else the modulator will drift away from quadrature. This is accomplished by adjusting V_N to compensate for the charging and discharging of the various equivalent resistor /capacitor pairs. If, however, V_N is changed such that it also causes V_1 to change, such as a power outage on the bias control loop electronics, then all of the resistor/capacitor pairs can be 'reset' to some new value. The longer term time constants are also 'reset' and information is lost as to the true long term drift behavior of the modulator. This has a number of consequences, one being that the equipment for this test needs to be on some kind of an instantaneous electrical power backup. On passively biased devices, this is not a issue as long as the DC electrodes stay shorted.

One of the problems with the constant phase test is how to test the modulator for its other parameters, like S_{11} or S_{21}. Normally, this is done by disconnecting the modulator from one measurement system and hooking up the next measurement system. If the bias loop cannot be disturbed, this disconnecting is not an option. An easy way to get around this is to have a second set of samples not in a constant phase test. Generally, this should be done with a constant voltage on the bias electrodes. Then, every 500 or 1000 hours, a set of intrinsic and extrinsic parameter measurements can be made to get the data.

The RF electrodes also take some special consideration, but both modulator types are handled the same way. GR-468 stipulates that the modulator shall be subjected to the maximum modulation rate during the test. This maximum rate is either a 2.5 Gb/s or a 10 Gb/s digital signal which are simulated in the lab with a $10^{23} - 1$ or $10^{30} - 1$ PRBS generator. At 10 Gb/s this is a quarter million dollar piece of equipment plus several thousands of dollars for a RF driver for each sample. There is another, less costly method that accomplishes the same intent, i.e., to stress the RF electrodes. An RF sine wave at the correct frequency and power level can be used to simulate the digital signal. The parameters must be selected so that power is properly divided between the waveguide and load resistors. The heat dissipation needs to approximate that of the digital signal, but this can be calculated for a particular design. Both the constant phase and the constant voltage must have the RF electrodes powered.

Data will now be presented for a number of accelerated aging programs on different types of modulators. It should be noted that these tests are very expensive, time consuming and continuously being refined.

2.4.1.1 Accelerated aging 2.5 Gb/s, X-Cut, active bias, non hermetic

Data will be presented for 2.5 Gb/s, X-Cut, active bias, non hermetic modulators in two constant phase long term aging tests at 85°C and 100°C respectively. Constant voltage tests are planned, but have not yet started. Two milliwatts of optical power in the 1550 nm band is launched into the polarization maintaining input fiber, the output fibers are monitored by a detector, 20 dBm of 900 MHz RF power is applied to the RF electrodes and ±12 volts is available at the DC electrodes to maintain quadrature for the life of the test. The test samples are mounted in an environmental oven. A computer acts as a datalogger and records bias voltage and detector current every 30 minutes. All electronics are on battery back up and the entire reliability lab including chambers is on generator back up with a 1 minute turn-on time. Eleven samples are in each test.

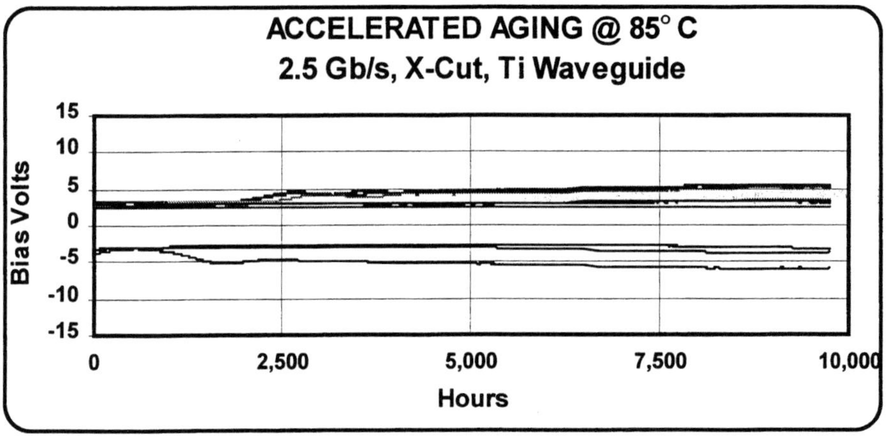

Graph 1. Bias volts vs. time @ 85°C

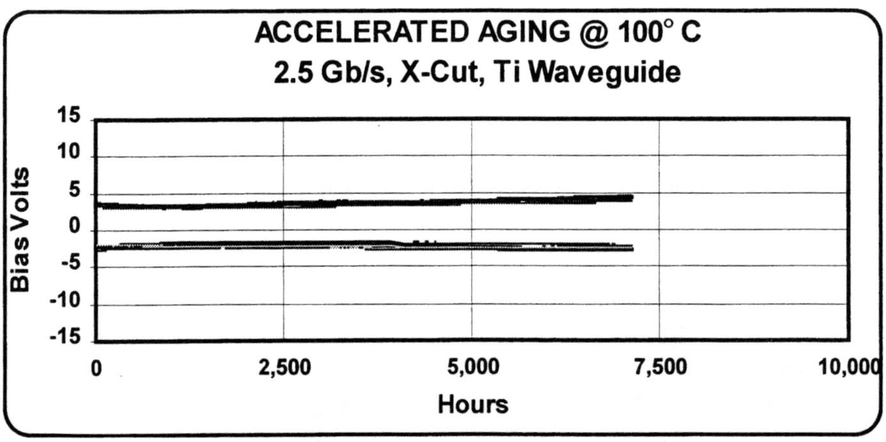

Graph 2. Bias volts vs. time @ 100°C

The pass/fail criteria, from a telecommunication system point of view, is when the drift of the modulator exceeds the rails of the bias voltage power supply, but this is a function of each system design. An impartial reliability test, however, needs a more precisely defined criteria. Industry practice has established this to be when the drift exceeds twice the initial bias voltage of a particular sample.

Graphs 1 and 2 show the bias voltage vs. time for each of the two tests while graphs 3 and 4 show the same data, normalized. Both sets of charts are needed to understand the behavior of the modulators. It is clear from these charts that all of the samples pass; the drift has not exceeded twice the initial voltage. Further analysis highlights the work of Korotky and Veselka[7] as the different time constants can easily be seen. What is very interesting to observe is the shape of the curves around 9000 hours. There is the start of a slight downward curvature strongly suggesting that, as this trend continues, these devices probably will never exceed twice the initial voltage and, therefore, never fail for bias drift.

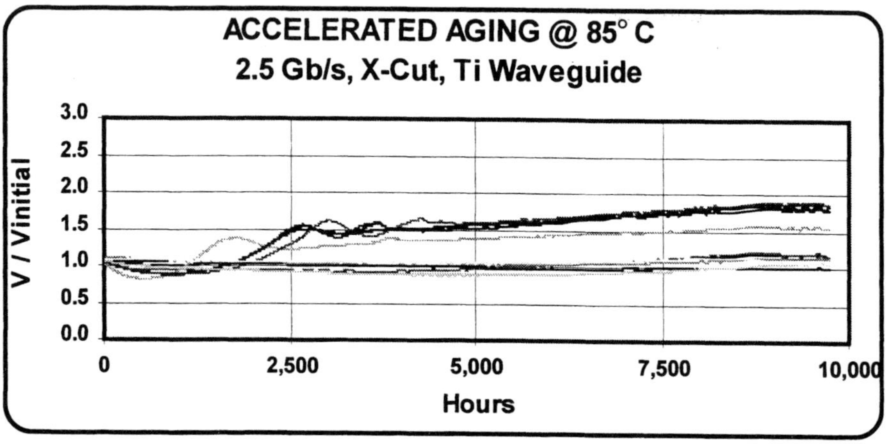

Graph 3. Normalized change in bias voltage vs. time @ 85°C

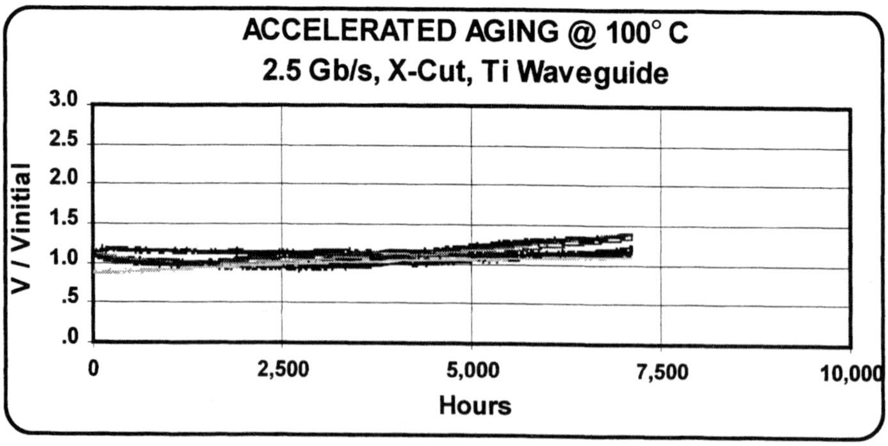

Graph 4. Normalized change in bias voltage vs. time @ 100°C

As a note, had the bias voltage been disturbed, these long time constants would not have been observed supporting the argument for the extreme measures taken to protect the test. It is also clear that very long test times reveal much information and should be performed. These tests will continue indefinitely.

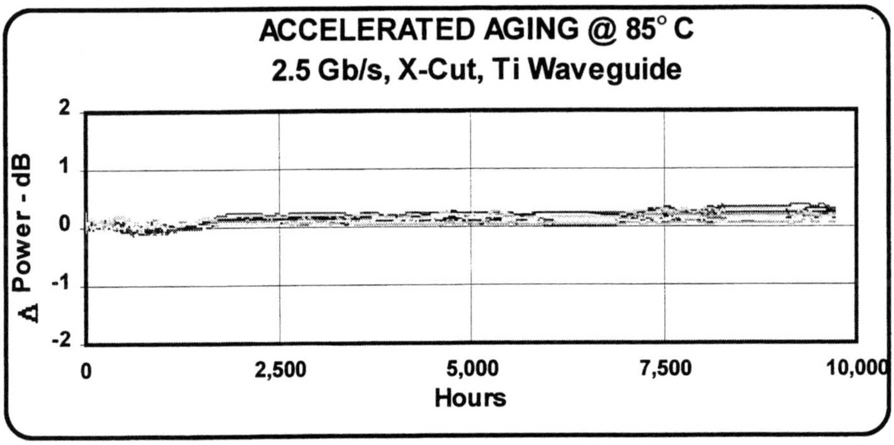

Graph 5. Change in detector power vs. time @ 85°C

Graph 5 is a plot of the change in detector power vs. time for the 85°C test. Typically, pass/fail criteria is 1 dB change in insertion loss, however, GR-468 suggests 3 dB. Clearly, loss is not a failure mechanism. In addition, much of the observed drift is believed to be due to drift in the source, not the samples. Unfortunately, there is no source reference being taken during this test. Similar data at 100°C is not available because this is above the maximum temperature specification for the device. Suffice to say that there is sufficient optical output power from all samples to maintain the quadrature loop.

2.4.1.2 Accelerated aging 2.5 Gb/s, X-Cut, passive bias, non hermetic

Data will be presented for 2.5 Gb/s, X-Cut, passive bias, non hermetic modulators in two constant phase long term aging tests at 85°C and 100°C respectively. The RF electrodes of the samples are powered with 20 dBm of 900 MHz RF sine wave. Measurements are taken about every 500 to 1000 hours. Eleven samples are in each test.

Graphs 6 and 7 show plots of the change in phase in degrees vs. time for both tests. Unlike the active biased devices, these curves appear to be straight lines. Move careful inspection and analysis, however, suggests that each is asymptotically approaching some limiting value. This makes sense. Going back to the Korotky and Veselka[7] model, V_N is now a direct short and its value is zero. This allows the charge to bleed off. In addition, the value of V1 should also be zero because the interferometer has been driven to quadrature by passive, non-voltage, techniques. As long as there are no voltage generating mechanisms, all the charge build up should eventually dissipate through the short at V_N and produce the observed asymptotic behavior.

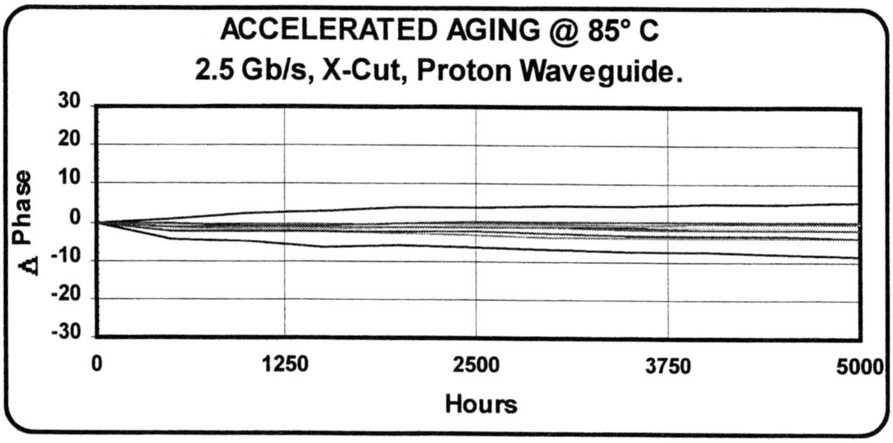

Graph 6. Phase vs. time @ 85°C

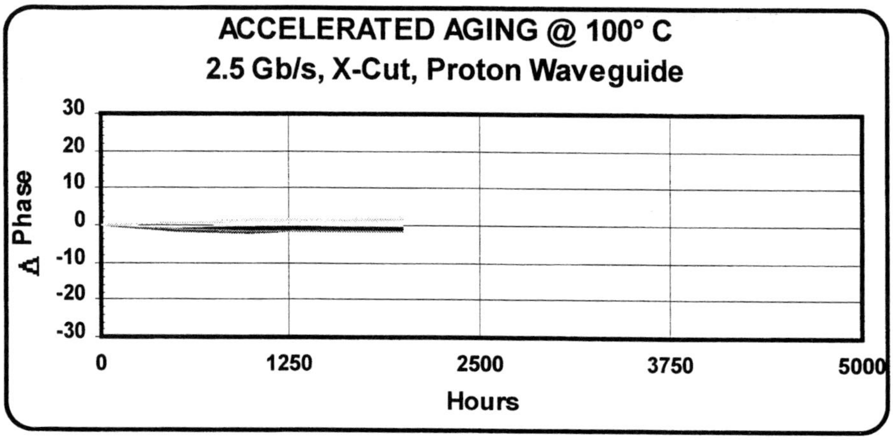

Graph 7. Phase vs. time @ 100°C

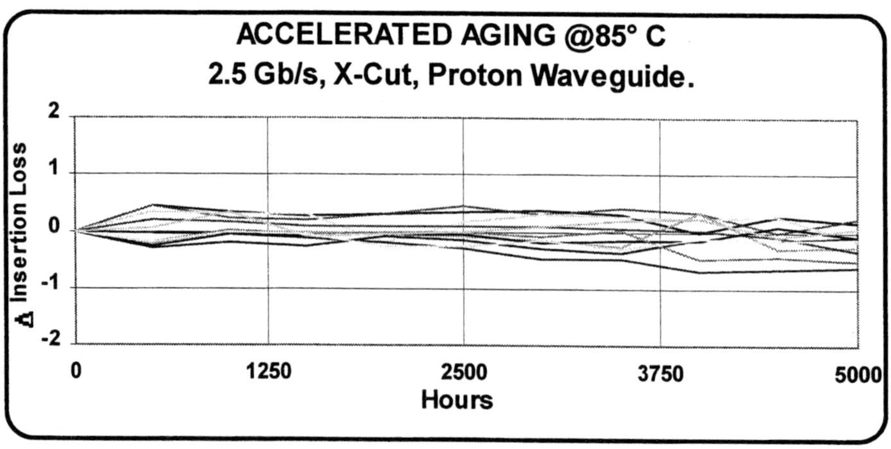

Graph 8. Insertion loss vs. time @ 85°C

Graph 8 shows the change in optical insertion loss vs. time at 85°C. This data does not exhibit any major trends. Most of the variances are probably due to set-up and measurement uncertainties especially since the trend seems to be that the average loss is slowly decreasing. These samples easily pass a 1 dB pass/fail criteria. Similar data at 100°C is not available as this is above the maximum temperature specification for the device.

Phase and phase drift are some of the many intrinsic parameters of the modulator. In order to establish a meaningful pass/fail criteria, a relationship between these intrinsic parameters and system performance needs to be established. Another way to approach this issue is to put a pass/fail criteria on an extrinsic measurement that integrates a number of the 'softer' intrinsic measurements into a meaningful extrinsic criteria. As stated earlier, the PRBS 'eye diagram' extinction ratio is just such a measurement. Graphs 9 and 10 show this extinction ratio for the samples in the 85°C and 100°C tests.

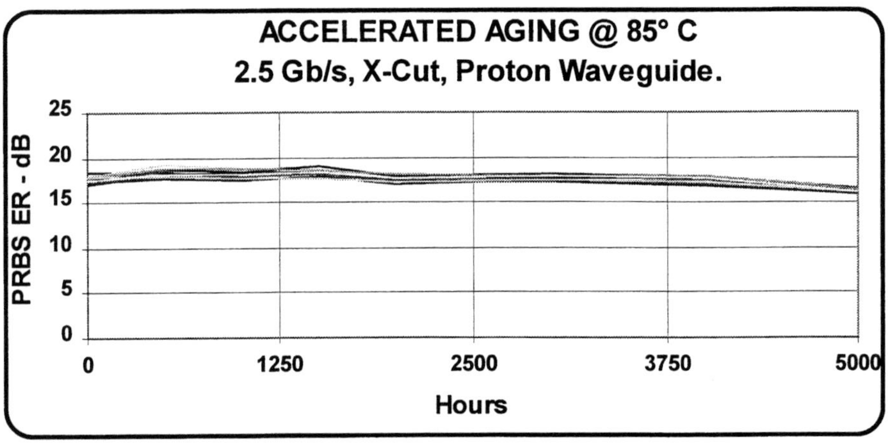

Graph 9. PRBS extinction ratio vs. time @ 85°C

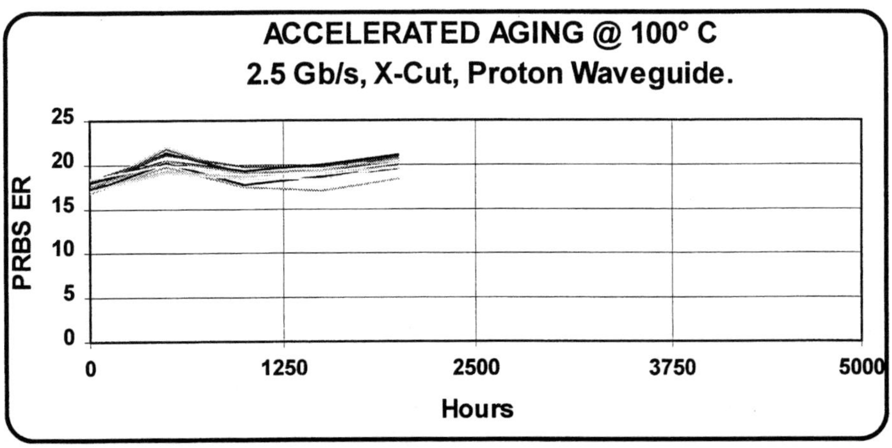

Graph 10. PRBS extinction ratio vs. time @ 100°C

In most transmission systems 12 to 14 dB is adequate to achieve the desired system bit error rates. Above about 16 dB, the measurement becomes very difficult to do because electrical noise starts to create high uncertainty in the value of the zero state. Examination of graphs 9 and 10 clearly show that all the samples have performed exceptionally well and clearly pass. In addition, there are no trends that suggest any failure modes.

2.4.1.3 Accelerated aging 10 Gb/s, X-Cut, active bias, non hermetic

Data will be presented for 10 Gb/s, X-Cut, active bias, non hermetic advanced design modulators in a constant phase long term aging test at 85°C. The samples used for this test were preliminary engineering prototypes because this produde had not yet been released to production. Two milliwatts of optical power in the 1550 nm band is launched into the polarization maintaining input fiber, the output fibers are monitored by a detector, there is no power on the RF electrodes and ±12 volts is available at the DC electrodes to maintain quadrature for the life of the test. There are 6 samples in this test

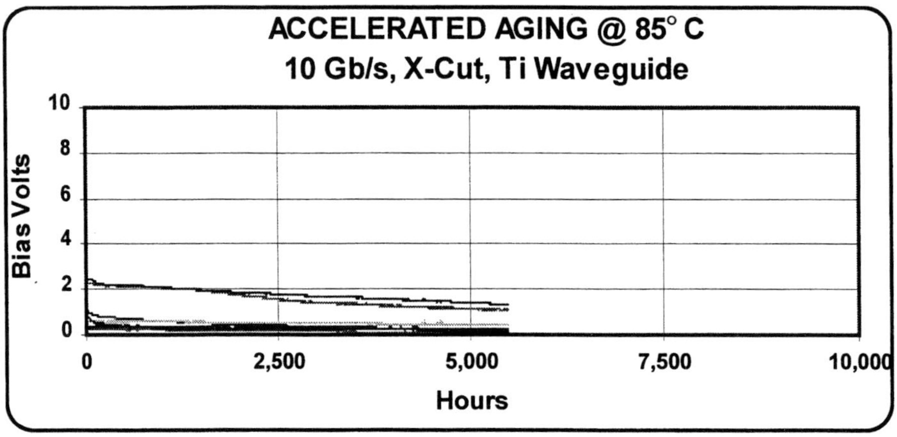

Graph 11. Bias voltage vs. time @ 85°C

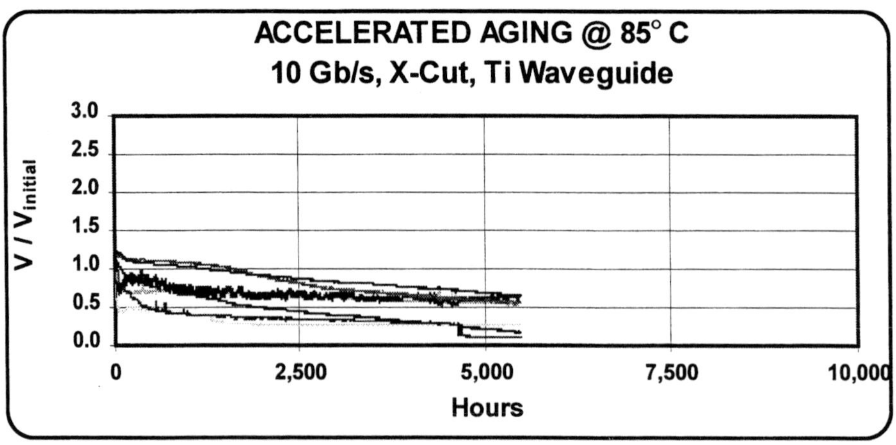

Graph 12. Normalized bias voltage vs. time @ 85°C

Graphs 11 and 12 show the bias voltage vs. time and change in bias voltage vs. time. As can be seen, the behavior is somewhat different than the 2.5 Gb/s, X-Cut, active bias, non hermetic devices in section 2.4.1.3. After an initial small upset, all of the voltages seem to be trending to zero or a value significantly below their initial value. A second observation is the relatively low initial values. This behavior is deliberate by design, however, much greater amounts of data are needed to verify the design concepts and manufacturing processes. Going back to Korotky and Veselka[7] and others, we know that drift rate seems to be proportional to voltage. Therefore, the lower the bias voltage, the lower the rate of drift. If a modulator could be designed such that its drift would trend to lower voltages, it should do so at ever lower rates explaining the observed asymptotic behavior.

Much more work is needed here, but it is in progress. Plans for this device include constant phase and constant voltage tests at 70°C, 85°C and 100°C of true production samples. The next generation of measurement equipment in being built which will include 2.4 GHz power on the RF electrodes. The first test will start up during the third quarter of 1999.

2.4.1.3 Accelerated aging 10 Gb/s, Z-Cut, active bias

A lot of work has been done on Z-Cut devices, however, not a lot has been published. Graph 13 shows work done by Nagata et al[15] of Sumitomo for the bias voltage performance of 12 devices at 85°C. This work was done as part of a qualification program very similar to GR-468. It was reported that this product passed all the tests including long term aging and hermeticity. Examination of the curves does show good performance. Using an exponential failure distribution and a conservative activation energy of 0.4 eV, the mean time to failure at 40°C was predicted to be 47.9 years.

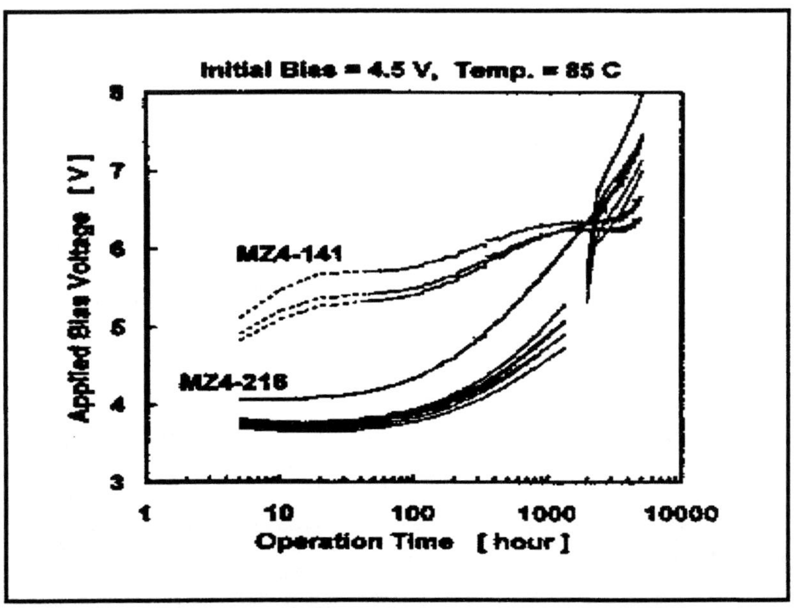

Graph 13. Bias volts vs. time @ 85°C, from Nagata, et. al.[15]

Graph 14 shows work presented in view graphs, but never published, by Minford[16] of Bell Labs also on long term aging studies of 10 Gb/s, Z-Cut, Ti waveguide hermetic modulators. He presents data on the drift of samples at three different temperatures, 85°C, 105°C and 125°C. From these curves, it is obvious that increased temperature shortens the time constants and presumably the time to failure. Current theory postulates that as temperature increases, charge mobility also increases and the drift curve moves to the left. Going back to Korotky and Veselka[7], it is a reasonable assumption the values of the resistors and capacitors can be highly dependent on temperature such that an inverse relationship exists between these time constants and temperature.

Minford then fitted this data to an Arrhenius relationship using a 1.1 eV activation energy and obtained the excellent results shown in Graph 15. Extrapolating the data back to a temperature of 50°C gives a 20 year time to double the bias voltage. As stated earlier, this is the generally accepted pass/fail criteria for modulators. All of this work attests to the excellent reliability of Z-Cut.

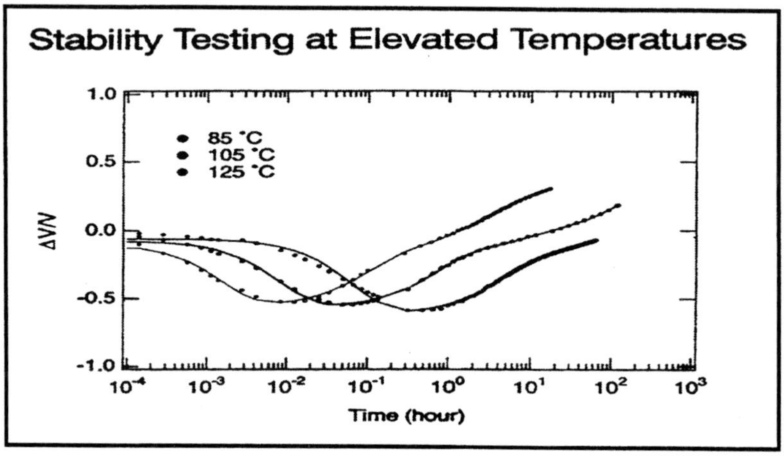

Graph 14. Bias drift vs. time at different temperatures from Minford[16]

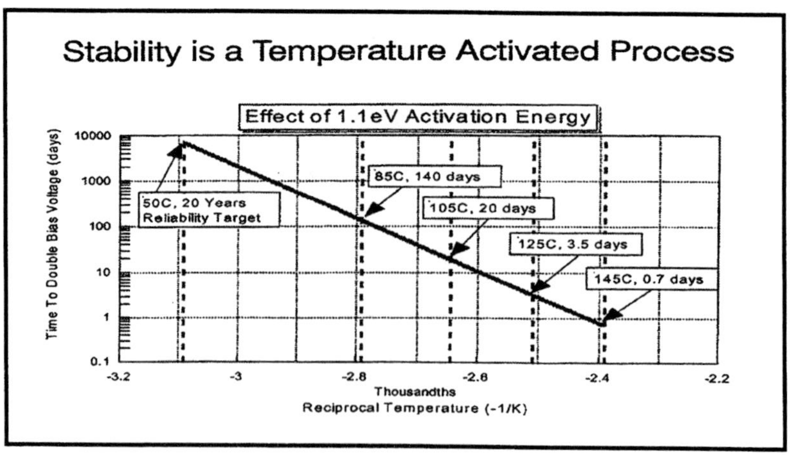

Graph 15. Lifetime predictions, from Minford[16]

Seino et. al. of Fujitsu has presented data on the reliability of Z-Cut modulators[17] over both temperature and humidity. Graph 16 shows his work for drift in percent of initial bias voltage vs. time at several temperatures. From this data, he was able to show that the activation energy for this drift mechanism was about 1.0 eV giving at reliability of 1 FIT at 15 years at 20°C. He also showed that at 80°C, the reliability was acceptable for 15 years in the field.

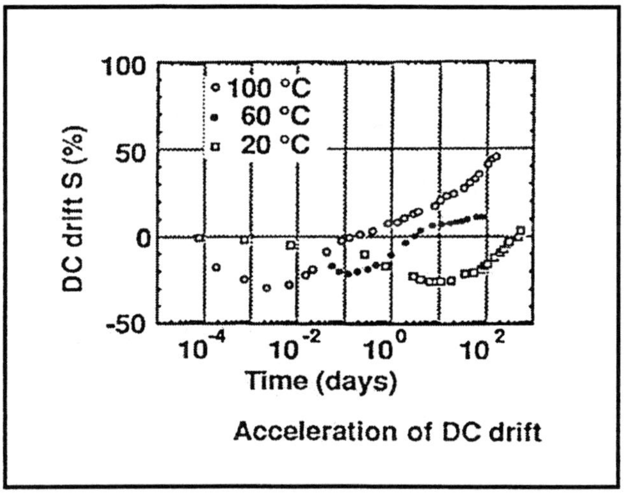

Graph 16. Bias drift vs. time, from Seino[17]

Seino et. al.[18] has also presented data comparing 2.5 Gb/s and 10 Gb/s Z-Cut modulators. Graph 17 shows this work for drift in percent of initial bias voltage vs. time at 100°C.. Notice that the 10 Gb/s seems to drift at a higher rate, however, this may simply be due to the limited number of samples. He also confirms his earlier results and shows that the real activation energy for drift is closer to 1.4 eV. This would increase his earlier reliability predictions suggesting even better field performance.

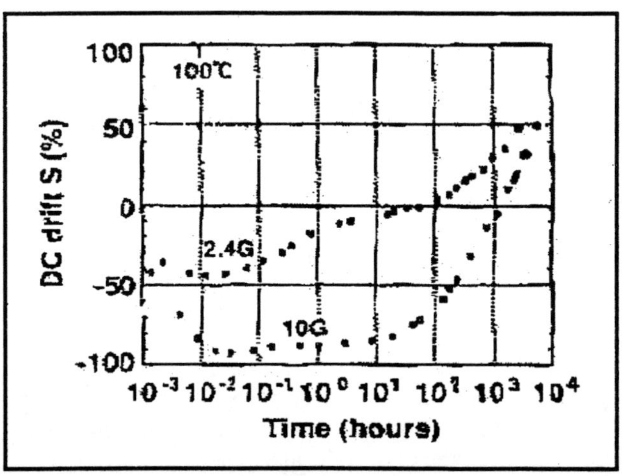

Graph 17. Comparison of 2.5 Gb/s and 10 Gb/s, from Seino[18]

2.4.2 High temperature storage

The high temperature storage test is performed in accordance with the conditions stated in Table 24 of GR-468. The specifics of this test are a 2000 hour soak at the maximum storage temperature of the modulator which typically is 85°C. In the programs outlined in this paper, the long term aging test is done at 85°C rather than the GR-468 minimum requirement of 70°C. In this case, the first 2000 hours of the long term aging test is substituted for the high temperature storage test.

Parameter	2.5 Gb/s, X-Cut Proton Waveguide		10 Gb/s, X-Cut Ti Waveguide	
	Before	Δ	Before	Δ
S_{11} (dB)	Pass		Planned	
S_{21} (GHz)	Pass			
Vπ RF (V)	3.2	.03		
Vπ DC (V)	6.5	.09		
Phase	0	−1.1		
Loss (dB)	3.2	0		
ER DC (dB)	37	−.2		
ER PRBS (dB)	18	−.5		

Table 7. Comparison of high temperature storage test results

The results from a number of high temperature storage tests on different modulator types is shown in Table 7. Examination of the results shows that little change occurred and all samples meet the requirements. Again, no quantitative data is available for Z-Cut, however, it is reasonable to assume that Z-Cut will also pass the test.

2.4.3 Low temperature storage

The low temperature storage test is an optional test and not usually performed unless a specific customer requires it. It is conducted in accordance with the conditions stated in Table 24 of GR-468 at the minimum advertised storage temperature for 2000 hours. Typically this temperature is -40°C. There are very few failure mechanisms that come into play for lithium niobate modulators in a long term cold storage test that are not brought out by the thermal cycling test. As such this test has not been performed on a large number of different modulator types. The test set up in the data includes a nitrogen purge on the changer to prevent excessive ice build up.

The results from a low temperature storage test on 2.5 Gb/s, X-Cut, proton waveguide devices are presented in Table 8. As can be seen, there was very little change in the performance of the modulators. Another test is being planned for 10 Gb/s, X-Cut, Ti Waveguide devices, however, the results are also expected to be good.

Parameter	2.5 Gb/s, X-Cut Proton Waveguide		10 Gb/s, X-Cut Ti Waveguide	
	Before	Δ	Before	Δ
S_{11} (dB)	Pass		Planned	
S_{21} (GHz)	Pass			
Vπ RF (V)	No data			
Vπ DC (V)	6.5	0.0		
Phase	0	−0.7		
Loss (dB)	3.2	−0.1		
ER DC (dB)	32	0.6		
ER PRBS (dB)	No data			

Table 8. Comparison of low temperature storage test results

2.4.4 Temperature cycling

The temperature cycling test is performed in accordance with MIL-STD 883, Method 1010 with modifications. The ramp rate must be ≥ 10°C / minute with end point dwell times ≥ 15 minutes. The two extremes are -40°C and +70°C. The test is run for 100 cycles and the samples are measured. The test then continues for an additional 400 cycles before a final measurement is made. Pass/fail criteria applies only to the first 100 cycles, however, the results of the additional 400 cycles must be reported. If a failure is observed in the 400 cycle portion, then a failure analysis must be performed and a corrective action issued. This test can either be a biased or a non-biased test, however, non-biased is more appropriate for lithium niobate due to its pyroelectric properties.

Parameter	2.5 Gb/s, X-Cut Ti Waveguide		2.5 Gb/s, X-Cut Proton Waveguide		10 Gb/s, X-Cut Ti Waveguide	
	Before	Δ 100 Φ / Δ 500 Φ	Before	Δ 100 Φ / Δ 500 Φ	Before	Δ 100 Φ / Δ 500 Φ
S_{11} (dB)	Pass		Pass		Planned	
S_{21} (GHz)	Pass		Pass			
Vπ RF (V)	No data		No data			
Vπ DC (V)	7.6	.02 / no data	6.7	0.0 / 0.0		
Phase	n.a.		0	−3.5 / −3.1		
Loss (dB)	4.1	0.1/ no data	3.0	.05 / .08		
ER DC (dB)	35	−0.1/ no data	39	−0.5 / −0.5		
ER PRBS (dB)	21	−1 / no data	No data			

n.a. = not applicable

Table 9. Comparison of thermal cycle test results

Temperature cycling is one of the more important tests because it stresses all aspects of the packaging designs. It is also one of the more difficult tests for hermetic devices to pass. At a ramp rate of 10°C per minute for 100 cycles, fairly high thermal stresses are generated. Thermal expansion and mechanical design problems will quickly be identified. Fiber and epoxy integrity are checked as well. The constant expanding and contracting stresses the solders in the seal can cause enough creep to break the hermetic seal between the fiber surface and the solder or between the solder and the package.

The results from a number of thermal cycling tests on different modulator types is shown in Table 9. Examination of the results shows that little change occurred and all samples meet the requirements for both 100 cycles and 500 cycles. Again, no quantitative data is available for Z-Cut, however, it is reasonable to assume that Z-Cut will also pass the test.

2.4.5 Damp heat

The damp test is performed in accordance with MIL-STD 202[19], Method 103 with a modified set of conditions and times. This is one of the tests where non-hermetic and hermetic modulators are treated quite differently. Non-hermetic modulators are tested for 2000 hours at 85°C and 85% relative humidity with an additional 3000 hours for information. These are the same conditions as for passive components in an uncontrolled environment per Telecordia GR-1221-CORE, Issue 2, "Generic Reliability Assurance Requirements for Passive Optical Components[20] . It is a very severe test for any component, but it represents Telecordia's conservative approach as non-hermetic laser diodes start to enter the marketplace. The test is waived for hermetic components unless there is exposed epoxy. Then the duration is only 500 hours. The test is meant to be passive, however, some of the tests have been conducted with a bias voltage. It is not clear what the best test parameters should be for modulators. Further work is needed to properly quantify the relationship between damp heat test conditions and field reliability.

Parameter	2.5 Gb/s, X-Cut Ti Waveguide		2.5 Gb/s, X-Cut Proton Waveguide		10 Gb/s, X-Cut Ti Waveguide	
Bias Volts	2 V		0 V		0 V	
Conditions	40° / 95% RH		40° / 95% RH		85° / 85% RH	
	Before	Δ	Before	Δ	Before	Δ
S_{11} (dB)	9.0	.06	Pass		In progress	
S_{21} (GHz)	Pass		Pass			
$V\pi$ RF (V)	3.16	−0.1	3.2	−.06		
$V\pi$ DC (V)	No data		3.2	6.7		
Phase	n.a.		6.7	0		
Loss (dB)	3.7	−.05	3.1	.09		
ER DC (dB)	30	1.0	37	−0.3		
ER PRBS (dB)	17	−1.2	16	−0.5		

n.a. = not applicable

Table 10. Comparison of damp heat test results

The results from a number of damp heat tests on different modulator types is shown in Table 10. As can be seen, one of the tests was run with a bias voltage. In addition, two of the modulator types were run under the old TR-468 specification with 40°C and 95% relative humidity for 56 days. Examination of all the results shows that little change occurred and all samples meet the requirements. Again, no quantitative data is available for Z-Cut, however, Z-Cut, being hermetic does not have to do this test unless there is exposed epoxy.

2.5 Other tests

Bellcore GR-468 has two final tests, internal moisture and ESD (electrostatic discharge) Sensitivity.

2.5.1 Internal moisture

Internal moisture is for hermetic designs only. It is performed in accordance with MIL-STD 883, Method 1018 which calls our for an residual gas analysis (RGA) of the device's internal atmosphere after a high temperature preconditioning. The requirement is ≤ 5000 ppm of H_2O. The material selection and manufacturing procedures to pass this test are well established from the electronics industry. Passing this test, however, does require careful design and attention, but it is, presumably, routinely passed.

2.5.1 ESD threshold

The very last test in GR-468 is ESD threshold which is preformed in accordance with Section 4.2.3 of Telecordia TR-NWT-000879[21]. The minimum requirement is 500 volts. Six samples need to be tested. In general, lithium niobate is not a static sensitive component, but the test is still being called out.

Parameter	2.5 Gb/s, X-Cut Ti Waveguide		2.5 Gb/s, X-Cut Proton Waveguide		10 Gb/s, X-Cut Ti Waveguide	
Volts	4,000 V		4,000 V		4,000 V	
	Before	Δ	Before	Δ	Before	Δ
S_{11} (dB)			Pass			
S_{21} (GHz)			Pass			
$V\pi$ RF (V)	No data		3.2	0.0		
$V\pi$ DC (V)	7.4	0.0	6.6	0.0	Planned	
Phase	n.a.		0	–0.25		
Loss (dB)	6.8	0.3	3.1	.05		
ER DC (dB)	34	–5	37	–1		
ER PRBS (dB)	No data		No data			

n.a. = not applicable

Table 11. Comparison of ESD test results

The results from a number of ESD sensitivity tests on different modulator types is shown in Table 11. All tests were performed at 4000 volts, well above the 500 volt minimum. Examination of the results shows that little change occurred and all samples meet the requirements. Again, no quantitative data is available for Z-Cut, however, it is reasonable to assume that Z-Cut will also pass the test.

4. FIELD DATA

As stated earlier, sufficient field data is now available to start to confirm and corroborate engineering and laboratory reliability study predictions. All of the above qualification and reliability data suggests that modulators should be extremely reliable devices. The long term aging studies detailed above in section 4.1.2 have yet to prove any random or wearout failure mechanisms that will affect expected lifetimes. All the Telecordia requirements have been met by a wide number of device types and manufacturers. All of this qualification and reliability data predicts that modulators, regardless of type, will have very good field performance.

Lithium niobate modulators have been commercially available for over 10 years starting with deep space and military sensor programs. The quantities were relatively small and field performance was generally unavailable although it was always believed to be very good. The telecommunications industry did not start deploying modulators until some time later. The last four years have seen explosive growth with many tens of thousands now installed. Table 12 outlines actual field experience.

FIELD DATA	2.5 Gb/s, X-Cut, Ti Waveguide	2.5 Gb/s, X-Cut, Proton Waveguide
In Service Device hours	> 100,000,000	> 300,000,000
Reported in service failures	0	2
FIT rate at 60% Confidence	9	10
FIT rate at 95% Confidence	30	21

Table 12. Field failure rates

These calculated failure rates assume that field failures are random and best described by an exponential distribution. The confidence intervals are determined by using a chi-squared estimator. FIT stands for failures in time per 10^9 device hours. As can readily be seen, the field failure rates for the listed devices are extremely low. Presumably Z-Cut is similar.

5. CONCLUSIONS

Lithium niobate modulators have turned out to be extremely reliable components in spite of the industry's early concerns. Field experience has confirmed and even exceeded the optimistic predictions of many qualification and reliability studies. The high degree of success and reliability that modulators have experienced will help the market acceptance for other integrated circuit technologies and products.

ACKNOWLEDGEMENTS

The author would like to acknowledge the team of all employees at JDS Uniphase Electro-Optics Products Division who have participated in the development of this technology to make it what it is.

REFERENCES

1. Ryan Hankin Kent, Inc., *WDM and Optical Networks: Technology and Markets,* RHK, San Francisco, 1999.
2 Crystal Technologies, "Product literature", CTI Bulletin 201, Mountain View, California.
3 INSPEC, *Properties of Lithium Niobate*, The Institution of Electrical Engineers, London, 1989.
4 H. Matthews, *Surface Wave Filters: Design, Construction and Use*, John Wiley & Sons, New York, 1977.
5 private discussions.
6 L. Hutcheson, *Integrated Optical Circuits and Components*, Marcel Dekker, Inc., New York, 1987.
7. S. Korotky and J. Veselka, "An RC network analysis of long term Ti:$LiNbO_3$ bias stability," *Journal of Lightwave Technology*, **14**, December 1996, pp. 2687-2697.
8 H. Nagata, N. Mitsugi, T Sakamoto, K Kiuchi, "Applied voltage induced fatigue of lithium niobate waveguide", Applied Physics Letters, January 1996.
9. TR-NWT-000468, *Reliability Assurance Practices for Optoelectronic Devices in Interoffice Applications*, Issue 1, Bellcore, December 1991.
10. GR-468-CORE, *Generic Reliability Assurance Requirements for Optoelectronic Devices Used in Telecommunications Equipment*", Issue 1, Bellcore, December 1998.
1.1 D. Williams, A. Akhtar, D. Attanasio, D. Maack and G. McBrien, "2.5 Gb/s PRBS extinction ratio eye diagram measurements of biased and bias free™ lithium niobate OC-48 modulators", pp. 175-178, Symposium on Optical Fiber Measurements, NIST, Boulder, 1998.
12. T. Clark, "Methodologies for extinction ratio measurement of high speed laser transmitters", page 18, Passive Component Measurement Considerations, Hewlett Packard Lightwave Division, Santa Rosa, 1999.
13. *General Specification for Microcircuits*, Military Specification MIL-M-38510F, October 1983.
14 *Test Methods and Procedures for Microcircuits*, Military Standard MIL-STD-883E, December, 1997.
15. H. Nagata, N. Mitsugi, K Kiuchi, J. Minowa, "Lifetime estimation for hermetically packaged 10 Gb/s LiNbO3 optical modulators", pp. 4-5, Engineering and Laboratory Notes in Optics and Photonics News, November 1996.
16 W. Minford, *The Taming of LiNbO3*, slide presentation, FIO III, St Martin Vesubie, France.

17. M. Seino, T. Nakazawa, Y. Kubota, M Doi, T. Yamane, H. Hakogi, "A low DC-drift Ti:LiNbO3 modulator assured over 15 years", post deadline paper PD3, OFC92, 1992.
18. M. Seino., T. Nakazawa, M. Doi, S. Taniguchi, "The long-term reliability Estimation of Ti:LiNbO3 for DC drift", paper PD1-8, IOOC95, 1995.
19. *Test Methods for Electronic and Electrical Component Parts*, Military Standard MIL-STD-202F, April 1980.
20. GR-1221-CORE, *Generic Reliability Assurance Requirements Passive Optical Components*", Issue 2, Bellcore, January 1999.
21. TR-NWT-000870, *Electrostatic Discharge Control in the Manufacture of TelecommunicationEquipment*, Issue 1, Bellcore, February 1991.

MEMS for optical switching: technologies, applications, and perspectives

Lih Y. Lin* and Evan L. Goldstein

AT&T Labs-Research, 100 Schulz Drive, Red Bank, NJ 07701

ABSTRACT

Micro-electro-mechanical-systems (MEMS), due to their unique ability to integrate electrical, mechanical, and optical elements on a single chip, have recently begun to exhibit great potential for realizing optical components and subsystems in compact, low-cost form. Recently, this technology has been applied to wavelength-division-multiplexed (WDM) networks, and resulted in advances in several network elements, including switches, filters, modulators, and wavelength-add/drop multiplexers. Due largely to the exploding capacity demand arising from data traffic, the transmission capacity demanded of and available from WDM networks is anticipated to increase rapidly. For managing such networks, optical switching is of particular interest due to the fact that its complexity is essentially immune to steady advances in the per-channel bit-rate. We will review various micromachined optical-switching technologies, emphasizing studies of their reliability. We then summarizing recent progress in the free-space MEMS optical switch we have demonstrated.

Keywords: MEMS, optical switching, optical crossconnect, WDM, free-space, micro-mirror, optical network, reliability.

I. INTRODUCTION

Long-haul optical fiber networks worldwide are experiencing swift growth not only in aggregate demand, but also in the bit rates of individual services. The advent of optical amplifiers and wavelength-division-multiplexing (WDM) have completely solved the associated transmission challenge, without requiring network-operators to engage in massive new deployments of optical fiber. Moreover, the essentially 'opaque', transponder-based long-haul network architecture[1] that is beginning to emerge solves the problems of accumulating performance-impairments, performance-monitoring and fault-localization, multi-vendor interworking, and piecewise upgradability. Still, substantial challenges remain: the swiftly increasing volumes of traffic referred to above need not only to be transported and provisioned; they need, perhaps most critically, to be restored in the event of link and equipment failures. This in turn imposes a strong need to move to coarse units of restoration. Whereas much of the current long-haul network restores at the DS-3 (45 Mb/s) level, there is an emerging need to restore in units of wavelengths (2.5 Gb/s and beyond).

* Correspondence: Email: lylin@research.att.com; Telephone: 732 345 3249; Fax: 732 345 3037

One natural means of accomplishing this is to deploy mesh-based networks whose optical crossconnects (OXC's) reconfigure for purposes of restoration and provisioning. The chief technological challenge lying in the path of this approach, however, is the development of practical optical-switching technologies around which to build an optical crossconnect.

Optical switching has been an active area of research for nearly two decades. Various optical switching technologies, employing both free-space interconnection and integrated optics, have been demonstrated, but all technologies fall well short of the roughly 500-port systems that are expected to be needed in core networks within roughly five years. Furthermore, optical switch fabrics will need to achieve low insertion loss, short switching time (on the order of one millisecond), low crosstalk, low polarization-dependence, and low wavelength-dependence. It is moreover desired that they be transparent to bit-rate in the interval of roughly 2.5 Gb/s to 40 Gb/s. These requirements impose stringent challenges on all optical switching technologies.

The emergence of this new application is, however, beginning to stimulate new device technologies. One emerging technology that appears to offer a promising future is that of optical MEMS (Micro-Electro-Mechanical Systems). Optical MEMS have the fundamental advantage of offering the benefits of free-space interconnection (such as low loss and crosstalk, and low polarization- and wavelength-dependence), together with the advantages of integrated optics (including compactness, optical pre-alignment, and low cost). Therefore, it offers the possibility of achieving high port-count in a small, low-cost system with excellent optical quality. In this paper, we review various optical MEMS switching technologies and their performance. We then focus on the free-space micromachined optical switches with free-rotating mirrors that we have demonstrated, and discuss their ability to incorporate advanced optical-networking functionality within the switch fabric itself.

2. VARIOUS OPTICAL MEMS SWITCHES

Optical MEMS switches have been demonstrated in both waveguide[2,3] and free-space configurations. Free-space approaches feature the advantage of good optical quality. Here, we focus on this category. Various free-space optical MEMS switches have been demonstrated; they include modulating On/Off switches[4-7], small 2×2 switches with four tapered single-mode fiber placed close to each other[8, 9], and potentially large-scale $M \times N$ switches[10-14]. We summarize here a few examples of free-space optical MEMS switches, focusing on both performance and reliability, without any claim of exhaustive coverage of this fast-growing area.

2.1. Reflective Modulating Switch

A reflective modulating switch, formed by attaching a Fabry-Perot intensity modulator in front of a single mode fiber, has been demonstrated by the University of Neuchâtel[7] (see Fig. 1). The Fabry-Perot modulator consists of two polysilicon mirrors with 1.6-μm initial separation. The device operates in the wavelength window between 1260 nm and 1360 nm. The spectral response of the Fabry-Perot modulator is tuned by applying a voltage between the two polysilicon mirrors. At 1310 nm, the device achieves less than 3-dB insertion loss, 10-dB On-Off ratio with 84-V extinction voltage, and a maximum modulation rate of 2.88 Mb/s. The reliability of this device has also been studied. A packaged device shows no observable degradation after thermal cycle tests between -40°C and +80°C. In the absence of high air humidity, the fracture strength of polysilicon is well above the operating stress of this device, and slow crack growth is not observed. However, the combination of high electric field with humid air can lead to mechanical breakdown, chiefly in the form of membrane rupture.

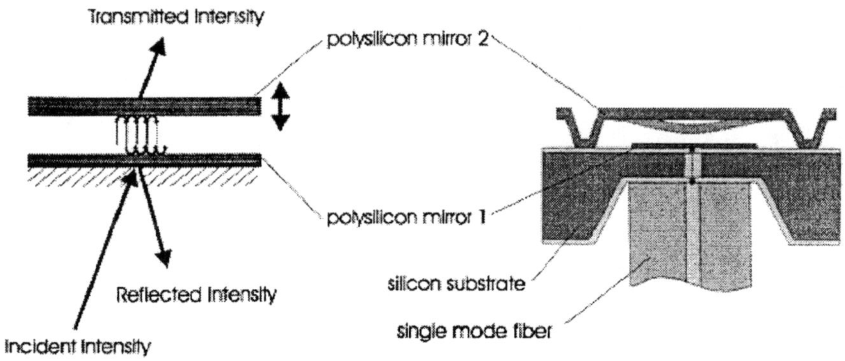

Fig. 1 Reflective MEMS Fabry-Perot Modulator[7].

2.2. Fiber-optic 2 × 2 Switch with Sliding Micro-mirrors

Figure 2 shows a 2× 2 free-space fiber optic switch demonstrated by UCLA[8]. The switch consists of an out-of-plane sliding micro-mirror driven by integrated scratch drive actuators. The micro-mirror is made of a 1.5-µm polysilicon plate with a 0.5-µm gold coating. Tapered single-mode fibers are brought close to each other (80-µm spacing) to reduce the insertion losses. The insertion losses have been characterized to be 1.3 dB and 1.9 dB for the transmission and reflection states, respectively. The switch achieves a rise time of 6 ms and fall time of 15 ms. The robustness of the assembled three-dimensional micro structure has been investigated by vibration testing. The hinged micro-mirror is oriented such that the mirror plane faces the axis of the vibration to obtain the worst scenario. Error-free operation at 100 MHz was observed with vibration intensity up to 89 G and vibration frequencies from 200 Hz to 10 kHz.

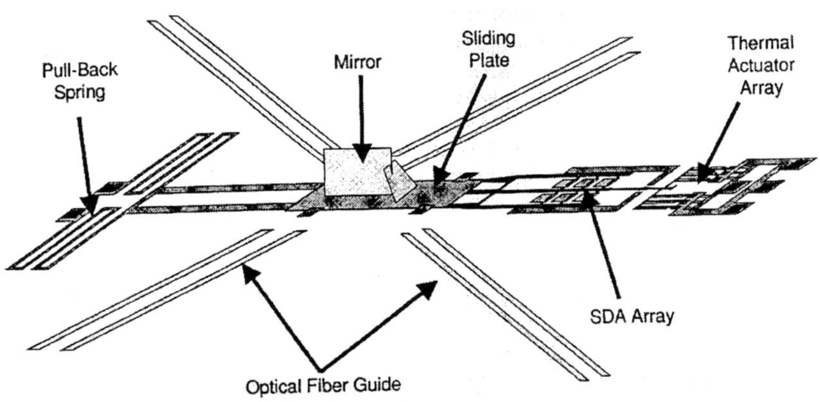

Fig. 2 2× 2 free-space MEMS fiber optic switch with sliding micro-mirrors[8].

2.3. Switch Matrix with Electrostatic Torsion Mirrors

Building a large-scale switch fabric from 2 × 2 switches is unlikely because of the high loss through each 2 × 2 switch. Collimated optical beam inputs are desired for building a large-scale switch matrix. Previous art includes the electrostatic torsion-mirror approach (Fig. 3), demonstrated by University of Tokyo[10]. The silicon torsion-mirror has a dimension of 300 μm × 600 μm, supported by a thin polysilicon beam (16-μm wide, 320-μm long, and 0.4-μm thick). The mirrors are independently actuated by electrostatic force impressed by an applied bias on the order of 100 ~ 150 V. The collimated optical beams are achieved by attaching ball lenses in front of single-mode fibers. Less than 7.66-dB loss was obtained for an optical path of 1 cm. A release process utilizing dry etching results in an improved yield of greater than 80%. A 3% drift of the resonant peak frequency was observed after the device was operated 40 million cycles at 25°C and 43% humidity. No mechanical failure was found after the cycling test.

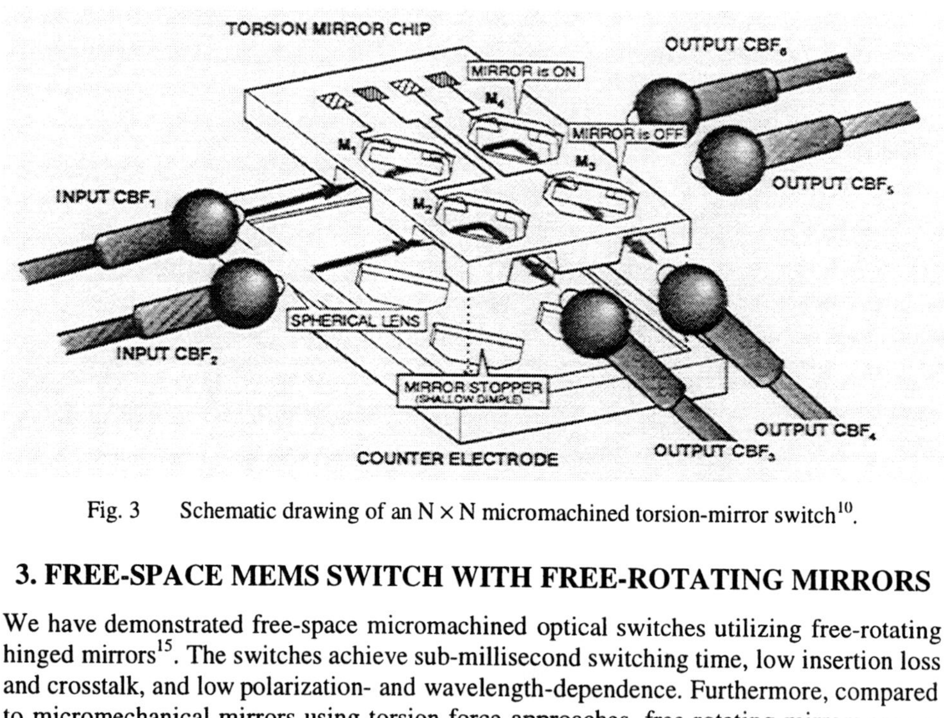

Fig. 3 Schematic drawing of an N × N micromachined torsion-mirror switch[10].

3. FREE-SPACE MEMS SWITCH WITH FREE-ROTATING MIRRORS

We have demonstrated free-space micromachined optical switches utilizing free-rotating hinged mirrors[15]. The switches achieve sub-millisecond switching time, low insertion loss and crosstalk, and low polarization- and wavelength-dependence. Furthermore, compared to micromechanical mirrors using torsion force approaches, free-rotating mirrors appear to be more suitable for OXC applications because the long switch-holding times of such applications require long-term stability and retrievability.

In this paper, we first review the design and performance of the switch. We then present a theoretical study of the device's expandability to large port-count. Finally, we describe the device's potential, due to its free-space nature, to be integrated with other micro-mechanical components to implement enhanced optical-networking functionality.

3.1. Device Design

Figure 4 contains a schematic drawing of a matrix free-space micromachined optical switch. Light is coupled into the switch by fiber collimator arrays. The free-rotating mirrors are integrated monolithically on a silicon chip by surface-micromachining techniques. By rotating a selected mirror with micro-actuators, the collimated light is switched to the desired output port. A schematic drawing of the micro-actuated mirror is shown in Fig. 5. The mirror is pivoted on the substrate by micro-fabricated hinges [16]. Figure 6 shows the fabrication process of the micro-hinges. The surface-micromachining process consists of two polysilicon structural layers and two phosphosilicate glass (PSG) sacrificial layers. The hinge design allows three-dimensional structures be assembled after release etching of the sacrificial material.

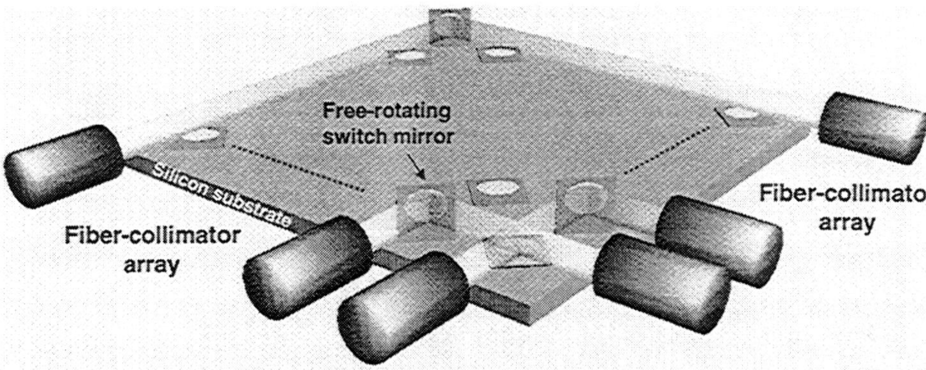

Fig. 4 Schematic drawing of the matrix free-space micromachined optical switch.

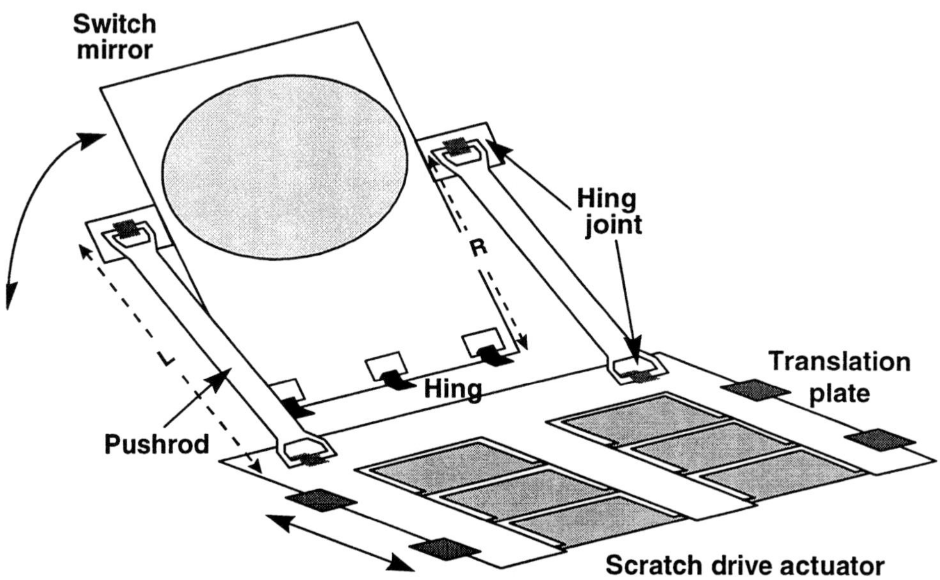

Fig. 5 Schematic drawing of the micro-actuated free-rotating switch mirror.

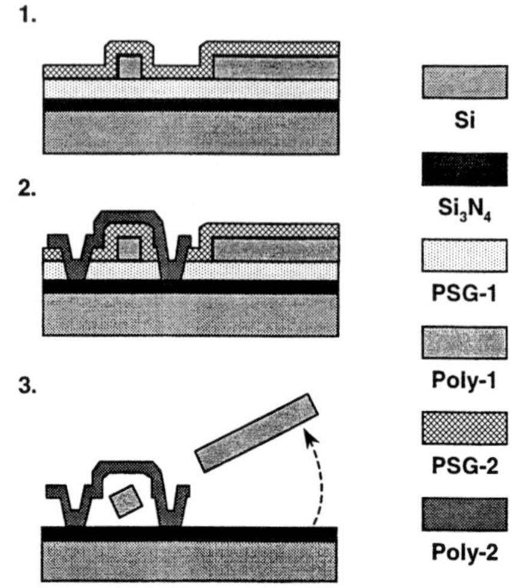

Fig. 6 Fabrication process of micro-hinges.

The translation plate, guided by a set of guiding rails, is actuated by arrays of precision scratch-drive actuators (SDA's)[17]. Pushrods connect the mirror with a translation plate through hinge joints. This design converts the translation of the micro-actuators into rotation of the mirror efficiently. This is one element of the device's speed. This design also reduces the area of the actuated mirror. The SDA consists of a flexible L-shaped polysilicon plate. It moves forward in steps of 10's of nm in response to applied electrical pulses through a combination of electro-static force and friction between the tip of the SDA and the substrate, as shown in Fig. 7. The stiction between the SDA plate and the substrate often causes low yield for the SDA's. We have designed SDA's with tip-edge to reduce liquid forming between the SDA plates and the substrate, as shown in Fig. 8. The combination of this structure with isopropyl alcohol rinse after release results in very high yield for the SDA's. No stiction was observed after the release etching.

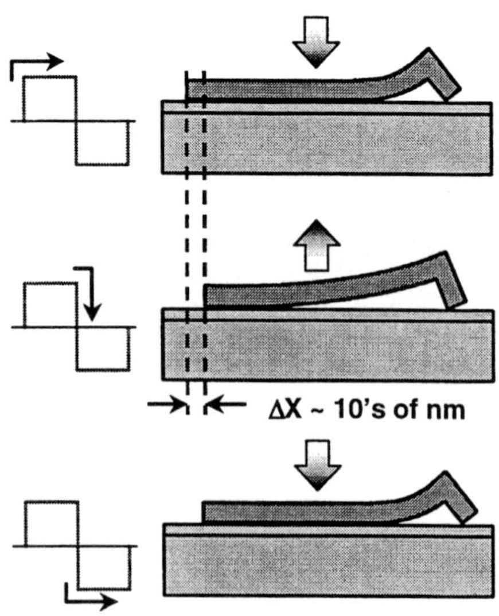

Fig. 7 Working principle of the scratch drive actuator (SDA).

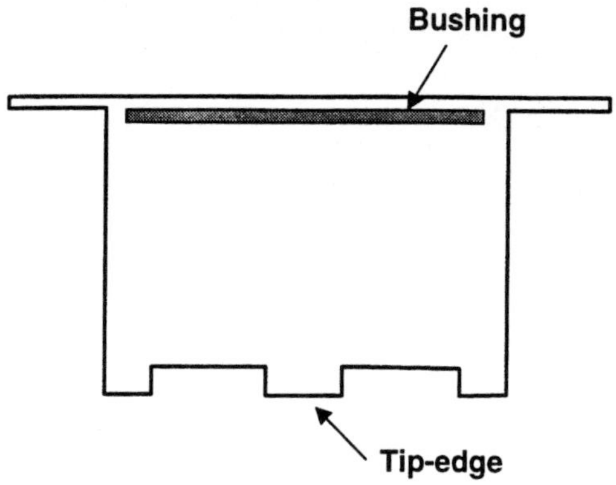

Fig. 8 SDA with tip-edge reduces stiction problem.

The device has been fabricated using the MCNC MUMPs (Multi-User MEMS Processes) fabrication process[18]. Figure 9 contains a scanning-electron micrograph (SEM) of an 8 × 8 switch. The complete switch fabric, exclusive of coupled fibers, occupies a chip area of 1 cm × 1 cm. The SEM of a micro-mirror is shown in Fig. 10. Figure 11 shows the SEM of close-up of the pushrod.

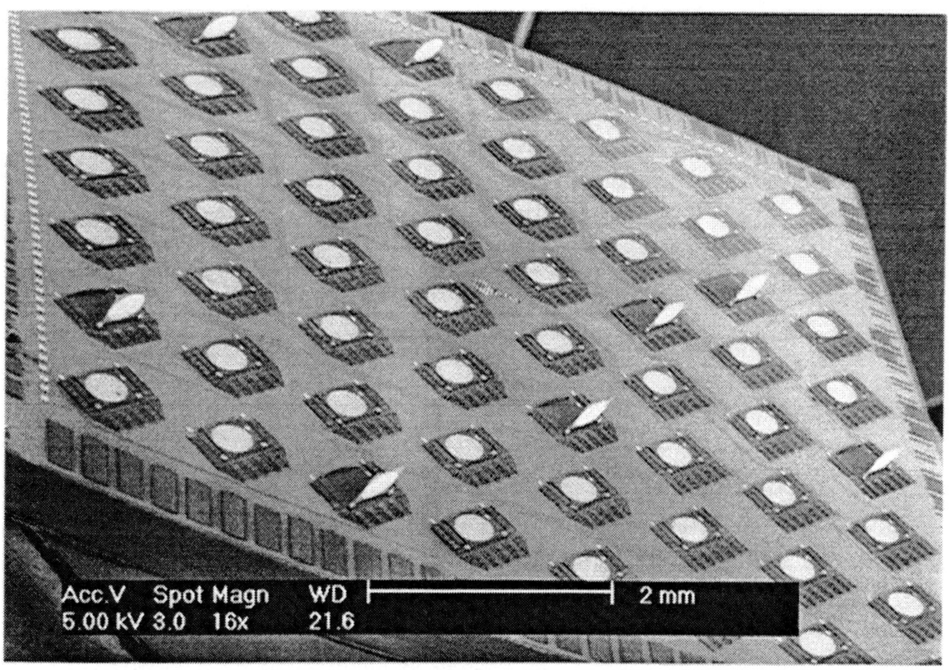

Fig. 9 SEM of an 8 × 8 free-space MEMS optical switch.

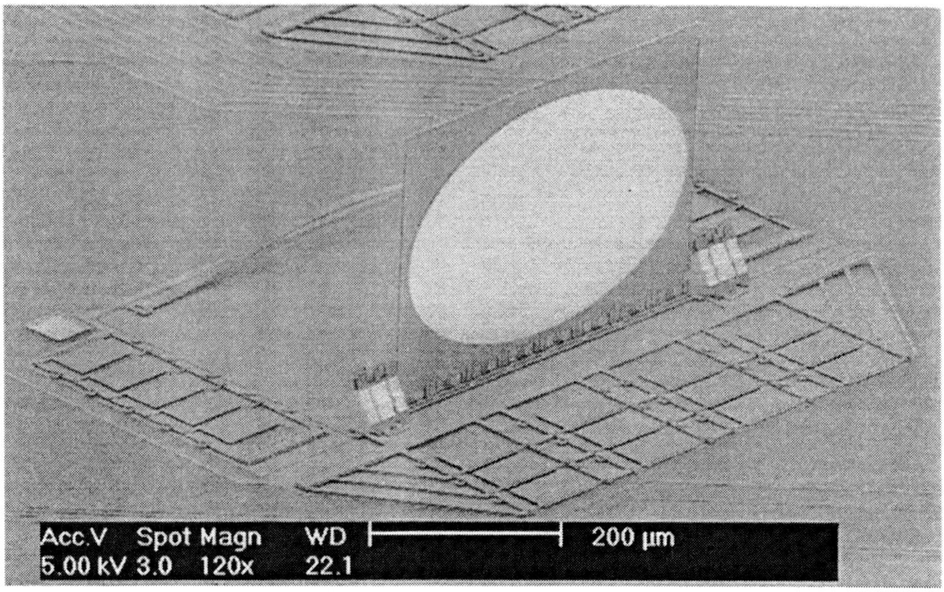

Fig. 10 SEM of the micro-actuated switch mirror.

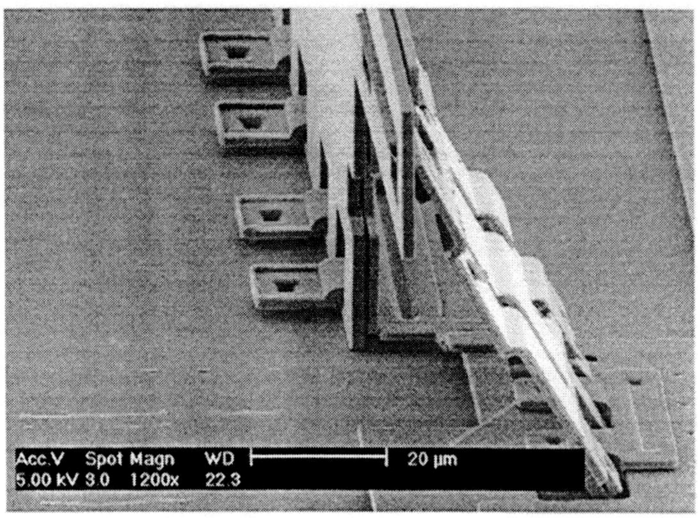

Fig. 11 Close-up of the pushrod.

3.2. Performance of the Switch

The free-rotating mirror achieves sub-millisecond switching times when the scratch-drive actuators are driven with a ± 100 V square wave at 500 kHz. The measurement result is shown in Fig. 12. The switching time is 500 μs for rotating the mirror from the OFF to the ON position. There is an additional 200- μs delay between the application of the electrical bias and the onset of measurable optical-switching action. In the first demonstration, the mirror is rotated down by pulling the translation plate back with polysilicon springs, resulting in an ON → OFF switching time of 560 μs. This measured switching speed matches well with the requirements of provisioning and restoration in core-transport networks. Indeed, since the software execution times for a mesh-based system would greatly exceed one millisecond, there is no value to be gained by striving for device-switching speeds greater than those demonstrated here. The switching curve of the device shows an extinction ratio of greater than 60 dB.

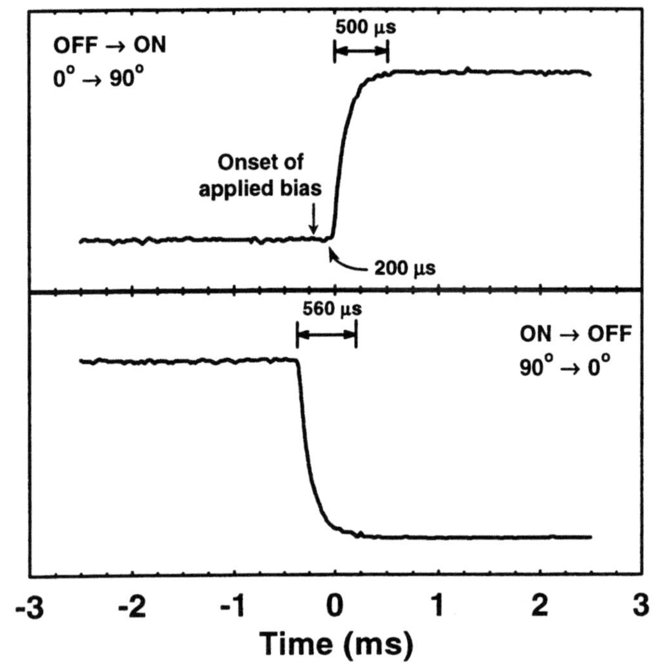

Fig. 12 Switching response of the micro-actuated mirror.

By employing fiber collimators for input and output coupling, very-low insertion losses have been achieved for switches with various port-count configurations. The results will be shown in Sec. 4. The crosstalk of the switch is characterized by switching the light to one output, and measuring the received optical power at the adjacent channel. Unlike guided-wave approaches, these devices exhibit very low crosstalk (< -60 dB) between adjacent channels without resorting to dilated architectures.

In addition to large extinction ratio, low crosstalk, and strict bit-rate transparency, another substantial advantage of the free-space MEMS optical switch is its low polarization-dependent loss (PDL). Figure 13 shows PDL as a function of the wavelength of incident light. The device's PDL was observed to be smaller than the measurement system's noise floor of about 0.3 dB. Because of the device's superior optical quality, no measurable power penalties for the various switching paths are observed. This is an inherent virtue of free-space interconnection approaches.

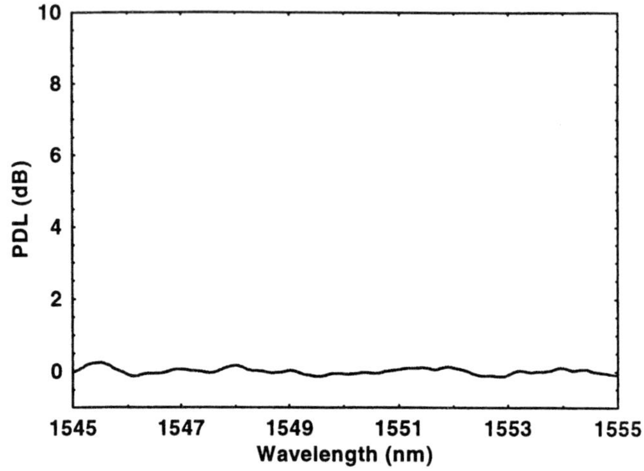

Fig. 13 Polarization-dependent loss (PDL) versus wavelength of the free-space MEMS optical switch. The device's PDL is below the noise floor of the measurement system.

One pays for the excellent optical quality of free-space interconnection with tight angular-alignment tolerances, as will be shown in Sec. 4. In the free-space MEMS optical switches with free-rotating mirrors, the moving distance of the translation plate, and therefore the mirror rotation angle, is roughly determined by the number of bias pulses applied to the SDA's. The maximum mirror rotation angle (90°) is further defined by the dimension of the designed ground electrodes for the translation plates. With the additional mechanical enhancement structure, angular precision better than 0.1° has been demonstrated[19].

4. SCALABILITY ANALYSIS

In this section, we analyze loss versus port count expected for free-space matrix-type switches. We also analyze the effect of angular misalignment in the switch mirrors and describe means of optimizing fiber-coupling efficiency. The analysis is based on Gaussian-beam theory[20]. The analysis and simulation results described here are generall applicable to optical matrix switches with free-space beam propagation.

First, it is assumed that the optical beam that emerges from the input fiber has a minimum Gaussian-beam waist at the point of emergence, characterized by the 1/e half-beam-width w_0 at the facet of the input fiber. A new variable given by the ratio of the switch-mirror radius R divided by the beam waist w_0 is defined. This dimensionless quantity $R/w_0 = a$ is a variable that can be optimized by adjusting the value of the switch-mirror radius R.

The coupling efficiency through the free-space MEMS optical switches is obtained by calculating the overlap integral of the wave functions of the optical beam and the fiber mode at the receiving fiber facet. The optical beam diverges as it propagates through free space due to the finite beam waist, thus limiting the coupling efficiency. Although Gaussian beams with larger beam diameters exhibit smaller divergence angles, the mirror size required increases with beam diameter. As a result, the propagation distance of the optical beam also increases for a given device port-count. Therefore, there is an optimal size of the Gaussian beam because of the compromise between these two competing factors. In particular, the pitch p between adjacent mirrors is geometrically related to the mirror radius R by $p = 3R + 800$ (in µm) for the free-space MEMS optical switch design described here.

Figure 14 shows the simulation results for various mirror radii and beam sizes, assuming that $a = 1.5$, i.e. that the mirror is 1.5 times as large as the beam waist. The simulation is also verified by experimental measurement. The measurement was done by retracting input and output fiber collimators to emulate the optical paths for switches larger than 8 × 8. The variable i labeling the abscissa in Fig. 14 represents the total propagation distance of the optical beam through the switch, measured in number of mirror pitches or, roughly, rows plus columns. The distance between the input/output fiber facet and the closest mirror is assumed to be $p/2$. Therefore, $i = 63$ is the number of pitches corresponding to the longest path in a 32 × 32 switch. The coupling loss in Fig. 14 decreases rapidly as the mirror/beam size increases, suggesting that beam divergence is the dominating factor in this region. Nevertheless, the effect saturates as the mirror radius increases beyond 400 µm.

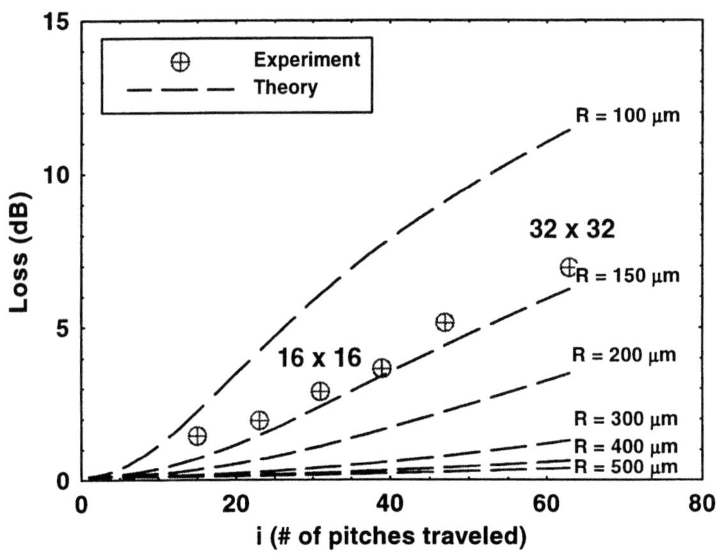

Fig. 14 Loss versus number of pitches traveled for various mirror radii, assuming $a = 1.5$, where a is the ratio of mirror radius to Gaussian beam waist.

The minimum loss for the longest optical path through a 512 × 512 or 1024 × 1024 switch fabric can therefore be obtained from the simulation. For illustration, a three-stage Clos switch-fabric architecture[21] is used as a means of constructing large-port-count switches using smaller constituent switch modules. Assuming the input port count of each switch module in the first stage is n, in order to achieve the property of strict non-blocking, the output port count of the switch module needs to be greater than $2n - 1$. This is also the required input and output port count for the switch module in the second stage.

The switch module in the third stage is a mirror image of that in the first stage. Therefore, to construct a 512 × 512 switch fabric, one can employ 32 layers, with 16 × 32, 32 × 32, and 32 × 16 switch modules comprising the three stages. For a 1024 × 1024 switch fabric, 45 layers of 23 × 45, 45 × 45, and 45 × 23 switch modules may be used. With these assumptions, using the approach described above, it can be shown that large-scale switch fabrics with insertion losses below the standard long-reach interface loss budget (25 dB) can be achieved with suitable mirror sizes.

The results described above assume a perfectly aligned optical system. To characterize imperfectly aligned systems, the dependence of coupling efficiency on the angular error in mirror alignment may be captured by performing a coordinate-system transformation on the wave function of the propagating Gaussian beam. Figure 15 shows the additional loss versus mirror-angle variation for various mirror radii, assuming that the traversed number of pitches is fixed at $i = 63$. The value of a has been optimized individually for different mirror radii R. The system's sensitivity to angular misalignment is seen to decrease rapidly as the optical beam size decreases. For R = 150 μm, for example, the misalignment penalty can be held to values < 3 dB if the mirror-angle variation is held smaller than 0.16°. The theoretical model, based on Gaussian-beam theory and coordinate transformation, has also been verified by experimental measurement, as shown in Fig. 16. A commercial mirror is used for this measurement. The distances between the input/output fiber collimators and the mirror are 5 cm.

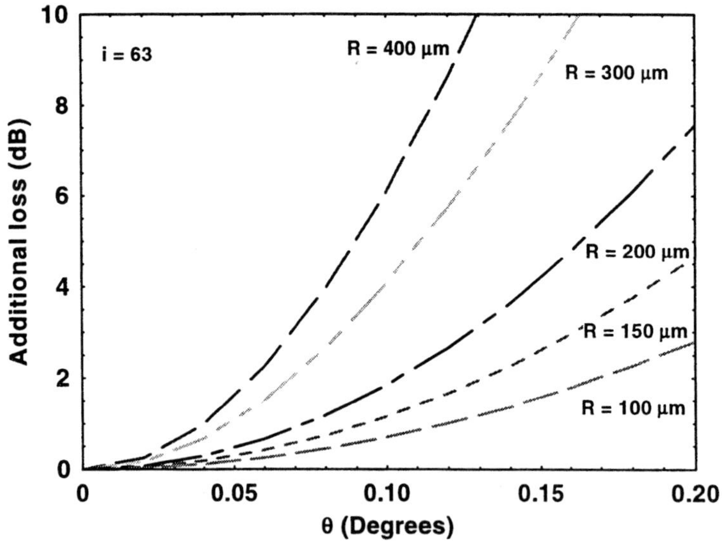

Fig. 15 Additional loss due to mirror-angle variation for the longest path in a 32 × 32 switch fabric. The values of a have been optimized for different mirror radii.

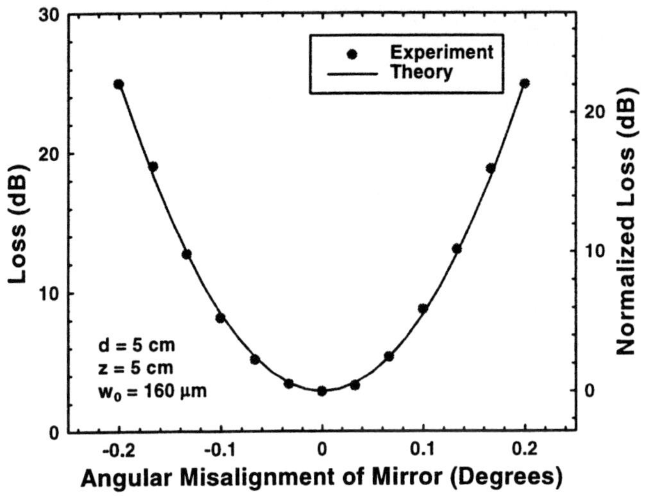

Fig. 16 The theory on the effect of angular misalignment agrees well with experiments.

5. FUNCTIONAL ENHANCEMENT

The free-space nature and micromachining fabrication process endow this technology with unique advantages of integrating various optical networking functionalities with the switch fabric easily. In this section, we describe various functionalities that have been demonstrated by this technology.

5.1. Bridging in Optical Networking

Any optical switching or crossconnection network element used for core-transport network restoration must provide the critical function of bridging. This is illustrated in Fig. 17. Under normal operation (Fig. 17(a)), traffic traverses the service link. Upon link failure (Fig. 17(b)), it is switched to the restoration path, sustaining an interruption of duration *T*. Upon repair, traffic must be *bridged* to *both* service and restoration links (Fig. 17(c)), in order to avoid a second interruption due to imperfectly synchronised head- and tail-end switching. What is critical is that the above bridging capability be incorporated into the optical switch with sufficient simplicity to avoid exacerbating the already-challenging scaling demands incumbent on it.

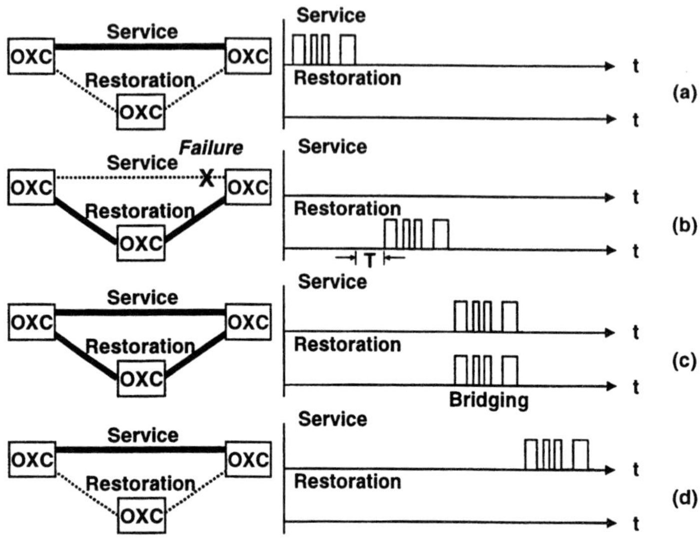

Fig. 17 Schematic illustration of the bridging operation in optical-crossconnect- (OXC-) based restoration.

We have demonstrated a means of accomplishing this in particular for the free-space MEMS optical switches[22]. By integrating into the switch fabric a few simple additional micro-optical elements, advanced functions like bridging can be incorporated at little cost in insertion loss or, ultimately, in scalability. Figure 18 contains a schematic drawing of the free-space MEMS optical switch with bridging. The switch fabric consists of arrays of free-rotating switch mirrors and beam-splitters, together with a single fixed mirror, all monolithically integrated on a single silicon chip. The any-one-to-any-two bridging function is provided jointly by the left column of switch mirrors, the fixed mirror in the lower-left corner, and the beam-splitters in the bottom row. To accommodate multiple simultaneous bridging operations, multiple copies of these elements could be integrated on a substrate.

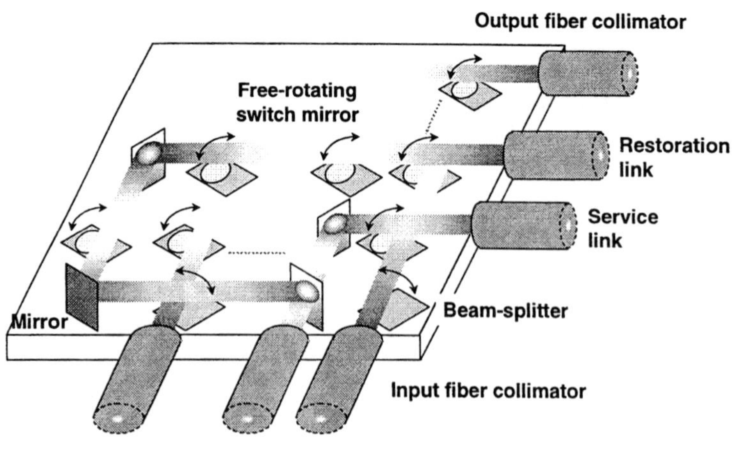

Fig. 18 Schematic drawing of the free-space micromachined optical switch incorporating the bridging function.

The device's operation is most readily apparent from Fig. 19. This is a top-view photograph of the FS-MOS, with array elements configured to execute bridging. The corner mirror (A) has been assembled and is held perpendicular to the substrate by precision side-latches[23]. Switch mirrors B and C and beam-splitter D have also been rotated up, perpendicular to the substrate. During normal operation, only switch mirror C is rotated up, providing a light path from the input to the output service link. Upon service-link failure, A, B, and D rotate up, thus directing traffic to the restoration link. With C upright as well, traffic is bridged onto both service and restoration links. Finally, to carry out recovery, A, B, and D are lowered, with C left upright, so that traffic again flows on the service link only.

The switch fabric with integrated micro-mirrors and beam-splitters was monolithically fabricated via the same surface-micromachining process as described in Sec. 3. The beam-splitter was fabricated using the same process as the switch mirrors, except that it consists of a 1.5-μm polysilicon plate without any reflective gold coating.

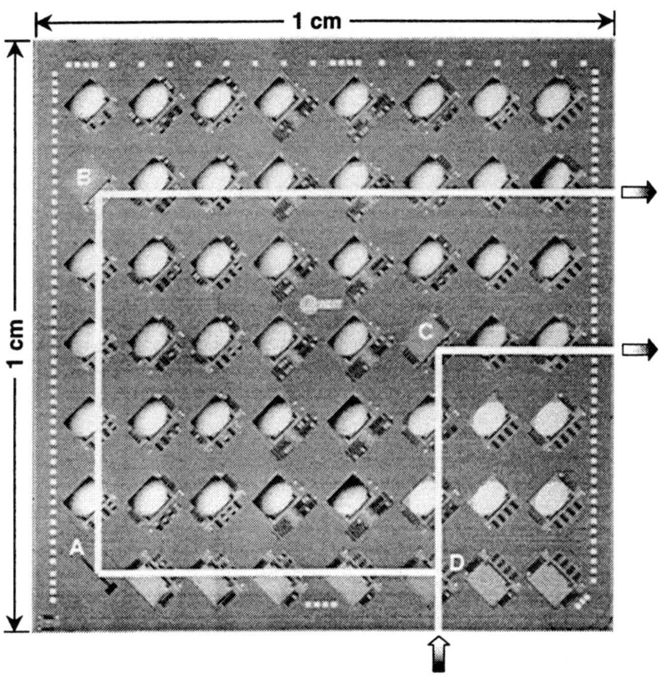

Fig. 19 Top-view photograph of the free-space micromachined optical switch configured to execute bridging. The switch fabric occupies a 1 c × 1 cm Si chip.

Among the chief determinants of scalability in these devices is loss. We have shown previously that low insertion loss can be achieved in these switches. Here, we determine the effect upon loss of the bridging structure. To characterise this, we measured insertion losses through beam-splitters and mirrors, for both service and restoration paths, by coupling light at 1550 nm into the switch fabric using a fiber collimator. Losses along the optical paths were then measured with a large-area detector. The measured insertion losses along the beam-splitter's transmission and reflection paths were 4.15 dB and 2.25 dB, respectively. The service path's switch mirror adds an additional 0.1-dB loss. Along the restoration path, the loss from beam-splitter to output is 0.3 dB. The total excess loss for the two paths is 0.31 dB, of which 0.09 dB results from scattering and absorption in the beam-splitter itself.

5.2. Integrated Signal Monitoring

Performance-management and fault-localization have become increasingly important issues in high-capacity optical-fiber networks. In the event of failures, fast automatic restoration is increasingly required. Current networks rely on transponders in the optical transport systems (OTS) (Fig. 20(a)) to detect faults; thus, communication through multiple interfaces is required before OXCs can respond to the failure. If the OXCs themselves detected faults, restoration would be greatly facilitated (Fig. 20(b)). Nevertheless, this has not been demonstrated by any optical switching technology. By integrating three-dimensional micromechanical structures into the free-space MEMS switch fabrics, such functions can be easily accomplished[24].

Figure 21 shows the schematic drawing of the free-space MEMS optical switch with signal-monitoring functionality realized by surface micromachining technology. At the outputs of the switch fabric, integrated 45° beam-splitters reflect a small portion of light onto hybridly integrated photodetectors. High-speed photodetectors solder-bonded on microwave coplanar transmission lines can monitor the signal states and quality. They can also verify the connection paths of the switch fabric by detecting the header information in the signal. To detect loss-of-signal, low-speed photodetectors are sufficient. Figure 22 shows the SEM photograph of the detector/beam-splitter module. The photodiode with 2.5 GHz bandwidth has an efficiency of 90%. After being integrated with the beam-splitter, the whole module has an efficiency of 11% at wavelength 1550 nm. The high-speed operation of the integrated photodetector has also been verified. Figure 23 shows the eye diagram at 2.5 Gb/s. Error-free signal has been obtained by bit-error-rate measurement.

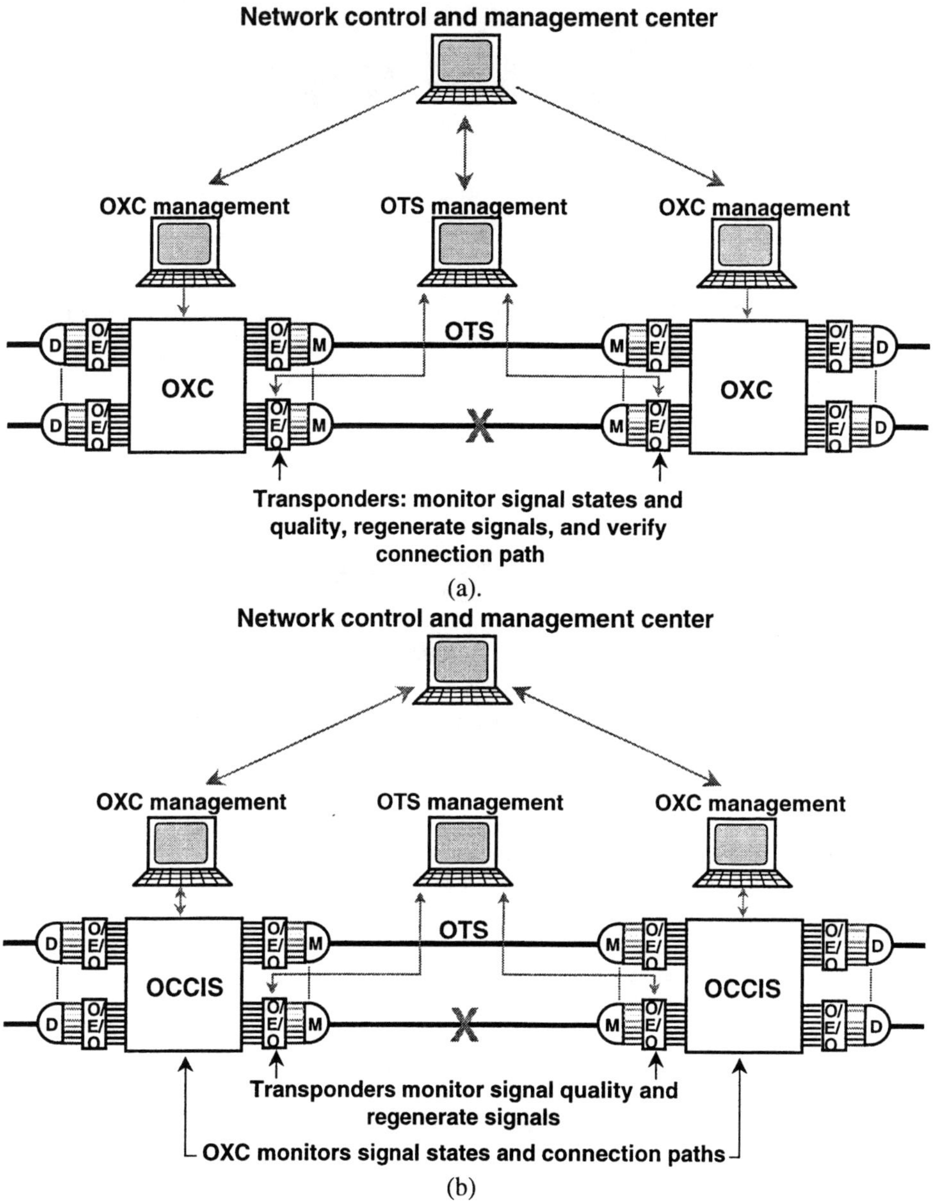

Fig. 20 (a) Monitoring signal states and connection paths by optical transport systems (OTS) results in extra restoration time. (b) Optical crossconnect integrated systems (OCCIS) can monitor signal states and connection paths within the crossconnect itself, thus facilitate fast network restoration.

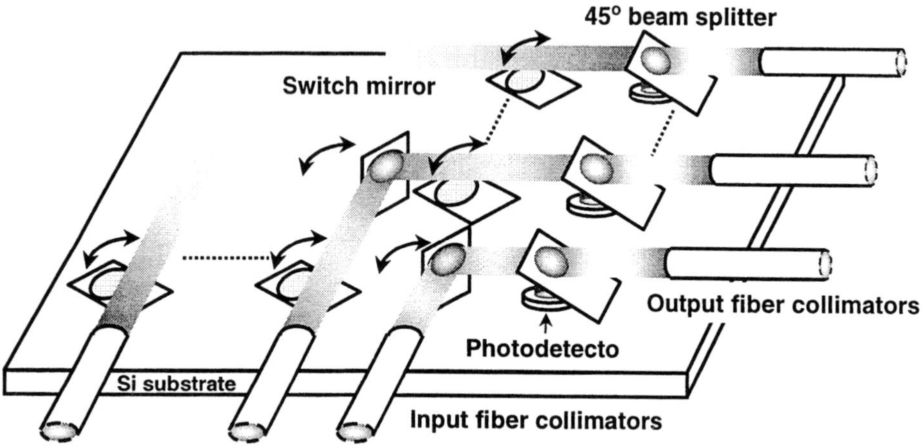

Fig. 21 Schematic drawing of the free-space MEMS optical crossconnect with signal-monitoring.

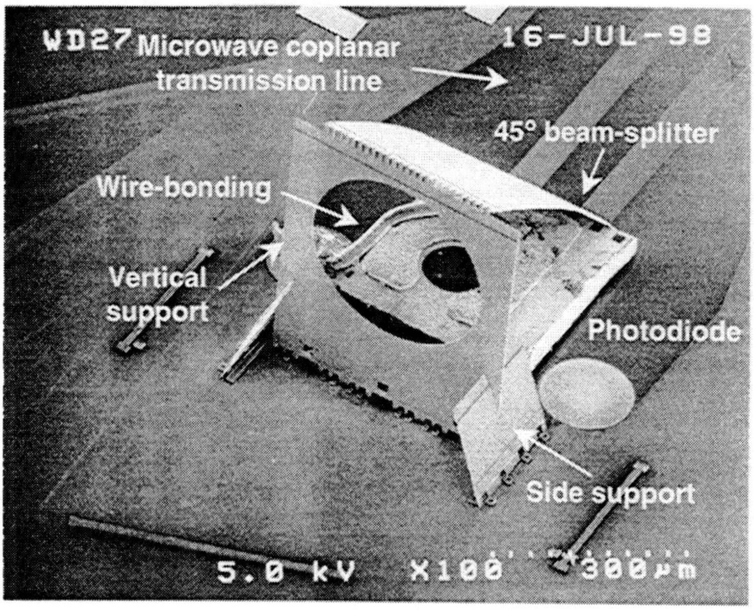

Fig. 22 SEM photograph of the photodetector/beam-splitter module.

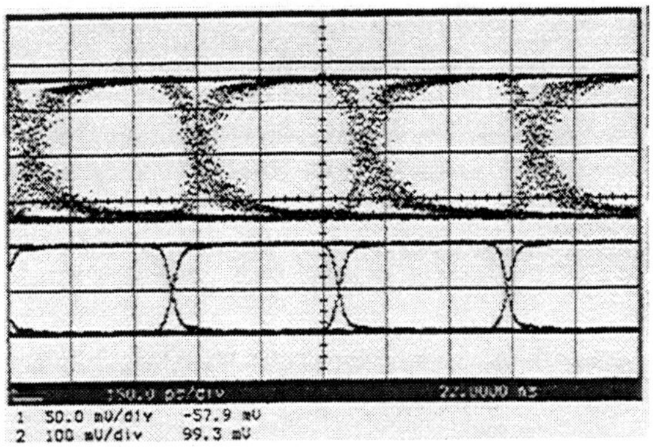

Fig. 23 Eye-diagram of the integrated photodetector at 2.5 Gb/s.

5.3. Connection-symmetry in Bi-directional Networks

Due to the connection-symmetry inherent in the conventional core-transport networks, OXCs deployed in such networks will also be connection-symmetric. This is illustrated in Fig. 24 for an OXC of nodal degree 2. Since all supported connections are bi-directionally symmetric, all switching is done in latched pairs. This is a feature whose substantial lurking economics can be naturally incorporated into a free-space MEMS OXC's[25].

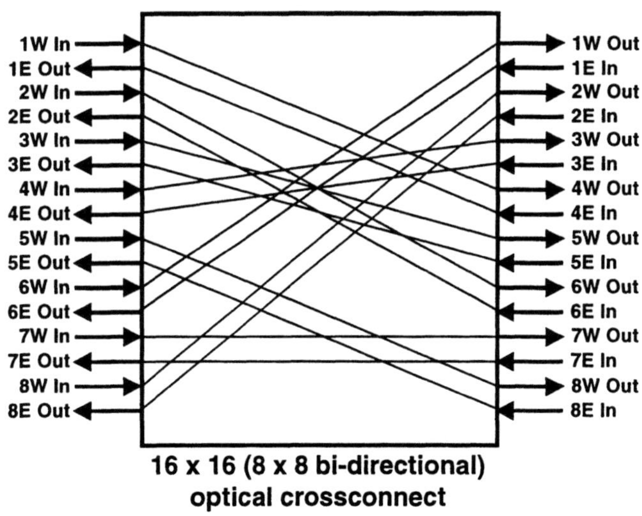

Fig. 24 An optical crossconnect supporting symmetric connections.

By reflecting signals in symmetric pairs from the two sides of free-rotating micro-mirrors, symmetric connections can be supported. Figure 25 shows the top-view photograph of a matrix switch with connection-symmetry of nodal degree two; here, the size and complexity of a 16 × 16 switch fabric is essentially reduced to that of an 8 × 8 conventional matrix switch. Nodal degree means the number of adjacent offices a network node is connected to. For higher nodal degree, optical *polygon* switches are required to avoid blocking. Figure 26 shows an optical hexagon switch for nodal degree 3. When B2 IN is connected to C2 OUT, C2 IN is automatically connected to B2 OUT.

This approach can in principle be extended to octagon and decagon switches for offices of higher nodal degrees. The current long-haul network, however, is sparse, with an average nodal degree of 2 to 3.

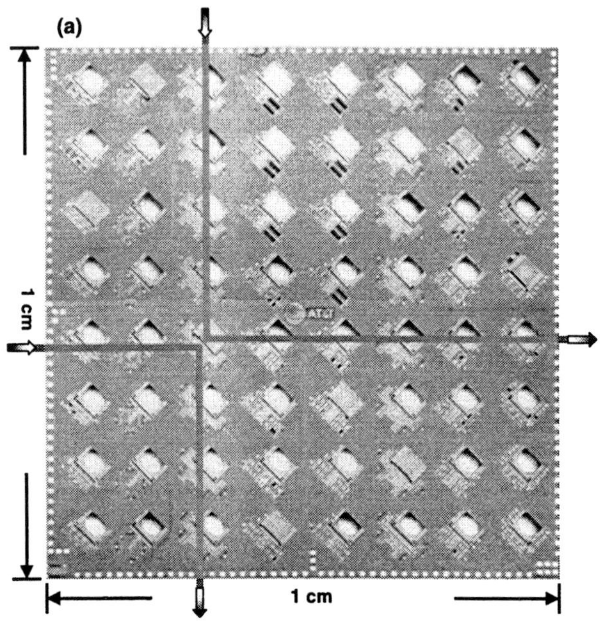

Fig. 25 Top-view photograph of a free-space MEMS matrix switch with connection-symmetry for nodal degree 2.

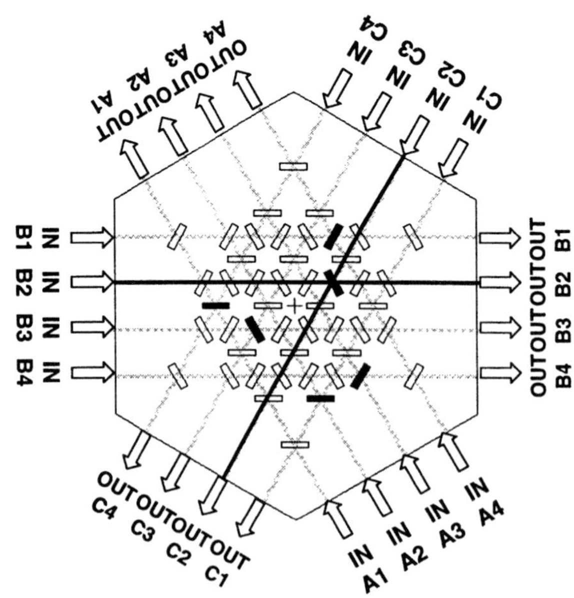

Fig. 26 Free-space MEMS hexagon switch with connection-symmetry for nodal degree 3.

6. CONCLUSION

MEMS technologies have begun to be demonstrated as underlying platforms for various optical network elements. Optical switching, which is of particular interest in future WDM networks, appears to be a particularly fertile area for application of MEMS technology. This is chiefly due to the promising optical quality of free-space MEMS switches, the inherently low cost of the devices, and their capacity to integrate advanced network functions, such as those described above. Although the technology has made swift progress in terms of functionalities and preliminary reliability data, it appears that demonstration of long-term reliability and cost-effective packaging are the chief challenges still facing it.

ACKNOWLEDGMENTS

The authors would like to thank Robert Tkach, Leda Lunardi, Adel Saleh, and Jane Simmons for many helpful discussions, and JDS Fitel for collaboration on packaging issues.

REFERENCES

1. E. L. Goldstein, J. A. Nagel, J. L. Strand, and R. W. Tkach, "Multiwavelength opaque optical-crossconnect networks," in *Lightwave*, 1998, pp. 92-95.
2. J. E. Fouquet, S. Venkatesh, M. Troll, D. Chen, H. F. Wong, and P. W. Barth, "A compact, scalable cross-connect switch using total internal reflection due to thermally-generated bubbles," in IEEE LEOS Annual Meeting, Orlando, FL, 1998.
3. H. Togo, M. Sato, and F. Shimokawa, "Multi-element thermo-capillary optical switch and sub-nanometer oil injection for its fabrication," in 12th IEEE International Conference on Micro Electro Mechanical Systems, Orlando, FL, Jan. 17-21, 1999.
4. J. A. Walker, K. W. Goosen, and S. C. Arney, "Fabrication of mechanical antireflection switch for fiber-to-the-home systems," *J. Micro-Electro-Mech. Syst.*, vol. 5, no. 1, 1996.
5. J. E. Ford, J. A. Walker, V. Aksyuk, and D. J. Bishop, "Wavelength add-drop switching using tilting micromirrors," *J. Lightwave Technol.*, vol. 17, no. 5, pp. 904-911, 1999.
6. C. R. Giles, B. Barber, V. Aksyuk, R. Ruel, L. Stulz, and D. Bishop, "Reconfigurable 16-Channel WDM DROP module using silicon MEMS optical switches," *IEEE Photonics Technol. Lett.*, vol. 11, no. 1, pp. 63-65, 1999.
7. C. Marxer, M.-A. Grétillat, N. F. de Rooij, R. Bättig, O. Anthamatten, B. Valk, and P. Vogel, "Reflective duplexer based on silicon micromechanics for fiber-optic communication," *J. Lightwave Technol.*, vol. 17, no. 1, pp. 115-122, 1999.
8. S. S. Lee, E. Montamedi, and M. C. Wu, "Surface-micromachined free-space fiber optic switches with integrated microactuators for optical fiber communication systems," in 1997 International Conference on Solid-State Sensors and Actuators (Transducers'97), Chicago, IL, June 16-19, 1997.
9. C. Marxer and N. F. de Rooij, "Micro-opto-mechanical 2 x 2 switch for single-mode fibers based on plasma-etched silicon mirror and electrostatic actuation," *J. Lightwave Technol.*, vol. 17, no. 1, pp. 2-6, 1999.

10. H. Toshiyoshi and H. Fujita, "Electrostatic micro torsion mirrors for an optical switch matrix," *J. Microelectromechanical Systems*, vol. 5, no. 4, pp. 231-237, 1996.
11. H. Laor, "MEMS mirrors application in optical cross-connects," in IEEE LEOS Summer Topical Meetings: Optical MEMS, Monterey, CA, July 20-22, 1998.
12. B. Behin, K. Y. Lau, and R. S. Muller, "Magnetically Actuated micromirrors for fiber-optic switching," in Solid-State Sensor and Actuator Workshop, Hilton Head Island, SC, June 8-11, 1998.
13. R. T. Chen, H. Nguyen, and M. C. Wu, "A low voltage micromachined optical switch by stree-induced bending," in 12th IEEE International Conference on Micro Electro Mechanical Systems, Orlando, FL, Jan 17-21, 1999.
14. L. Y. Lin, E. L. Goldstein, and R. W. Tkach, "Free-space micromachined optical switches for optical networking," *IEEE J. Selected Topics in Quantum Electronics: Special Issue on Microoptoelectromechanical Systems (MOEMS)*, vol. 5, no. 1, pp. 4-9, 1999.
15. L. Y. Lin, E. L. Goldstein, and R. W. Tkach, "Free-space micromachined optical switches with submillisecond switching time for large-scale optical crossconnects," *IEEE Photonics Technol. Lett.*, vol. 10, no. 4, pp. 525-527, 1998.
16. K. S. J. Pister, M. W. Judy, S. R. Burgett, and R. S. Fearing, "Microfabricated hinges," *Sensors and Actuators A*, vol. 33, pp. 249-256, 1992.
17. T. Akiyama and H. Fujita, "A quantitative analysis of scratch drive actuator using buckling motion," in IEEE Workshop on Micro Electro Mechanical Systems, Amsterdam, the Netherlands, Jan. 29 - Feb. 2, 1995.
18. MCNC, http://mems.mcnc.org.
19. L. Y. Lin, E. L. Goldstein, and R. W. Tkach, "Angular-precision enhancement in free-space micromachined optical switches," submitted to *IEEE Photonics Technology Letters*, 1999.
20. Joseph T. Verdeyen, *Laser Electronics*, 2nd ed: Prentice-Hall, 1989.
21. C. Clos, "A study of non-blocking switching networks," *Bell Sys. Tech. J.*, vol. 32, pp. 406-424, 1953.
22. L. Y. Lin, E. L. Goldstein, and R. W. Tkach, "Free-space micromechanical optical crossconnect with bridging functionality for optical-layer restoration," in European Conference on Optical Communication, Madrid, Spain, Sep. 20-24, 1998.
23. L. Y. Lin, S. S. Lee, K. S. J. Pister, and M. C. Wu, "Micro-machined three-dimensional micro-optics for integrated free-space optical system," *IEEE Photonics Technol. Lett.*, vol. 6, no. 12, pp. 1445-1447, 1994.
24. L. Y. Lin, L. M. Lunardi, and E. L. Goldstein, "Optical cross-connect integrated system (OCCIS): A free-space micromachined module for signal and switching configuration monitoring," in IEEE LEOS Summer Topical Meeting: *Optical MEMS*, Monterey, CA, July 20-22, 1998.
25. L. Y. Lin, E. L. Goldstein, J. M. Simmons, and R. W. Tkach, "High-density micromachined polygon optical crossconnects exploiting network connection symmetry," *IEEE Photonics Technol. Lett.*, vol. 10, no. 10, 1998.

Advanced Optical Systems and Networks

Association of submarine cable reliability and network protection for very high availability transoceanic transmission networks

Steven Desbrulais, Vincent Lemaire, Loic Le Gall, Christophe Mathieu
Submarine Networks Division, Alcatel
Villarceaux, F-91625 Nozay cedex, France

Abstract: To match the bandwidth requirements of our fast-growing communication world, submarine networks have to provide ever greater transmission capacity. Thus, almost 2 Tbit/s will soon be the capacity available in submarine networks. A high level of availability is achieved owing to a very reliable submarine cable and to powerful network protection mechanisms.

1. Introduction

Optical Wavelength Division Multiplexing (WDM) technology and its tremendous potential for transmission and routing applications has been studied in most of the telecom research laboratories for many years. Over the last 10 years, the submarine telecommunications world has been urged to introduce the latest technology, offering the maximum possible connectivity and capacity. But at the same time, this tremendous increase of cable capacity has compelled customers to demand even greater network availability. In the following paper we present the methodology used to combine the high reliability of the submerged plant with the use of the latest technology of network protection equipment to offer networks with a very high availability for SDH or IP traffic, including terminal protection, Network Protection Equipment, Optical Path Switch equipment or IP routers. After a description of the reliability of the submerged plant equipment, we describe different network architectures focusing on protection mechanisms.

2. Submarine link reliability

2.1 Traditional Submarine Systems

The traditional optical submarine system comprises a single link between two terminal stations. The link comprises a cable containing a number of pairs of fibres, into which are spliced repeaters at regular intervals to amplify the signal. The repeaters are powered by Power Feed Equipment (PFE) at the terminal stations and the Submarine Line Terminating Equipment (SLTE) provides the interface between the submarine system and the terrestrial network. The system is shown schematically in fig. 1. Each double arrow represents bi-directional information flow along a number of fibre pairs.

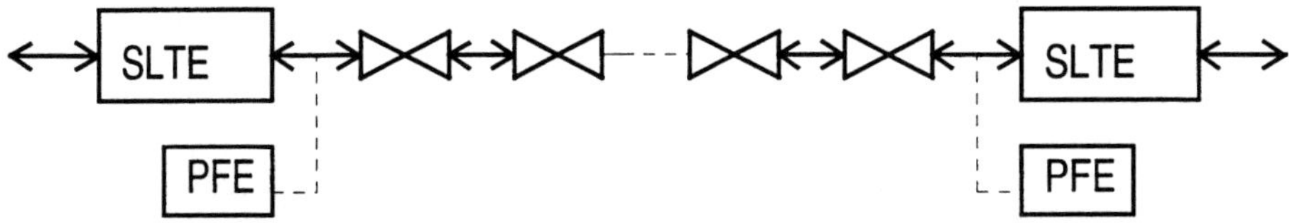

Figure. 1. Traditional submarine system components. The submarine link is terminated by the Submarine
Line Terminating Equipment (SLTE) and is powered by the Power Feed Equipment (PFE).

From this basic configuration, a number of key points are pertinent. Firstly, the system, which has a design life of 25 years, can be divided into repairable and non-repairable elements. The terminal equipment, comprising the PFE and SLTE, is clearly repairable, since it is readily accessible. For this reason, it's dependability is assessed in terms of unavailability. By contrast, the submarine plant is considered to be non-repairable and so its dependability is quantified in terms of reliability. In practice, if the submerged plant fails, it is, of course, repaired. However, if it were classified as a repairable system and the unavailability combined with that of the terminal equipment, it would be found that the unavailability of the SLTE is dwarfed by that of the submerged plant. The distinction is nevertheless cogent, since a single failure of the submerged plant can result in an outage of several weeks and repairs are very costly. By contrast, a failure within the terminal station is repairable within a matter of hours. As a non-repairable system, the reliability of the submerged plant is quantified in terms of the expected number of ship repairs.

It is also worth noting that most of the terminal equipment in traditional links is duplicated so that its effective failure rate is typically of the order of 2000 FITs per fibre pair, this being predominantly due to the unduplicable elements. By contrast, the failure rate of a repeater is of the order of 10 FITs per fibre pair, so for a long haul system comprising 100 repeaters, say, the effective failure rate of the terminal equipment and the failure rate of the submerged plant are comparable. Exceptions to this are links with few or no repeaters. However, short links are always deployed in shallow water areas and are therefore susceptible to trawler and anchor damage. Such extrinsic faults are considered further in section 2.3.3 below.

2.2 The Development of Submarine Networks

The first submerged optical system to be installed was UK-Belgium in 1986. This used regenerative technology, each repeater converting the optical signals to the electrical domain to regenerate the signal before re-launching into the next fibre span. Whilst such point to point links were adequate initially, it soon became necessary to introduce submerged Branching Units (BUs) to accommodate 3 or more landing points. These are passive units with respect to the signal and only provided fibre routing. They do, however, include switching elements for the power feed so that, in the event of a failure in one branch, it is still possible to power the remaining two branches from the PFE. Such branching units were first used on PTAT-1 in 1990. A further level of sophistication was introduced in TAT-9 in 1992. In this system, 3 submerged multiplexed BUs were provided to allow active signal routing. The last submarine regenerative system to be installed was CANTAT-3 in 1994. From this point onwards, the transition to optically amplified systems was made.

The first system for which optically amplified technology was developed was TAT-12/13, installed in 1995/96. TAT-12 and TAT-13 comprise 4 point to point links to form an Atlantic ring network, but no submerged BUs were employed. With the advent of optically amplified systems, the electrical submerged multiplexors were no longer applicable. However, with the introduction of WDM systems, it became feasible to allow wavelength routing in addition to fibre routing. Such WDM BUs were first introduced in SEA-ME-WE 3, which was commissioned this year.

Contemporary submarine networks use Dense WDM technology with up to 40 wavelengths per fibre pair, each carrying a 10Gb/s signal[1,2]. These Terabit systems, which are currently being introduced, will be the mainstay of international communications and their reliability is paramount. During the progression of the technology indicated above, various strategies have been employed to assure the reliability of the submarine plant and the availability of the terminal equipment. With regard to the SLTE, it is no longer practicable to duplicate the equipment due to cost and space considerations. However, redundancy is still possible by reserving some of the wavelengths to allow for transmitter or receiver failures. In the next section, the strategies which can be used to assure the reliability of modern links will be explored.

2.3 Strategies Assuring Link Reliability

The basic building blocks for a link within a submarine network are shown in fig.2. In addition to the WDM BUs, DWDM systems also require equalization due to the finite bandwidth of the EDFAs. Since there are very few Equalizer Units (EUs) in a system and they are passive elements, their impact on reliability is small. There are also very few BUs in a system, but since they constitute submarine network nodes, their high reliability is vital.

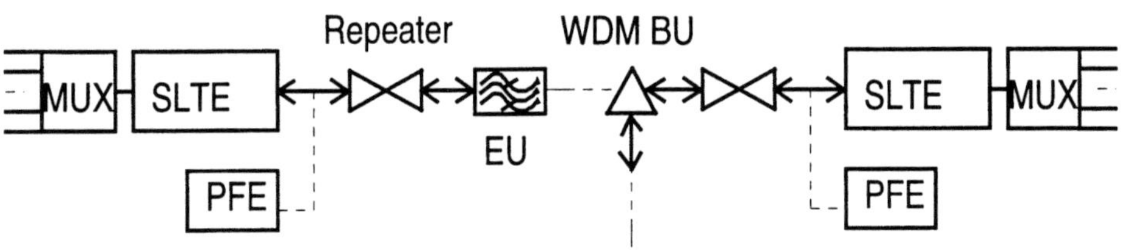

Figure. 2. Submarine link components. In addition to the components shown in fig. 1, the links also include
WDM BUs and EUs. The multiplexors are also shown which convert inputs such as STM-1 tributaries up to 10Gb/s.

2.3.1 Reliability of Submarine Elements

The key element whose reliability must be maximized within the submarine plant is the repeater. Functionally, the repeater must extract power from the cable to power the internal electronics and perform both signal amplification and supervisory functions. The reliability diagram of a repeater is shown in fig. 3.

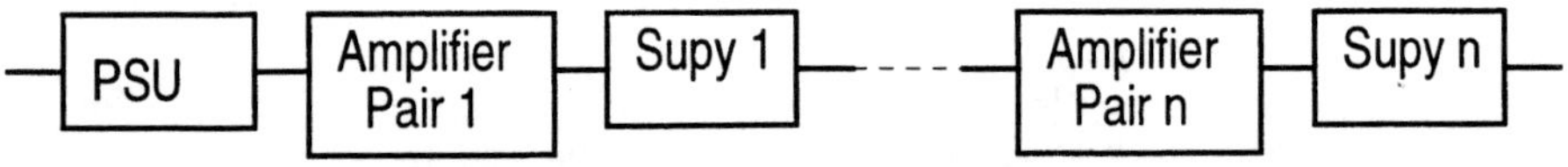

Figure. 3. Repeater Reliability Diagram showing the Power Supply Unit, Amplifier Pairs and Supervisory elements.

The amplifiers and supervisory elements are shown in series suggesting that the failure of any one element causes a repeater failure. Although the repeater functionality is compromised if any of these elements fail, in practice, if the supervisory fails, the repeater will continue to function but it will not be possible to alter or monitor any of its operating parameters remotely. Also, if an amplifier pair should fail, the remaining amplifier pairs will still be operable. The only element which would cause a total system failure is the Power Supply Unit (PSU). For this reason, the PSU employs massive redundancy. The effective failure rate of the PSU has been estimated to be only 0.2FITs. There has never been a PSU failure in any of our repeaters or BUs. Whether there has ever been a component failure within a PSU is unknown since such failures are, by design, hidden from external observation.

Every component which is used within the submarine plant must be most rigorously qualified. In addition to resilience to temperature and humidity extremes, bump, vibration and other robustness tests, those components which are expected to have measurable failure probabilities are extensively life tested, usually under suitable chosen accelerated conditions. Furthermore, a number of screening tests are employed at all levels of build to route out any possible early life failures.

Of all the components used within the submerged plant, the most notorious is the laser, which is used to pump the EDFAs. Failure rates for pumps are typically 30 FITs for 1480nm lasers and 100 FITs for the recently introduced 980nm pumps. Since failure rates of the order of 10 FITs are required per amplifier pair, redundancy schemes are mandatory. Fig. 4 shows a schematic representation of an amplifier pair used within current repeaters which demonstrates the high degree of redundancy employed [3].

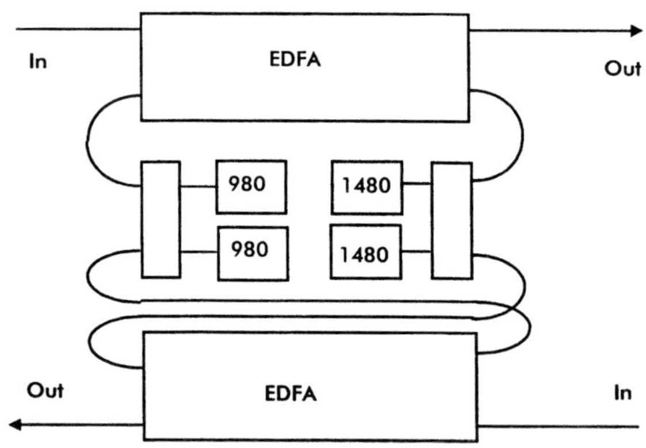

Figure 4. EDFA Pump redundancy scheme. 980nm lasers are shared for co-pumping and 1480nm lasers are shared for counter-pumping

A total of 4 pumps are used per fibre pair, but not all of these need be working for perfect or adequate operation. Both pairs of lasers are shared between both erbium doped fibres so that if one pump of a pair fails, the output of the remaining pump is increased and can still be used for amplifying signals in both directions. The 980nm pumps are used in the co-propagation direction in order to minimise the noise figure and the 1480nm pumps provide power amplification.

Reliability performance

The reliability diagram of the amplifier pair is shown in fig. 5. The first level of redundancy is provided by pump sharing at both the 980 and 1480nm wavelengths and the second level is provided by using independent pump control elements and pump pairs. An amplifier pair fails under the following conditions:

All four pumps fail
Two pumps of the same pair fail and the pump control circuit for the other wavelength fails
The amplifier pair common circuitry fails
Any combination of the above

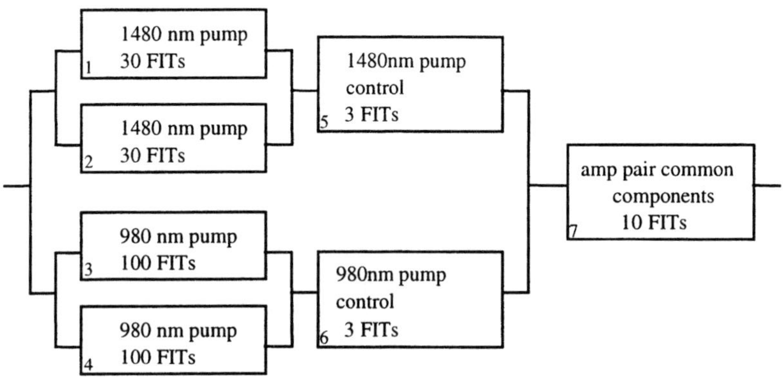

Figure 5. Amplifier pair reliability diagram illustrating the 2-level redundancy scheme

In the following analysis, expressions are developed for the failure probability of the amplifier pair. In the world of submarine networks, the design life, as mentioned above, is 25 years. The failure probabilities F used below are therefore 25 year failure probabilities. These are related to an equivalent average failure rate (AFR) λ_{eff} using the expression

$$\lambda_{eff} = -\frac{1}{t}\ln(1-F)\times 10^9 \qquad \text{FITs} \tag{1}$$

where *t* is 25 years, expressed in hours.

Similarly, the failure rates for the individual components are rarely constant in time. The failure rates quoted are AFRs, derived from failure probabilities.

The failure probability F_T of the amplifier pair is given by

$$F_T = 1 - R_{com}(1 - F_P) \tag{2}$$

$$F_P = \prod_{i=1}^{2}\left[1-\left(1-F_i^2\right)R_c\right] \tag{3}$$

where F_1 is the failure probability of a 980nm pump, F_2 the failure probability of a 1480nm pump, R_c is the reliability of the pump control circuits, R_{com} is the reliability of the amplifier pair common components, and F_P is the failure probability of the combined pumps and pump control elements. The failure probability F_T implicitly contains all the failure mechanisms identified above. From the failure rates given in fig.5, it is found that $F_p << 1$, demonstrating that the amplifier pair failure rate is dominated by the common components. Since there will in general be some system degradation in the event of partial failures, e.g., less than 4 effective pump failures, it is necessary to establish the probability of these partial failures. The general form of F_P is modified in this case and is given by

$$F_P = \sum_{s=1}^{smax}\prod_{j=1}^{6} X_{j,s} \tag{4}$$

where the index *j* references the first 6 elements within the reliability diagram above. Index *s* represents a single failure microstate of the system, *smax* of which comprise the macrostate which constitutes the partial failure type identified in table 1 and X is either the failure probability F or the reliability R of the *j*th element appropriate for the microstate in question. Using (4) and (2) gives the partial failure probability.

Impact of partial EDFA failures on system performance

In assessing the reliability of a link in terms of the individual repeaters, it is clear that there must be an interrelationship between system performance and partial failures. For instance, how many partial failures can be withstood, and how severe can each partial failure be?

To assess the system performance in the presence of the failed and partially failed amplifiers, it is necessary to calculate the probabilities of each partial failure state. The system power budget can then be designed to accommodate them. The following table gives the EDFA performance degradation versus partial failure state probability for the pessimistic case in which the 980nm pump failure probability is 5% over 25 years (230FITs) to 95% UCL. Using such pessimistic 980nm pump failure rates is prudent since they represent a new technology to submarine networks and must therefore be introduced with caution.

The first and last states in the table are the fully working and totally failed states. Each state probability was calculated using (4) above. The 4 columns headed pump1 and pump2 refer to the availability of pump power to the EDF, and so the entry 'N' refers not only to the failure of a pump itself, but also to failures associated with the pump control or amplifier pair common components elements.

State Probability (%)	No. of failed pumps	1480 pump1	1480 pump2	980 pump1	980 pump2	NF penalty (dB)	O/P penalty (dB)
88.99	0	Y	Y	Y	Y	0	0
9.37	1	Y	Y	Y	N	0.5	0
1.14	1	Y	N	Y	Y	0	0
0.30	2	Y	Y	N	N	1.0	1
0.06	2	N	N	Y	Y	0.1	2
0.12	2	Y	N	Y	N	0.5	2
3.86E-4	3	N	N	N	Y	0.5	5
3.17E-3	3	Y	N	N	N	1.5	2
2.16E-4	4	N	N	N	N	Ship Repair	

Table 1. Probability of each externally observable operational state of an amplifier pair, categorised in terms of power available from each pump.

The expected number of amplifiers in each of these states in a system comprising 100 amplifier pairs (say) can be read directly from the table. For example, the expected number of amplifier pairs with 2 failed 980nm pumps over a 25-year period is 0.3. By summing the noise figure and output power penalties in the table, weighted by the corresponding state probability, the average degradation is found to be 0.058dB per amplifier pair. However, system simulations have shown that for a 100 repeater system, a realistic power budget apportionment can accommodate a margin of 0.125dB for each amplifier pair. Clearly, the system can easily accommodate the expected number of partial failures, indicating that the repeater design is extremely robust and performs well even in the case of pessimistic 980nm pump failure probabilities.

Other submarine elements

The same principles which are used to assure high repeater reliability are also applied to the WDM BU, passive BU and EU. Since the WDM BU is inherently complex due to the optics required to add and drop wavelengths, the failure rate is typically 5 times that of a repeater. The failure rates of the passive BU and EU are of the same order as a repeater. The cable and splices constitute the remaining submarine components. Providing the fibres and splices are proof tested, then it can be shown, using Weibull statistics, that their failure rates are negligible[4].

2.3.2 Reliability of Submarine Line Terminal Equipment

Throughout this evaluation of terminal equipment availability we have used the worst case performance figures. Where required, the transmission equipment is protected by a 1:N (N < 20) system where almost every unit is duplicated.

A common WDM MUX and protected amplifier feed the assembled WDM signal to the optical line. At the receive side a protected optical pre-amplifier feeds the line signal into the WDM DEMUX unit.

The figure below shows the reliability diagram of the path through the SLTE.

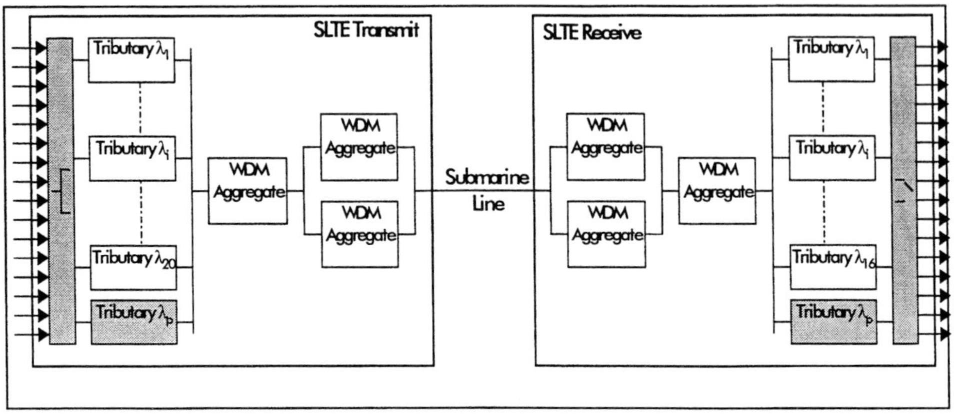

Figure 5a : Reliability diagram of the 10 Gbit/s SLTE

An outage would occur for any of the wavelengths if the unduplicated components failed, or if there were failures in both the working and standby paths. The outage can thus be calculated by summing the appropriate values.

	Unit
A	SLTE Tributary Tx Function
B	SLTE Duplicated Aggregate Tx
C	SLTE Tx non duplicated part
D	SLTE Tributary Rx Function
E	SLTE Duplicated Aggregate Rx
F	SLTE Rx non duplicated part
G	SLTE 20+1 Protection Coupling (Tx+Rx)
H	SLTE Protection Amplified Switch Matrix (Tx+Rx)

Table 2 : Transmission Equipment Reliability

The unavailability of an unidirectional path is given by the following formula :

Unprotected SLTE : $U_{L1} = A + B^2 + C + D + E^2 + F = 7.36\ 10^{-5}$, 38.7 min/year

20+1 protected SLTE : the SLTE transmit and receive tributary are switched together.

$U_{L2} = C_{20}^{2} (A + D)^2 + 20*(A + D)*H + B^2 + E^2 + C + F + G = 2.52\ 10^{-6}$, 1.33 min/year

2.3.3 Reliability of Power Feed Equipment

Just as a failure of PSU within a repeater can cause a link failure, so too can a failure of the PFE at the terminal stations. Again, redundancy is employed extensively to improve the link availability. There are several ways in which redundancy is employed over a link. Firstly, the power units themselves can be duplicated and secondly, if the system power requirement is not too large, double end feeding can be employed such that a failure of the PFE in one terminal station is compensated for by the PFE at the other end. This latter method, which uses dual stage redundancy, is illustrated in the reliability diagram of fig. 6.

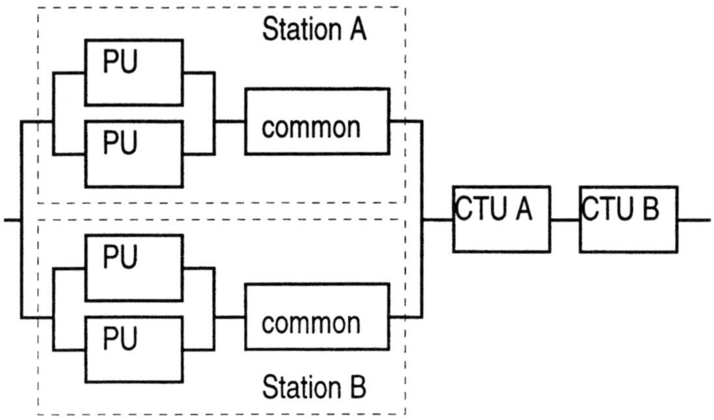

Figure 6. Reliability diagram of Power Feed functionality incorporating double end feeding. The diagram illustrates the 2-level redundancy scheme

At each station, there are two power units. Those elements of the PFE which are unduplicable are shown as 'common' .For double end feeding, the PFEs are effectively duplicated except for the Cable Terminating Units (CTUs) at each end. Both of these must be functioning for power to be delivered to the submarine plant. The unavailability of the configuration, assuming an adequate supply of spares is available at the terminal stations, is given by

$$\begin{aligned} \overline{A} &= 1 - \left\{1 - \left[1 - \left(1 - \overline{p}^2\right)c_o\right]^2\right\}c_t^2 \\ &\approx \left(\overline{p}^2 + \overline{c}_o\right)^2 + 2\overline{c}_t \end{aligned} \tag{5}$$

where p is the availability of a Power Unit

c_o is the availability of the PFE common

c_t is the availability of the Cable Terminating Units

and the bars represent unavailability, i.e., $a + \overline{a} = 1$.

For each element, the unavailability is given by

$$\overline{a} = \frac{\lambda}{\lambda + \mu} \approx \lambda t_r \tag{6}$$

where λ is the failure rate, $\mu = 1/t_r$ is the repair rate and t_r is the Mean Time To Repair (MTTR). Typical values for the element unavailabilities, assuming a 2 hour MTTR are $\overline{p} \approx 10^{-4}$, $\overline{c}_o \approx 4 \times 10^{-7}$ and $\overline{c}_t \approx 10^{-7}$. Substituting these values in (5) clearly shows that the unavailability of the power units and common elements have a negligible contribution to the PFE system unavailability. The combined failure rate of the two CTUs is 100FITs, so the probability of a power failure over the life of the system is only 2%.

2.3.4 Intrinsic Versus Extrinsic Failures

The reliability of the links discussed above are quantified in terms of the intrinsic reliability of their constituent components, or where available, by the analysis of field returns. The probability of extrinsic faults are far more difficult to quantify, yet they are frequently the most significant cause of failures. Submarine links are prone to failure due to trawler or anchor damage[5] and despite armouring and burying cables in shallow water areas, it is known that the number of submarine faults due to extrinsic failures typically exceeds those due to intrinsic failures by a factor of 4[6]. This immediately brings into question the viability of using different fibres within the same cable for restoration purposes and must be taken into account when designing network protection strategies. Terminal equipment is also susceptible to extrinsic failures due to human errors, but no data is available to quantify this.

In the next section, submarine networks are presented which are based upon the link reliability principles given above.

3. Architectures for a high availability submarine network.

3.1. overview

The evolution of the telecommunications networks from circuit oriented networks toward data-centric networks begins now to impact the core network architecture and therefore the submarine networks architecture.

While SDH (Synchronous Digital Hierarchy) network elements were, in the last few years, the main transmission equipment in the scope of the submarine networks architecture, we can now envisage to define architectures where IP routers (and ATM switches) and optical network elements play a more important role in the network, specially on the protection issues.

This doesn't mean that the SDH transfert mode will be removed from the transmission layers; direct mapping of data flows like IP packets or ATM cells onto WDM is not yet industrially mature but one can imagine to replace network protection mechanisms performed at the SDH level by mechanisms working at the optical level or even at the 'IP level' (layer 3).

This could lead to increase the number of possible submarine networks architectures where, beside the 'classical' combination of SDH and WDM technologies , we introduce optical routing and IP technologies in the submarine networks.

WDM technology allows a network operator to mix on its network all these new technologies by dedicating wavelength(s) to each different technology or to a combination of each technology.

For instance, some wavelengths could be dedicated to a typical circuit oriented network using SDH equipment for access and protection purpose while other wavelengths could carry IP packets and therefore use only IP routers or a combination of IP routers and optical switch for protection purpose.

The new paradigm of the submarine networks is illustrated in figure 7.

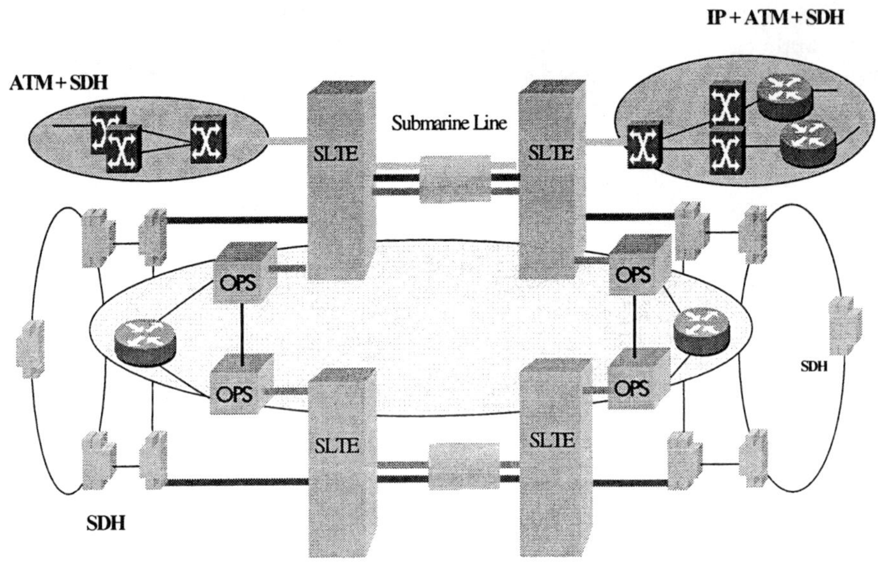

Fig. 7. submarine networks paradigm.

Each main approach to ensure network protection will now be detailed in the following chapters starting with SDH mechanisms , then presenting protection mechanisms operating at the optical level and finishing with the IP traffic handling.

3.2. SDH protection mechanisms

One of the key benefits of using SDH technology is the availability of powerful protection mechanisms.

Network protection mechanisms can be split into two main families, section protection mechanisms (operating at STM-1 and higher level) and path protection mechanisms (operating at VC-4 and lower levels). They are self-healing mechanisms: pieces of equipment do not need network management activation to re-route the traffic, by contrast with restoration.

3.2.1 Section protection

Section protection mechanisms aim at protecting multiplex sections connecting SDH ports at all bit-rates (STM-1 to STM-64) from various kinds of failures whether from components failures or from extrinsic activity such as cable cuts.

They are applicable to all SDH network topologies and have different characteristics and levels of complexity according to the topology supporting the protection scheme.

One of the main issues is the use of a protocol between network elements to perform coordinated switching actions when network self-healing mechanism has to be engaged. This protocol is transported by dedicated bytes in the overhead of the protection multiplex section.

3.2.1.1 Linear Trail protection

So called MS Linear Trail protection mode in SDH world and Line switching in SONET world applies to the multiplex section between two adjacent nodes. This is the basic application of section protection and two modes are possible: one plus one and one for one.

1+1 protection mode

The principle is to duplicate the signal on both transmission directions on the normal (working) and on the protection sections. The best signal is chosen at the end-node. When a failure occurs, all traffic between the two nodes is carried out by the protection resource. See figure 8.

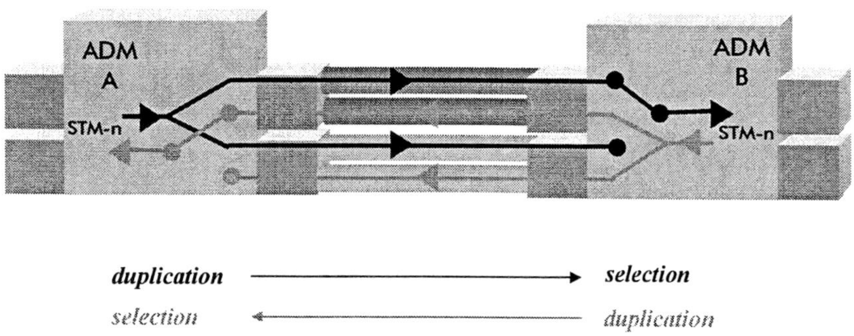

Fig 8. Linear trail protection; 1+1 mode.

There is a need for protocol only when the mechanism has to be implemented in a bidirectional way that aims at having both transmission directions on the same resource (working or protection) even if the failure occur in one direction only.

This protection can be either revertive or non-revertive. In the first case the signal goes back automatically to the working channels after failure repair while in the second case it remains on the protection channel even if normal resource is available again and until an operator requires the switch through the management system.

For maintenance purpose and thanks to the management system, the operator is able to force the switch from one resource to the other.

1:1 protection mode

An improvement of the former approach is achieved by using the protection resource only when necessary instead of permanently duplicating the signal on both normal and protection resources.

Pre-emptible extra-traffic could therefore be transported on the protection channel and revertive mode could be used for better availability of extra-traffic.

Protocol is needed in order to make both sides of the section switching.

See figure 9.

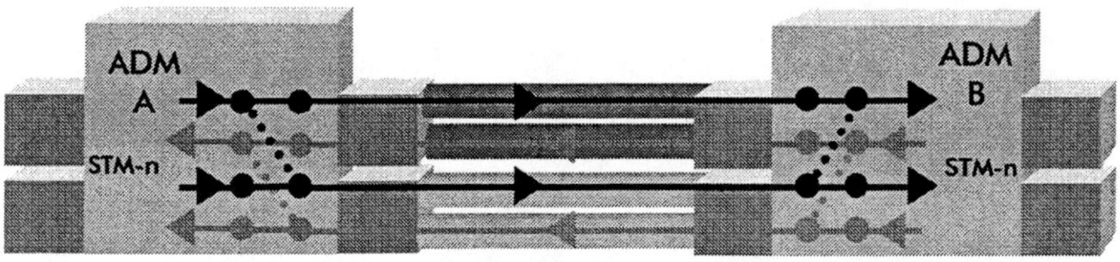

High priority traffic

High priority traffic

Extra traffic

Extra traffic

Fig 9. Linear trail protection: 1:1 mode

3.2.1.2 Section protection with ring topologies

Although it remains possible to use the classic Linear Trail protection one would rather use benefits of a ring topology by using MS-SPRing protection mode or its Transoceanic application. MS-SPRing is defined as two or four fibre Multiplex Section Shared Protection Ring in ITU and two or four fibre bidirectional Line Switching Ring in SONET world.

Those protection modes can be seen as an extension of the Linear Trail protection 1:1 mode. They use the same basic mechanisms (fault detection at multiplex section layer and protocol transported by dedicated bytes in the multiplex section overhead of the SDH frame) with some extensions defined hereafter.

Protection mechanisms are described specifically for four fibre rings in the document but remain true with two fibre rings.

As for Linear 1:1 protection mode resources at the multiplex section are physically duplicated (optical ports and fibre pairs) forming two stacked independent rings, one using the normal (or working) fiber pair and one the protection fibre pair all around the ring.

As for Linear 1:1 mode, preemptible extra-traffic is available on the protection fibre pair.

Two-fibre MS-SPRING mode is the same but instead of having two fibre pairs along the ring, only one is available and consequently the multiplex section payload is split into two parts; half of the AU-4s supports the normal traffic while the other half is reserved for protection purpose.

Thanks to the ring topology two kinds of switching can occur: span switching and ring switching.

Span switching is used when the multiplex section carrying the normal traffic is faulty (for instance in case of a STM-n optical interface defect). In that case all the traffic is switched to the protection multiplex section of the same span. This is illustrated in figure 10 :

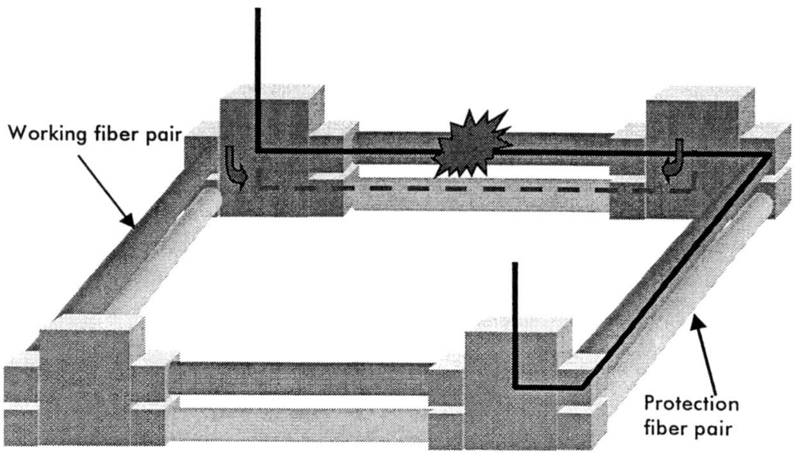

Fig 10: Section protection in ring configuration: span switch

Ring switching is used when both working and protection sections of the same span are not available (owing to a cable cut for instance). In such a case all the traffic, at both nodes adjacent to the failure, is looped back to the opposite direction of the failure. This is illustrated in figure 11:

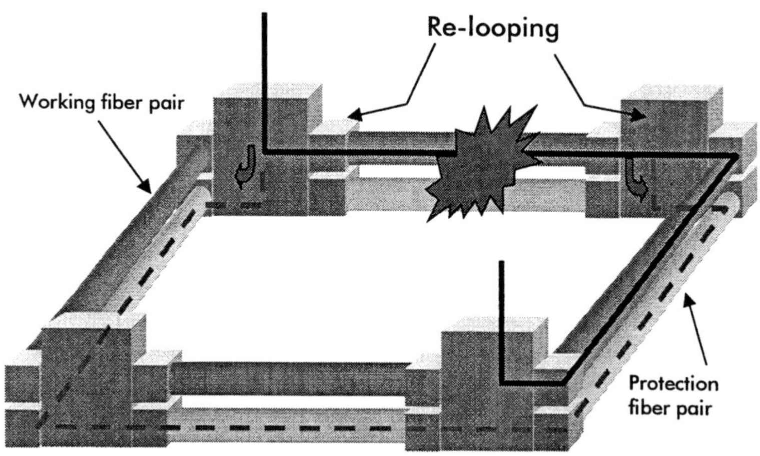

Fig 11: Section protection in ring configuration; ring switch

Switching time is 100ms plus protocol transmission delay.

One major drawback of this mechanism is that traffic looping at the nodes adjacent to the failure could lead to triple ocean crossing for the signal and therefore introduce unacceptable transmission delays.

An improvement of the four-fibre MS-SPRING protection mode has been defined in Annex A of ITU G.841 recommendation and is known as NPE mode.

The principle is to avoid loopbacks at the node adjacent to the failure by directly re-routing the signal between originating and terminating nodes. This is done on path (AU-4) basis and requires a little more time to protect the network against a failure (300ms for switching the overall traffic affected by a failure).

Compared to MS-SPRing, this protection mode requires less protection resource in the case of a ring switch, preserving by this way the extra-traffic and therefore providing a better usage of the bandwidth.

Span switching remains available within 50ms.

An example of a ring switch is given in figure 12.

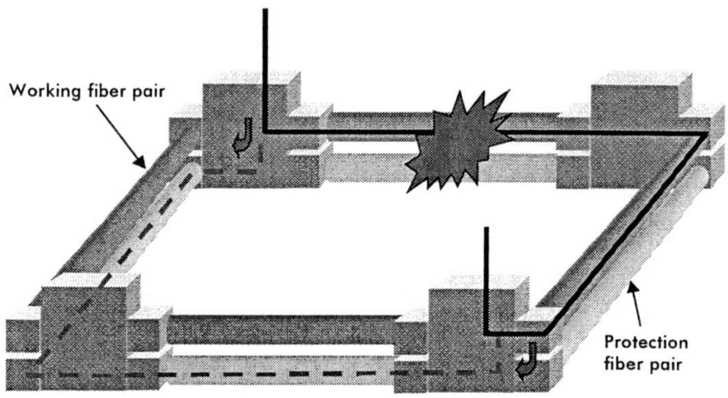

Fig 12. Section protection in ring configuration. Transoceanic application.

Pre-emptible extra-traffic is available with NPE protection mode and the management of this type of traffic in the ALCATEL implementation guarantees a hit of maximum 300 ms when corresponding bandwidth is not needed for protection purpose.

3.2.2. Path protection

This protection mode, called SNC-P in ITU standards and Path Switching in SONET applications, operates at VC-4 and lower levels (VC-3 and VC-12) and aims at using two different routes to link the termination nodes of a signal. It is applicable to every kind of network topology from linear to ring and meshed networks.

The principle is to duplicate the signal at the originating node on both directions of the ring and to select the best one at the terminating node. Both transmission directions operate independently.

Because no protocol is required between nodes this mechanism is very robust and fast (switching time is up to 50ms).

Figure 13 illustrates the principle:

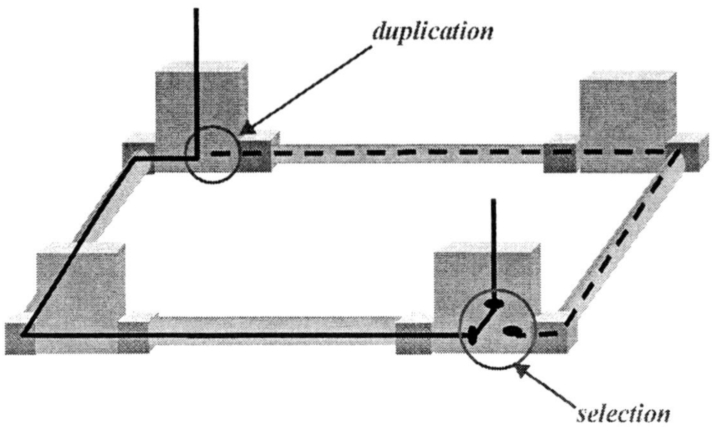

Fig 13: Path protection

Each of the described protection mechanism has advantages and drawbacks. The choice among them has to be made according mainly to the traffic distribution and the customer preferences.

A comparison, in terms of advantages and drawbacks, of Path protection and NPE protection mode is given in table 3:

	Advantages	Drawbacks comparing to each other
Section protection in ring configuration (NPE)	- Bandwidth usage. - Multiple (span) failures in a ring - Extra-traffic available	- switching time (ring switch: 300ms)
Path protection	- Simple, robust mechanism - Switching time (50ms) - Reduced number of hits in non-revertive mode.	- bandwidth usage (path duplication)

Table 3. Comparison of NPE and path protection mode.

Thanks to a powerful network management system ALCATEL systems upgrade from one protection mode to the other remains possible with the least number of hits as possible (for instance, two hits of 50ms for path protection to NPE migration) .

It is even possible to mix protection types in a single network (e.g. on a per couple of wavelength basis) hence allowing a telecom operator to adapt protection mechanisms to traffic patterns.

3.2.3. Protection of interconnected rings

They are many reasons for the SDH topology built over a submarine cable to provide multiple interconnected rings. The first one is obviously because the physical cable itself builds true interconnected rings.

Another reason could be the protection issue. It could be interesting to divide up long distance rings with a large amount of landing points in multiple rings, and leading to tolerance to multiple failures in the overall network.

Although single node interconnection remains possible, the network reliability can be enhanced using dual nodes interconnection and so called "Drop and Continue" feature avoiding the weakpoint of the interconnection node.

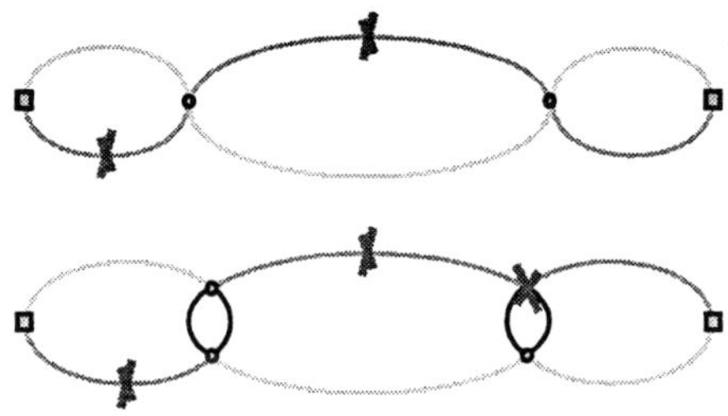

Fig 14: multiple rings interconnection

A typical application of protected, interconnected rings is the GEMINI project which is an interconnection of three rings, one for the submarine part and two for the terrestrial part of the network.

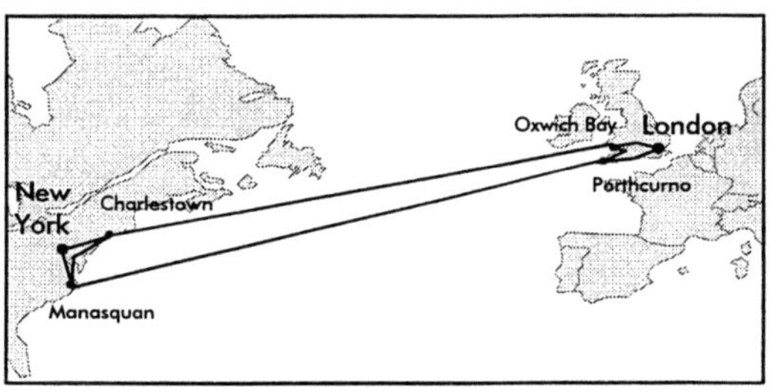

Fig 15: Dual node interconnection. GEMINI project

The Drop & Continue feature is applicable to rings configured in both path protection and NPE (4f MS-SPRING) modes.

The principle is to duplicate in an interconnection node the incoming signal from one side of the ring to the second interconnection node of the same ring and to the other ring.

In SNC-P rings, both interconnection nodes perform Drop & Continue allowing each interconnection node to select the best signal coming from one side or from the other side of the ring before sending it to the ring downstream. In this way the signal enters by two different points into the next ring and arrives from both sides to the terminating node which performs the selection of the best signal.

Figure 16 shows an example of a double failure in two rings providing SNC-P with Drop & Continue feature.

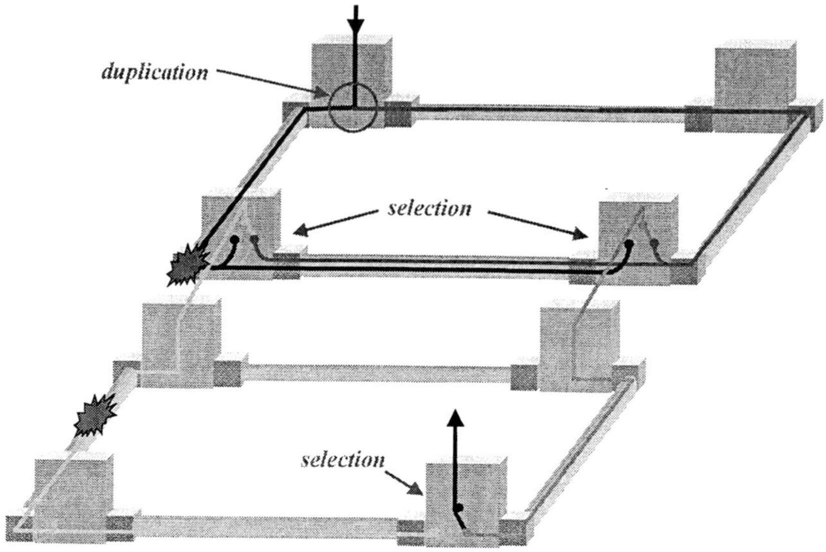

Fig 16: Dual node interconnection. SNC-P protection mode

With NPE protection mode the Drop & Continue feature as well as the path selection feature are only performed at the so called primary node. The other one (secondary node) just drops the signal towards the downstream ring. Primary and secondary nodes have to be chosen on a path (AU-4) basis on each interconnected ring and need not be adjacent. Primary and secondary nodes remain the same for each path after a NPE ring switch.

Figure 17 is an example of two dual node interconnected rings protected in NPE mode.

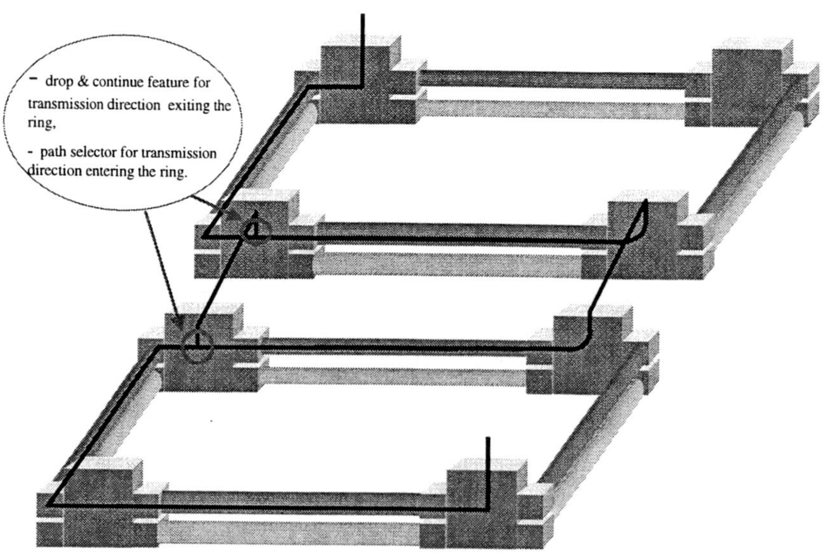

Fig 17: Dual node interconnection: NPE protection mode.

Dual node interconnection with Drop and Continue feature of two rings protected with different mode is possible; for instance one ring in SNC-P and the other one in NPE mode.

3.3. Network protection at the optical level

Although the SDH layer brings powerful protection mechanisms new possibilities are now emerging.

Instead of performing the protection switching at the electrical AU-4 level (155 Mbit/s) of the SDH, one possibility is to operate at higher bit-rate which is more accurate to the failures transmission networks want to protect against.

SDH network elements dedicated to the network protection could therefore be replaced by network elements operating at the optical level in the submarine stations.

This could lead to a dramatic saving of power and footprint in the already over-crowed submarine stations .

SDH network elements remain at the access of the network and the network protection is achieved with a combination of both SDH mechanism and optical level mechanism.

In addition to the considerable saving of space and power consumption, the main advantages of using optical technology in the submarine networks architecture are the transparency to the bit-rate and the possibility to have different types of equipment at the access: classical SDH Add drop multiplexer as well as IP routers and ATM switches which, for the next few years, will feature SDH interfaces at their high speed interfaces can be supported at the access of the network.

The very first target network topology that could benefit from an optical switching equipment is the so called "city to city" architecture where a submarine ring is inserted between two terrestrial backhaul networks forming by this way a typical trans-oceanic network.

This is illustrated in figure 18:

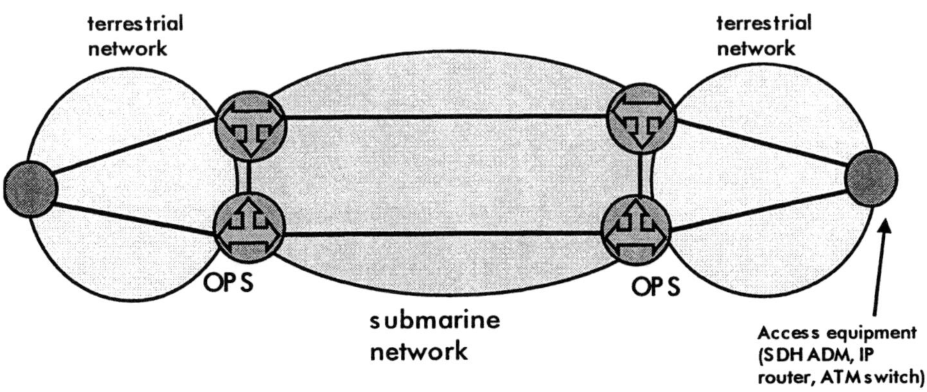

Fig 18. The OPS network element in the submarine network context.

The optical path Switch (OPS) takes place at each interconnection points (actually in the submarine stations) between terrestrial backhauls and submarine networks and replaces SDH network elements that were used only for protection purpose.

This piece of equipment operates on a wavelength basis and can be put in place of the equivalent SDH network element when the traffic carried by the wavelength has not to be dropped in the submarine station. This case is more and more common specially for operators entering the market.

The access equipment, at the edge of the terrestrial backhauls, can be either SDH equipment or IP routers and ATM switches provided that their physical interfaces is an SDH one.

In the case of an SDH Add Drop Multiplexer the detailed architecture is the following one:

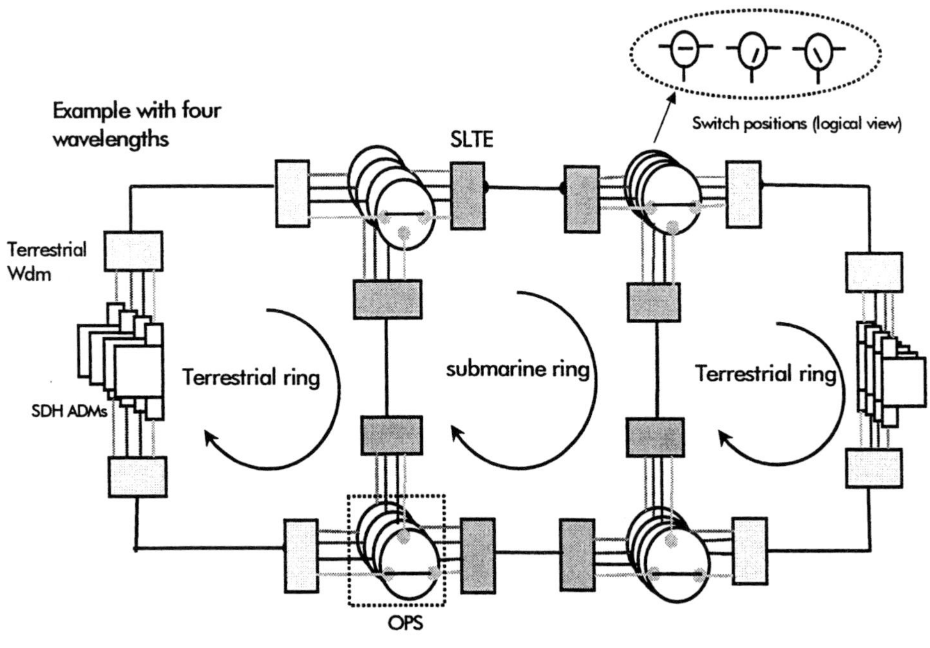

Fig 19: overall architecture.

Note that such an architecture doesn't lead to double the capacity of the inter-link between submarine stations on each side on the ocean. This is a key benefit comparing to classical SDH "drop & continue" mechanism which requires the duplication of this inter-link.

This is an important issue for new operators entering the market and who hold, in the first step of their deployment, a terrestrial network with a relative poor connectivity.

Protection of such a network is achieved with a combination of SDH mechanism (SNC-P at ETSI, Path Protection in the ANSI world) operating at AU-4 level and optical switching (OPS) at the optical channel (or wavelength) level.

This is illustrated in figure 20.

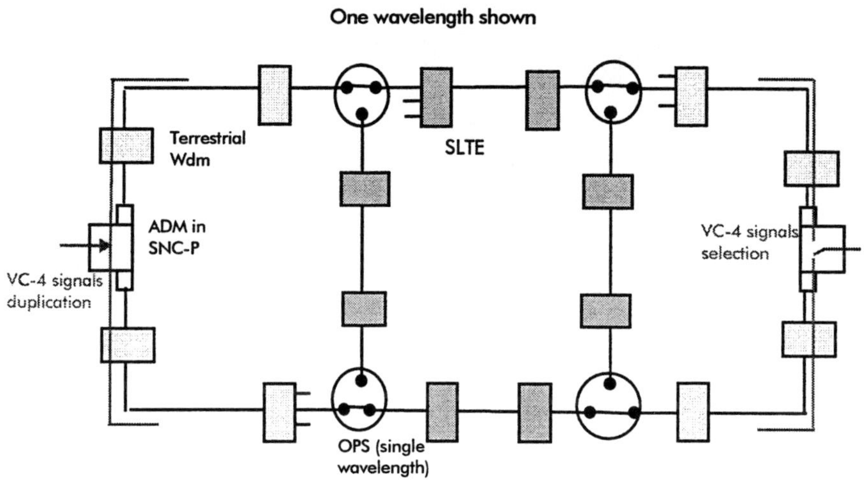

Fig 20. network architecture: failure free

In the case of a single failure , or multiple failures in the same "side" of the network, only the SDH mechanism is involved and no optical switching is necessary to recover the traffic. Switch is performed, if needed (the selector already may be in the good position) , at VC-4 level and traffic recovery time is therefore less than 50 ms.

This is illustrated in figure 21.

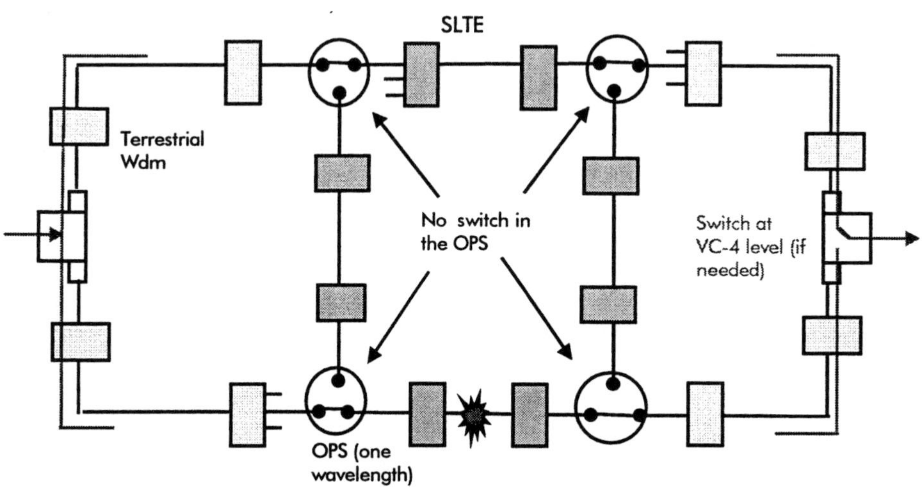

Fig 21. network architecture: single failure

If the first level of protection provided by SDH mechanism (or re-routing in IP routers) is not able to recover the traffic, switching at the optical level can be additionally experienced in order to recover at least one optical path.

The maximum traffic recovery time is therefore the one of the OPS switching time, which is 300 ms, plus the ADM switching time (if needed) which is 50ms.

This is illustrated in figure 22.

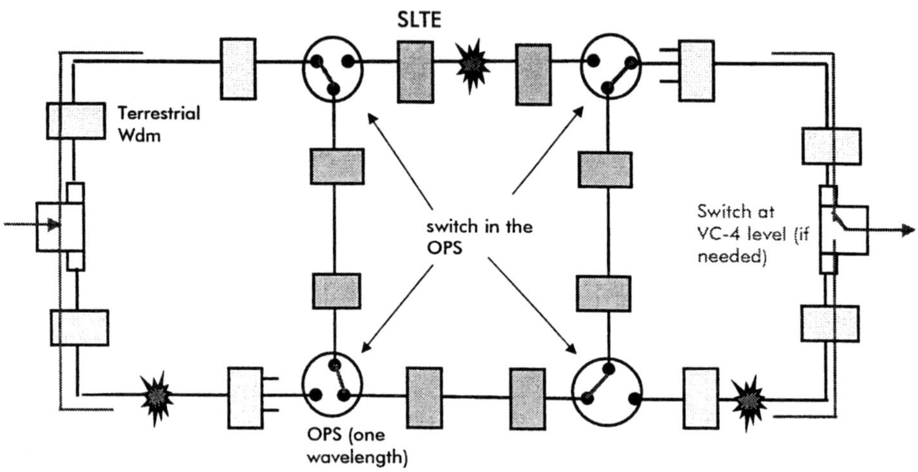

Fig 22. network architecture: three simultaneous failures

As seen in the previous example, when the traffic cannot be recovered with the only SDH protection mechanism because no optical path are available due to the failures positions in the network, the OPS switches must operate in a coordinate manner in order to recover one optical path within the fastest possible timeframe.

The only way to achieve the 300ms target time is to implement a self-healing mechanism based on a protocol that allow the OPS network elements present in the network to share all together the information on the status of each segment of the whole network they got locally.

This protocol is carried by the submarine part of the network in the available bytes of the FEC frame overhead.

Thanks to a direct link with each co-located terminal (terrestrial or submarine), the OPS network elements gets the status from all segments they are directly connected to.

Each OPS can individually be controlled by the management system in order to allow an operator to set the switches in the desired position.

The principle is illustrated in figure 23:

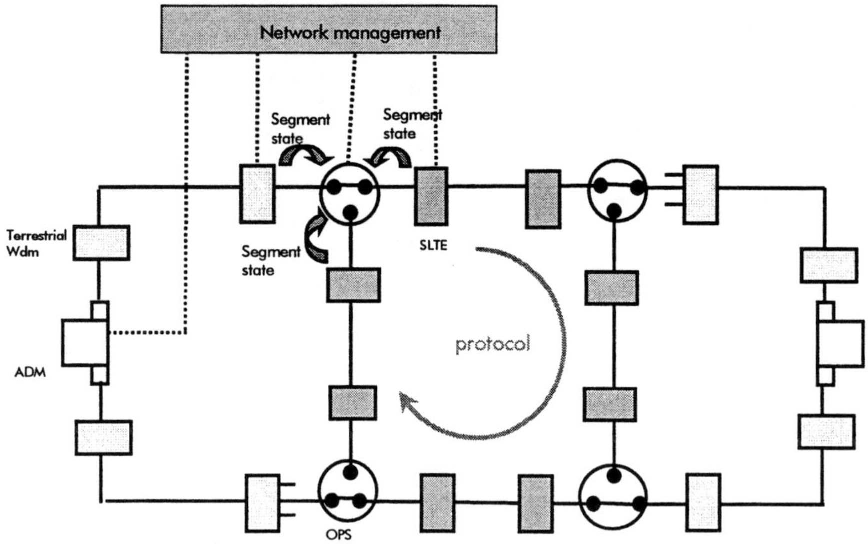

Fig. 23. overview of network operation principles

3.4 IP traffic handling

For several years, the main characteristic of the international traffic has been the dramatic increase of the data traffic owing to the expansion of applications such as the World Wide Web or Virtual Private Networks. And today, it is widely agreed that this change will continue in the next years leading to a new paradigm where data, and more specifically IP packets, will be the major part of the overall traffic.

As submarine systems are essential to interconnect the national backbone networks (for example, transatlantic submarine systems have to support the huge and ever increasing Internet traffic between Europe and US), it is currently intended to optimise the design of these systems in order to offer the most efficient end to end service for the data traffic.

Since new equipment, such as IP routers, are going to be integrated in the submarine networks, it is possible to extend the classical set of protection mechanisms (such as the well-known SDH protection mechanisms) used to offer a very good availability of the network by including the specific mechanisms provided by these new equipment.

3.4.1. Integration of IP routers in submarine networks

The main interfaces of the current generation of high-capacity IP routers (also called gigabit or terabit routers) are STM-4 and STM-16 SONET/SDH interfaces. As the current generation of submarine systems are based on 10 Gb/s WDM transmission equipment, it is necessary to use an ADM (Add/Drop Multiplexer) between the router and the submarine line terminal equipment (see figure 24). The ADM allows to multiplex several IP traffics (STM-4 or STM-16 traffic) in a STM-64 SONET/SDH frame which will be transported on the submarine link (refer to the figure below). In order to use efficiently the SONET/SDH payload, interfaces supporting the contiguous concatenation feature are used on both sides of the connection between the router and the ADM.

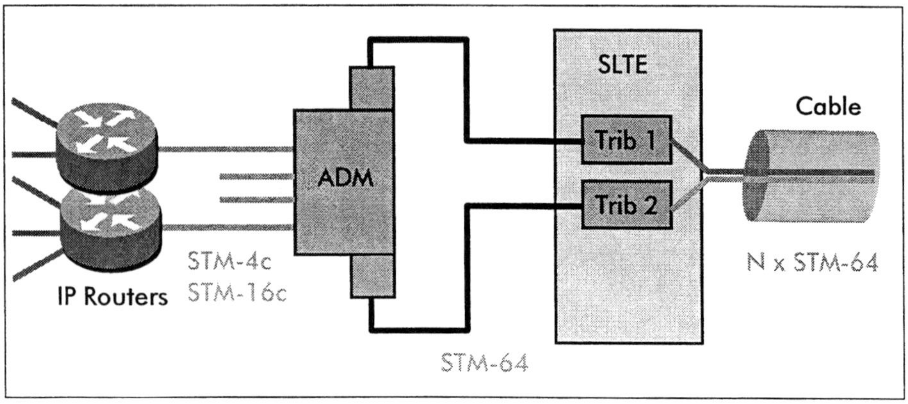

Figure 24. Connection of a router via an ADM

Because of the fast growth of the capacity transported in the backbone networks, STM-64 SONET/SDH interfaces will be provided soon with the next generation of high-capacity IP routers. A new solution will then be available to integrate these equipment in the submarine networks : it will be possible to connect directly the IP router to the tributaries of the 10 Gb/s submarine line terminal equipment (SLTE). This solution is described in figure 25.

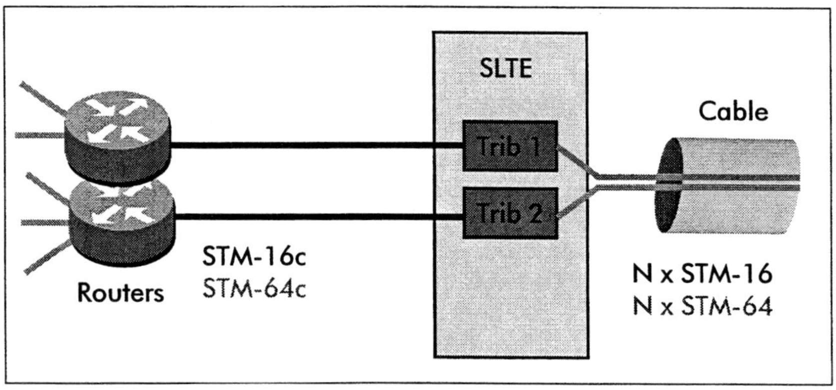

Figure 25. Direct connection of an IP router

3.4.2. Impact on the network availability

Since the architecture of the submarine networks is going to evolve in order to integrate new equipment such as IP routers, the set of "network mechanisms" used to guarantee a high availability of a network is going to be extended in order to integrate the mechanisms provided by these new equipment.

Concerning IP routers, three main mechanisms are available to improve the availability of a network.

The first mechanism is called "load sharing". With this mechanism, an IP router is able to share the traffic between several routes defined between two points. This mechanism can be applied when there are several parallel links between two adjacent routers or when there are several available paths between two distant routers (refer to figure 26). The result is the multiplication of the number of available routes between two points of the network. This set of available paths is managed automatically (owing to the IP routing protocols) and, in case of failure of one route, all the remaining routes are able to maintain the connectivity.

The "load sharing" mechanism is supported by the main routing protocols currently used in the IP world : OSPF (Open Shortest Path First) and BGP (Border Gateway Protocol).

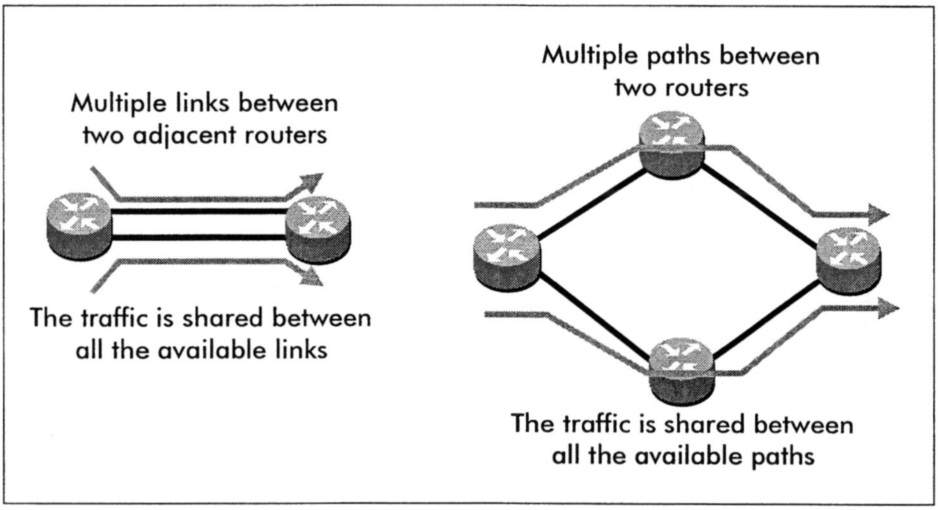

Figure 26. Load sharing between two routers

The second mechanism which can be used to offer a good availability of the network is the classical SONET/SDH APS (Automatic Protection Switching) mechanism. The high-speed interfaces provided on IP routers are SONET/SDH interfaces and most of them allow to support this well-known and fast protection mechanism. The APS is configured in a 1+1 mode (duplication of the traffic) and guarantees that the traffic is recovered within 50 ms after the detection of a failure.

The APS can be used with both solutions defined previously to integrate an IP router in a submarine network.

When the IP router is connected to an ADM, the connection between the router and the ADM can be protected by the APS.

When the IP router is directly connected to the SLTE, the APS can be used to protect the whole path between two distant routers.

Both cases are illustrated in figure 27.

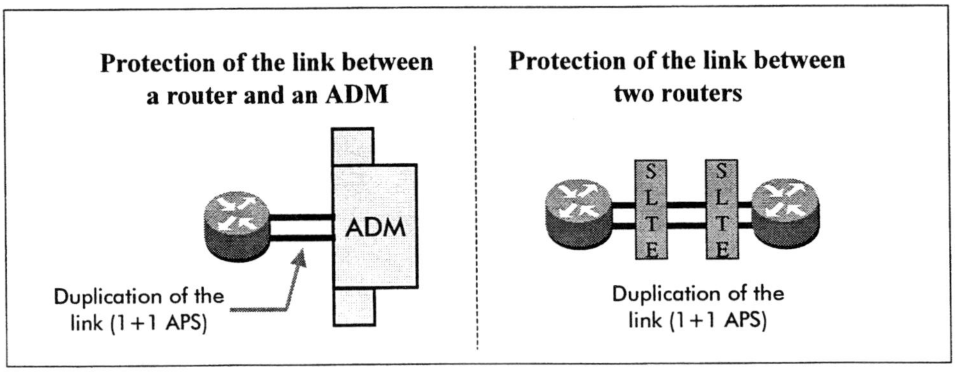

Figure 27. APS protection mechanism with routers

The third mechanism which allows to improve the availability of the network is the routing protocol. Indeed, when a path between two routers becomes unavailable after a failure, the routing protocol is able to discover a new path automatically (when there is still at least one available path between the two routers).

This protection of the network may be quite slow (especially compared to the fast SDH APS) since it may take some time before detecting a failure (the detection of some failures is based on the utilization of timers) and then it may also take some time to recalculate the routing tables. But the strengths of this mechanism are that it is automatic and very efficient in terms of utilization of the bandwidth.

All the routing protocols currently used in the IP world (OSPF and BGP) provide this automatic recovery of the traffic.

This mechanism is illustrated on figure 28.

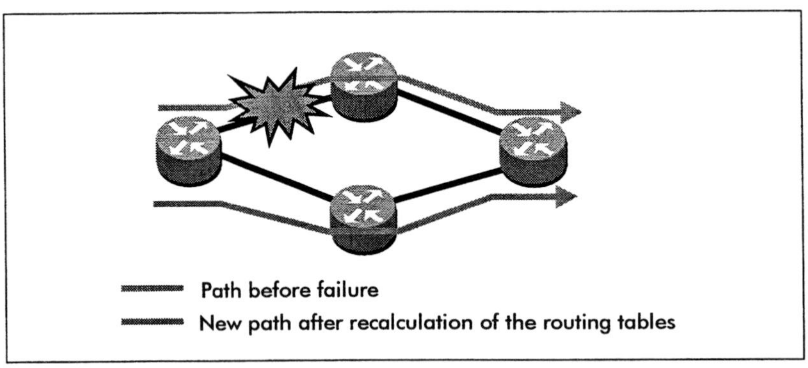

Figure 28. Automatic discovery of a new path in case of failure

3.4.3. Examples of IP-based configurations

In order to illustrate how the IP mechanisms described in the previous paragraph can be used to improve the availability of a network, the description of some new architectures for the global « city to city » network seems to be a good example.

A « city to city » network allows to interconnect two terrestrial stations which are separated by a sea or an ocean and located far from the coast (refer to figure 29). The network comprises three parts : one submarine network connecting two terrestrial backhauls. Since there is no need to drop some traffic in the submarine stations, the IP routers are just located in the terrestrial stations (cities) in order to optimise the cost of the overall solution. In the submarine stations, there are only optical equipment : WDM transmission terminals, optical switches...

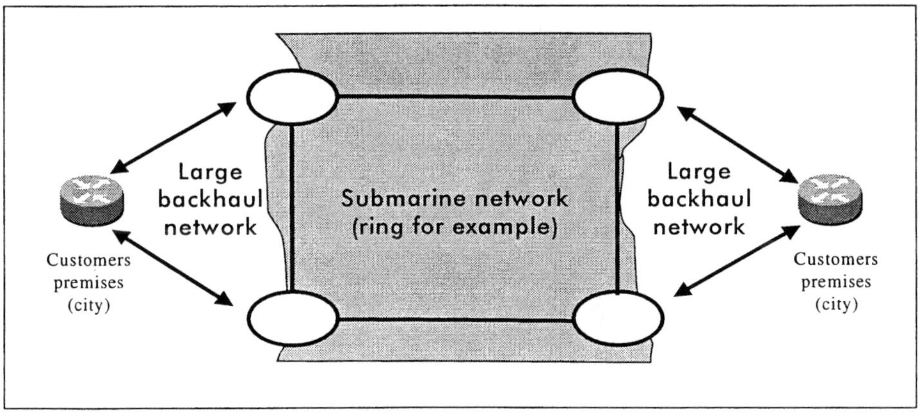

Figure 29. Description of a "city to city" network

A first available configuration for a « city to city » network is to build a submarine ring (4 stations connected by two transoceanic cables and two inter-station cables) and two connections between the submarine stations and the customers premises (in the city).

Owing to the properties of the routers, it is possible to connect the routers by several parallel logical links (one IP hop for each link) using various physical paths (this configuration is also called « route diversity »). The whole traffic is shared between all the available links. In case of failure (one submarine cable break for example), there are still some available paths in order to transport the whole traffic. The connectivity is maintained. The possible consequence of the failure may be a degradation of the Quality of Service.

This solution is described in figure 30.

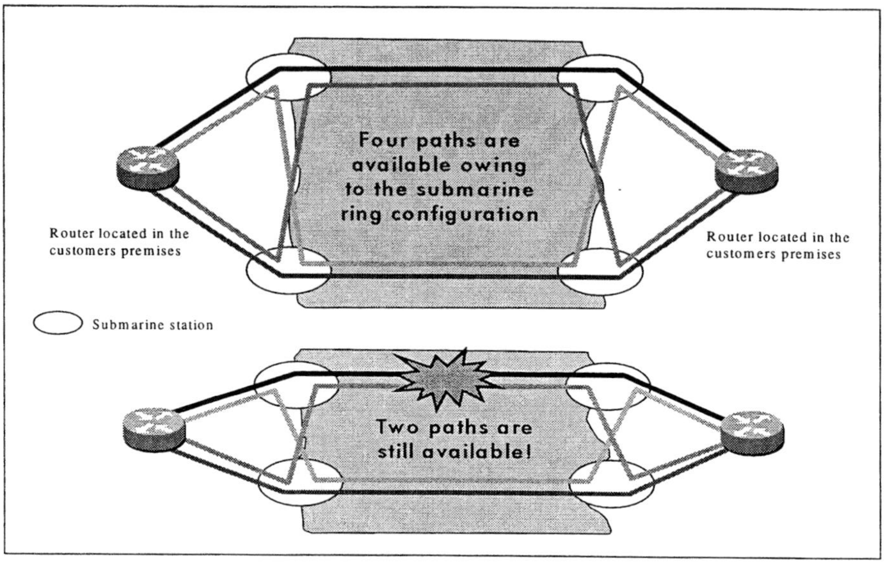

Figure 30. First configuration : route diversity

A second available configuration for a « city to city » network is to use three independent submarine cables (connecting six submarine stations) and to connect the customers premises (in the city) to each submarine station.

The routers are able to optimise the utilisation of all the available links, especially with the « load sharing » mechanism. And in case of failures (for example, one terrestrial link break during the repair of one submarine cable), the routing protocol is still able to find an available path to transport the whole traffic.

This solution is described in figure 31.

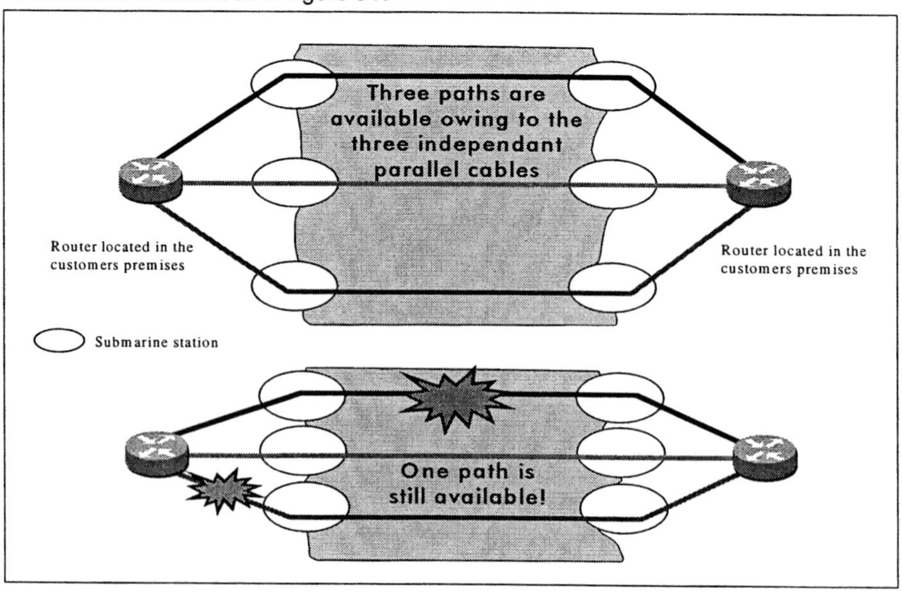

Figure 31. Second configuration : three cables

4. Conclusion.

In this article we have seen that the redundancy of the power feed equipment together with the duplication of the PFE at the submarine line end-points provide a well secured submerged plant power feed.

The repeaters, with their amplifiers redundancy, have a high level of availability.

The probability to get an equipment failure (a transmitter or receiver failure for instance) in the submarine terminals during the submarine cable repair time is not negligible. For that reason, the SLTE implements a 1:n protection mechanism that increases dramatically its availability.

Together with those previous equipment protection schemes, powerful network protection mechanisms are available to provide an unrivalled submarine networks availability.

Beside the classical SDH protection schemes, the OPS equipment is one of the most interesting protection dedicated equipment that can match the requirements, in terms of protection features as well as in terms of cost, footprint and power savings, of most operators acting in both classical telephony networks and data oriented networks.

References

[1] Vareille G., Pitel F., Uhel R., Bassier G., Collet J.P., Bourret G. and Marcerou J.F., "340Gb/s Straight Line Transmission Over 6380km with Full System Implementation Assessment", *Opt. Fib. Conf.*, San Diego, CA, paper PD18, Feb. 1999.

[2] Chesnoy J., "Global Undersea Networks: technical developments", *Opt. Fib. Conf.*, San Diego, CA, paper TuD1-1, Feb. 1999.

[3] Desbruslais S. R., Kendrick S., Lemaire V., Hazell N., "Superior Reliability of 980nm co- and 1480nm counter-pumped amplifier for long-haul submarine systems", *proc. ECOC'99*, Nice, France, paper MoD1, Sept 1999.

[4] Weideshorn S. M., and Evans A G., "Proof testing of Ceramic Materials – An Analytical basis for failure prediction", *Imt. J Fracture*, Vol. 10, no. 3 1974 pp379-392.

[5] Shapiro S., Murray J. G., Gleason R. F., Barnes S. R., Eales B. A., Woodward P. R., "Threats to Submarine Cables", *proc. Suboptic'97*, San Francisco, CA, pp742-749, May 1997.

[6] Desbruslais S. R., "Meeting the need for increasing reliability in the field of Submarine Systems", *proc. Suboptic'97*, San Francisco, CA, pp109-114, May 1997.

Reducing channel degradations in WDM systems and networks

Alan Eli Willner
Room EEB 538, Dept. of Electrical Engineering - Systems
University of Southern California
Los Angeles, California 90089-2565
willner@solar.usc.edu

ABSTRACT

Future higher-performance systems will be more complex than today's systems since the wavelength domain will be used to help route signals through different static or reconfigurable network paths. In these next-generation systems, several parameters may vary which would have deleterious effects in a WDM environment, including: variable insertion losses, channel addition and deletion (i.e., add/drop multiplexing), unstable laser power, non-uniform EDFA gain, non-uniform accumulation of dispersion and nonlinearities, and polarization mode dispersion that varies with fiber temperature. In order to ensure robust system operation, we discuss in this paper various dynamic schemes for compensating damaging effects so that these complex systems maintain high performance.

Keywords: optical communications, wavelength-division multiplexing, photonic networks, add/drop multiplexers

1. INTRODUCTION

The general research thrust that will be addressed in this paper is the dynamic compensation of channel-degrading effects in reconfigurable wavelength-division-multiplexed (WDM) systems [1].

Systems with simple designs are being deployed which can serve a critical need for upgrading overall capacity. However, future higher-performance systems will be more complex since the wavelength domain will be used to help route signals through different static or reconfigurable network paths. In theses next-generation systems, several parameters may vary which would have deleterious effects in a WDM environment, including: variable insertion losses, channel addition and deletion (i.e., add/drop multiplexing), unstable laser power, non-uniform Erbium-doped fiber amplifier (EDFA) gain, and non-uniform accumulation of dispersion and nonlinearities. In order to ensure robust system operation, it is imperative to provide dynamic schemes using feedback for compensating damaging effects so that these complex systems maintain high performance under many different conditions.

Some of the specific projects that we will highlight in this paper include:

(a) *Dynamic Dispersion Compensation in 10-Gbit/s Optical Systems using a Novel Tunable Nonlinearly-Chirped Fiber Bragg Grating*: The accumulation of dispersion for a given channel may significantly vary for reconfigurable systems in which a channel may originate locally or far away, requiring that the dispersion compensation techniques actively track the accumulated dispersion. We use a novel nonlinearly-chirped FBG for ~10-ms dynamic dispersion compensation for a 10-Gbit/s system. By using a single mechanical stretcher, we can actively tune the resonant wavelength of the FBG and dramatically change the generated chirp from 300 to 1000 ps/nm [2]. Moreover, we use a sampled nonlinearly-chirped FBG to provide dynamic dispersion compensation to many WDM channels simultaneously [3].

(b) Dynamic Power Equalization of WDM Channels: Robust systems require that the power differential among many WDM channels remains small. We use a wavelength demultiplexer and a parallel array of polarization-insensitive acousto-optic modulators to provide a fast (~0.1 μs) loss element for each individual WDM channel. All channel-power differentials can be independently reduced from >10 dB to <1 dB [4].

(c) *Compensation Of PMD Using A Grating Written Into A High-Birefringence Fiber*: Polarization mode dispersion (PMD) has recently emerged as one of the next critical hurdles in achieving high-performance optical transmission systems and networks. This situation has occurred since many fiber segments that were deployed throughout the 1980's have a PMD parameter that is as much as 100-1000 times greater than the more-optimized fiber deployed today. We use a tunable nonlinearly-chirped fiber Bragg grating (FBG) [5] written into a high-birefringence photosensitive fiber to compensate for varying amounts of polarization mode dispersion. We use this novel approach to compensate for an induced differential group delay and show good performance at 10 Gb/s signal with 175 ps of adjustability.

(d) *Reduction of Coherent Crosstalk in WDM Add/Drop Multiplexing (ADM) Nodes by Bit Pattern Misalignment*: Add/drop multiplexers are key subsystems in WDM networks. Coherent crosstalk has been shown to be a serious limiting factor in scaleable WDM networks. We analyze a method of reducing the effects of coherent crosstalk in ADMs by purposely introducing fiber delays on the order of bit times between different crosstalk paths [6].

2. DYNAMIC DISPERSION COMPENSATION USING A TUNABLE NONLINEARLY-CHIRPED FIBER BRAGG GRATING

Two of the most troublesome issues in ≥10-Gb/s optical fiber transmission are the accumulation of dispersion and nonlinearities. Dispersion-managed single-channel or wavelength-division-multiplexing systems have been demonstrated in which the dispersion accumulated in one span of fiber is negated by an optical element that provides the opposite dispersion. The nonlinearities are controlled by always maintaining some dispersion at any given point along the fiber span. The accumulation of dispersion in high-speed optical transmission systems can be compensated either with dispersion compensating fiber (DCF) [7] or a linearly-chirped fiber Bragg grating [8]. However, the accumulated dispersion for a data channel may vary in time for: (i) dynamically reconfigurable optical networks in which a given channel may originate locally or far away (see Fig. 1), and (ii) transmission systems having changing operating conditions (i.e., laser or modulator chirp). For such dynamic systems, dispersion compensation must be adjustable in order to dynamically track the accumulated dispersion. For the DCF, there is no method at present for tuning *in situ* its dispersion. For the linearly-chirped fiber Bragg grating, simply stretching the grating only produces a shift in the resonant wavelength range of the reflected band but does nothing to change the induced dispersion. A recent report used 21 separate stretching segments to asymmetrically stretch a uniform fiber Bragg grating to change the induced dispersion [9].

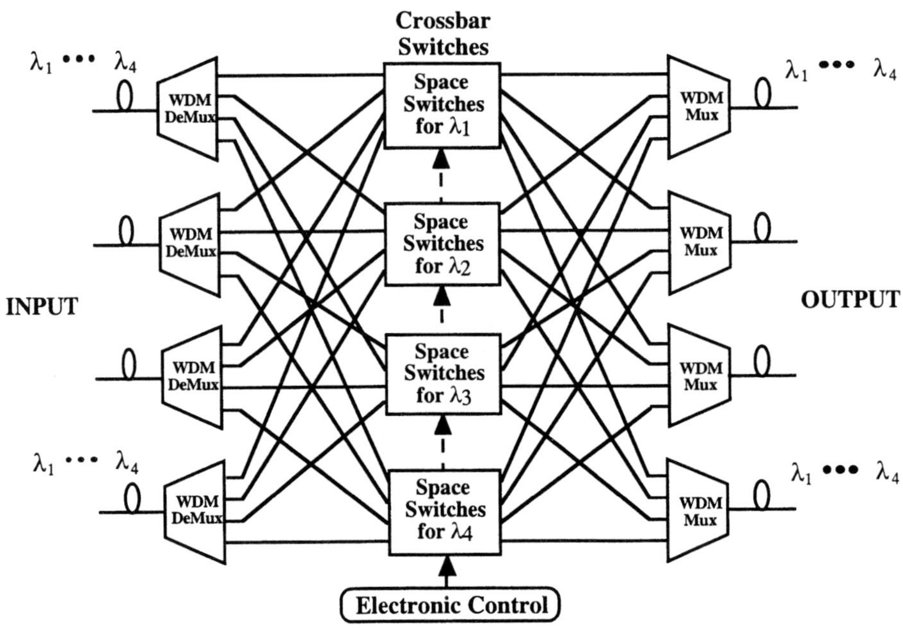

Fig. 1. A typical optical cross-connect for which a given channel at any output may have originated from any input.

Here, we demonstrate the tuning characteristics of a novel *nonlinearly-chirped* fiber Bragg grating in which a *single* piezoelectric transducer is used to tune the dispersion, and (ii) the first system demonstration of dynamic dispersion compensation for a 10-Gb/s optical channel. The time delay induced by this nonlinearly-chirped fiber Bragg grating changes with wavelength in a nonlinear fashion. We tune the resonant wavelength of the fiber Bragg grating and change the generated dispersion continuously from 300 to 1000 ps/nm for a given channel. By switching between two different lengths of single-mode fiber (SMF), the power penalty is reduced to <1 dB both for a link containing high dispersion (104 km, initial penalty = 3.5 dB) and for one containing low dispersion (50 km, initial penalty = 1.5 dB). Compensation is achieved in less than 2 ms, which is appropriate for circuit-switched networks.

The nonlinearly-chirped grating is mounted on a single piezoelectric transducer (PZT). Stretching the fiber Bragg grating with the piezoelectric transducer tunes the effective induced dispersion. Figure 2 shows a family of fiber Bragg grating-induced time delay vs. wavelength curves for different applied voltages to the piezoelectric transducer. For any given applied voltage, the dispersion (i.e., the *slope* of the curve at that wavelength) varies nonlinearly from 300 to 1000 ps/nm. Changing the applied voltage merely shifts the curve without altering its shape. For a given fixed channel wavelength, the dispersion varies with applied voltage from 300 to 1000 ps/nm. With ~1000 V applied, the reflected band is shifted by ~1.5 nm and the wavelength shift with voltage is linear. The inset in Fig. 2 is the reflected spectra under different stretching voltages. The bandwidth is ~1 nm and the reflectivity varies by ~0.7 dB.

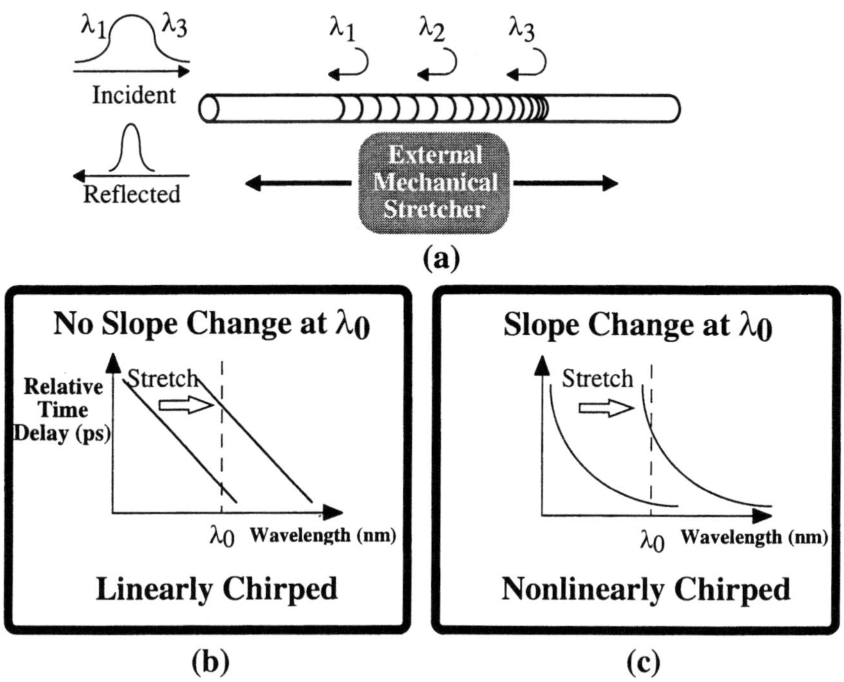

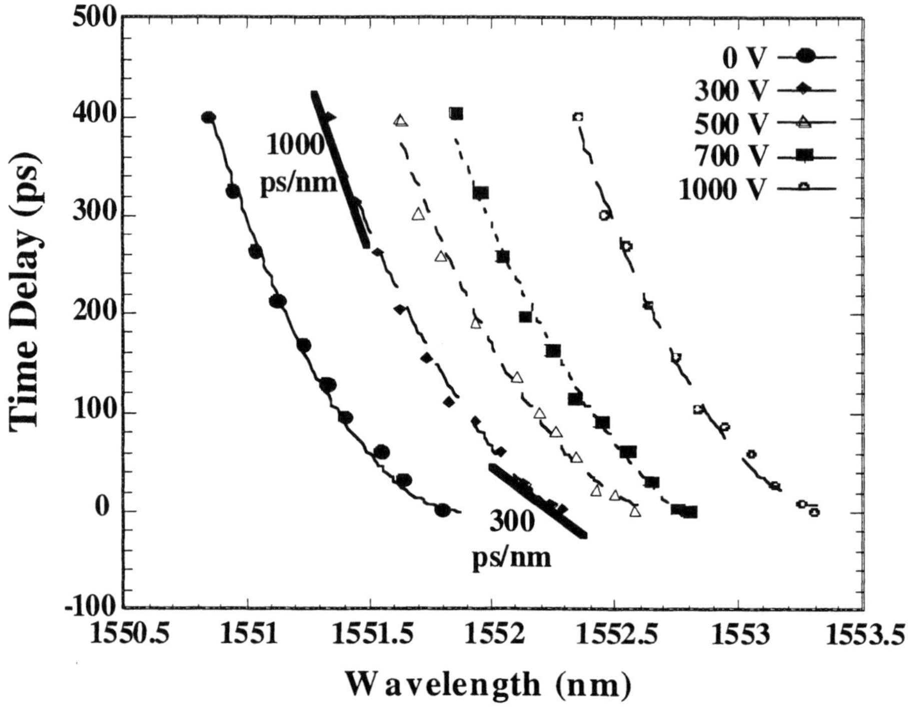

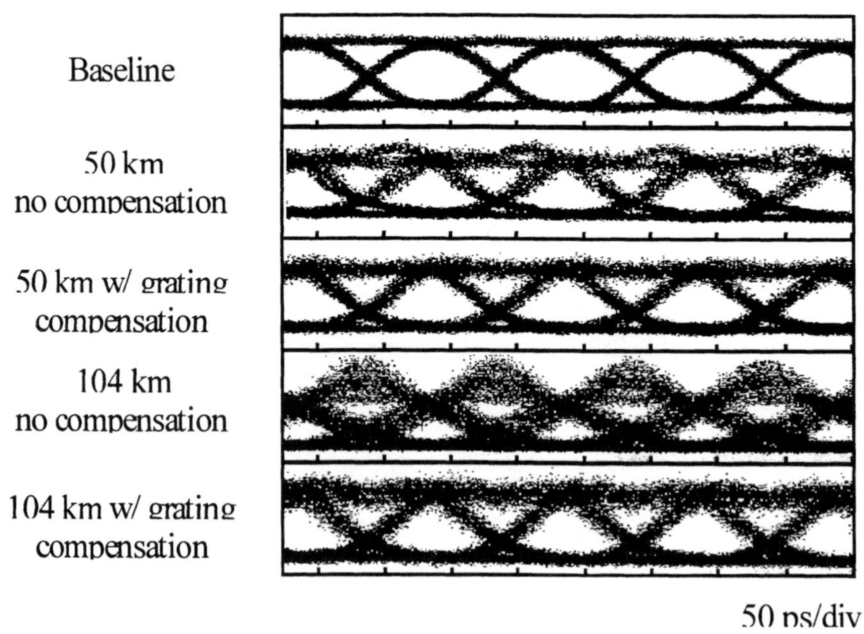

Figure 2. Curves of time delay vs. wavelength for different applied voltages of the piezoelectric stretcher. Inset: reflected spectra for different stretching voltages.

The operating wavelength is set at 1551.7 nm. The signal is modulated at 10 Gb/s and passes through either 50 or 104 km of conventional singlemode fiber by alternating between two parallel paths as controlled by two acousto-optic switches (AO). The dispersion value for the 1551.7 nm channel is 300, 500, and 1000 ps/nm for piezoelectric transducer voltages of 0, 300, and 500 volts, respectively.

By switching between 50 and 104 km of transmission fiber, we demonstrate dynamic dispersion compensation in a 10 Gb/s system. Figure 3 shows the eye diagrams under switching conditions for baseline and with and without compensation after both 50 and 104 km. Without compensation, the eye is fairly open after 50 km, whereas the eye is closing after 104 km. After tunable compensation, the eyes are all open.

(a)

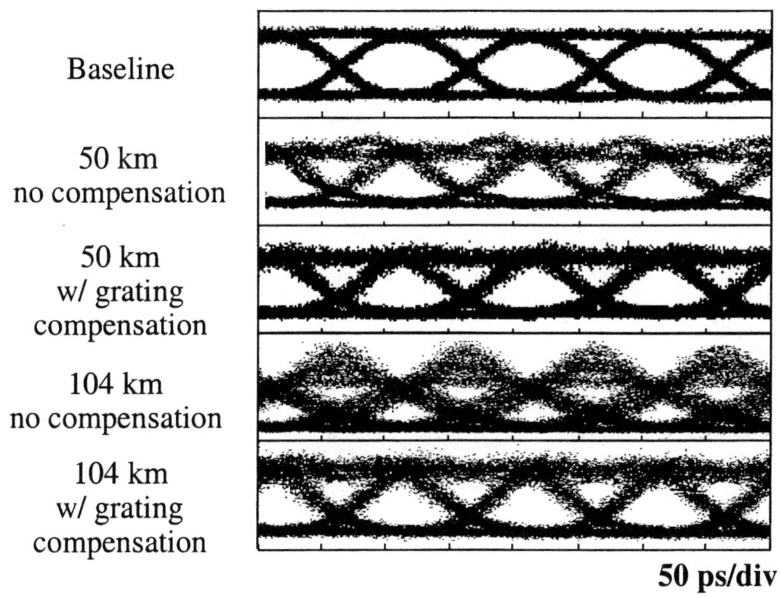

Figure 3. Eye diagrams after 0 km, 50, km, and 104 km with and without the nonlinearly-chirped fiber Bragg grating.

One disadvantage of using such a compensator in a WDM system is that each channel would need its own nonlinearly chirped Bragg grating. It would be highly desirable if a single grating could compensate the dispersion of multiple WDM channels. We extend the work on dynamic dispersion compensation to include the simultaneous compensation of several WDM channels using a single *sampled* nonlinearly-chirped FBG. Sampled gratings [10, 11] are fabricated by imparting a periodic modulation onto the refractive index of a fiber grating, causing the grating to have a multiplicity of spectral resonances.

We demonstrate adjustable dispersion compensation for three 10-Gbit/s WDM channels simultaneously. The dispersion of each wavelength channel is tuned in unison from -200 ps/nm to -1200 ps/nm by stretching the grating. Such a device would compensate the dispersion of a subset of transmitted wavelengths in a reconfigurable network for which the entire subset is switched to propagate along varying fiber distances. A single sampled nonlinearly-chirped grating is more attractive than using many gratings since: (i) a single grating is easier and cheaper to fabricate and package, (ii) temperature stabilization demands are reduced, (iii) exact channel spacing is ensured by the manufacturing process, (iv) insertion loss is significantly less than that of several concatenated gratings, and (v) all channels have almost the same reflectivity if a sinc-shaped sampling function is employed.

Figures 4(a) and 4(b) are the shifts of the reflected spectrum and the grating-induced time delay curves, respectively, for the 3 channels under different stretching conditions. All channels exhibit almost identical optical and time-delay characteristics. The reflectivity difference among the three channels is <2 dB and can be reduced by using a sinc-shape modulation of the sampled grating [20]. Within one wavelength reflection band, the dispersion changes smoothly from -200 ps/nm to -1200 ps/nm for different wavelengths. By uniformly stretching the grating, the dispersion varies smoothly from –200 ps/nm to –1200 ps/nm for a fixed wavelength within each band.

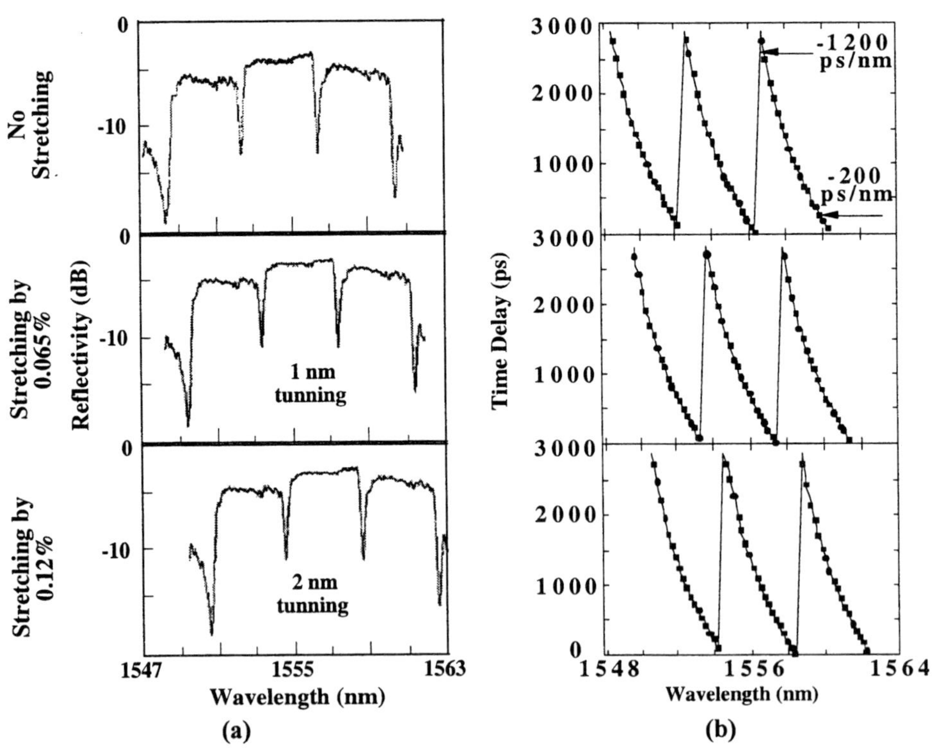

Figure 4. (a) Grating reflection spectrum under different stretching conditions, and (b) time delay vs. wavelength under different stretching conditions.

The experiment setup for tunable dispersion compensation in a WDM system is described as follows. Three WDM channels (1551, 1555, 1559 nm) are externally modulated at 10-Gb/s with a pseudorandom bit stream (PRBS) 2^{15}-1. The transmission distance for our two sets of data is 60 and 120 km. Figure 5 shows the eye diagrams at –20 dBm channel input power for the three WDM channels with and without the compensating grating at different distances. After 60 km, the eye diagrams for the 3 channels are fairly open without compensation, and the grating was tuned to provide a relatively small amount of dispersion compensation. The eye diagrams of the 3 channels after 120 km of propagation are fairly closed without compensation, and the grating was stretched to shift the resonance bands by ~2 nm to provide sufficient dispersion and open the eye diagrams. Near-error-free transmission is achieved for all channels.

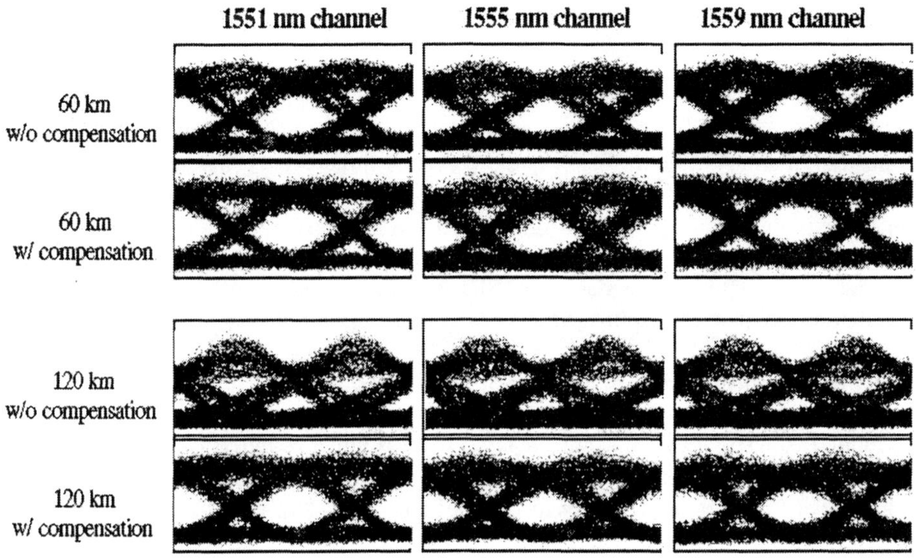

Figure 5. Eye diagrams for 3 WDM channels with and without sampled nonlinearly chirped fiber Bragg grating after 60 km and 120 km.

3. DYNAMIC EQUALIZATION OF WDM CHANNELS USING FIBER-GRATING-BASED MODULATORS

EDFAs and WDM are key enabling technologies for high-capacity optical fiber transmission systems and routing networks [12]. However, robust systems and networks require that the power differential between many WDM channels remains small so as to minimize interchannel crosstalk and ensure adequate EDFA gain for all channels. Several system parameters may change which would have deleterious effects on signal power, including: variable insertion losses, neighboring channel addition and deletion, variable insertion losses, unstable laser power, non-uniform EDFA gain, and μs-long gain transients in EDFA cascades. Furthermore, the non-uniform gain of the EDFA severely limits the usable bandwidth, and can be as small as ~5 nm after an EDFA cascade.

Passive channel equalization methods, such as long-period fiber grating [13] and end-to-end telemetry [14], have been demonstrated in long distance transmission. These techniques are adequate for static or slowly-changing links, but they cannot provide robust equalization for a dynamic WDM system. Since dynamic variations in the total input power into an EDFA will cause μs-long gain transients in chains of deeply-saturated EDFAs [15], the dynamic equalization module should respond faster than 1 μs so as to compensate any possible EDFA-induced channel power variation. Finally, the module should equalize many possible input channels independently to maximize the system robustness.

Here, we use a parallel array of acousto-optic (AO) modulators as part of a solution for dynamic WDM channel power equalization in which each AO modulator independently acts as a loss element for an individual WDM channel. The acousto-optic modulator has <μs rise/fall time and is polarization insensitive. When combining AO modulators with a dense wavelength division multiplexer (DWDM) [16-17] (i.e., an integrated frequency router) and a feedback control circuit, all channels can be independently equalized with channel spacings being as small or as large as can be accommodated by the DWDM. We demonstrate that this module equalizes the power of three WDM channels from 10 dB to 1 dB and responds in ~0.1 μs. Moreover, we insert this module into a recirculating fiber loop to actively equalize the non-uniform EDFA gain for the three channels over a transmission distance of 1,000 km. Since the DWDM passively filters each wavelength, the amplified-spontaneous-emission (ASE) between channels is suppressed, thereby enabling the gain in each amplifier to more effectively amplify signals and not noise.

The configuration of the module is shown in Figure 6(a). The module consists of: (i) a 16x16 DWDM with channel spacings of ~1 nm, (ii) two acousto-optic modulators, (iii) a 4x4 coupler, and (iv) a dynamic control circuit. Three input channels (i.e., λ_1, λ_2, and λ_3) are located at 1552.3, 1556.3, and 1560.3 nm, respectively. The 1556.3 nm channel is unmodulated and acts as a reference channel for our active equalizer. The other two channels are modulated by electro-optic (EO) modulators to simulate optical signal power variation. For our demonstration, we set the reference channel power to be several dB lower than those of the modulated channels. The three WDM channels are spectrally separated by the DWDM into three parallel paths. A variable attenuator is connected to the reference channel to ensure identical insertion loss for each optical path. The two modulated channels are each connected to an AO modulator loss element to dynamically control each channel's transmission and reduce the power level to be equal to the reference channel. After equalization, the separated WDM channels are passively combined by a 4x4 coupler; note that a second DWDM could provide the same function.

The feedback control loop can be described as follows. Three output ports of the 4x4 coupler are used to monitor the power levels in the three channels by connecting each output port to an optical filter and a PIN detector. The detectors' outputs, representing the relative powers of λ_1, λ_2, and λ_3, are input to a control circuit which compares the channel powers and generates appropriate RF acoustic signals to drive the AO modulators. The AO modulators then attenuate power of each channel to equalize all the output powers.

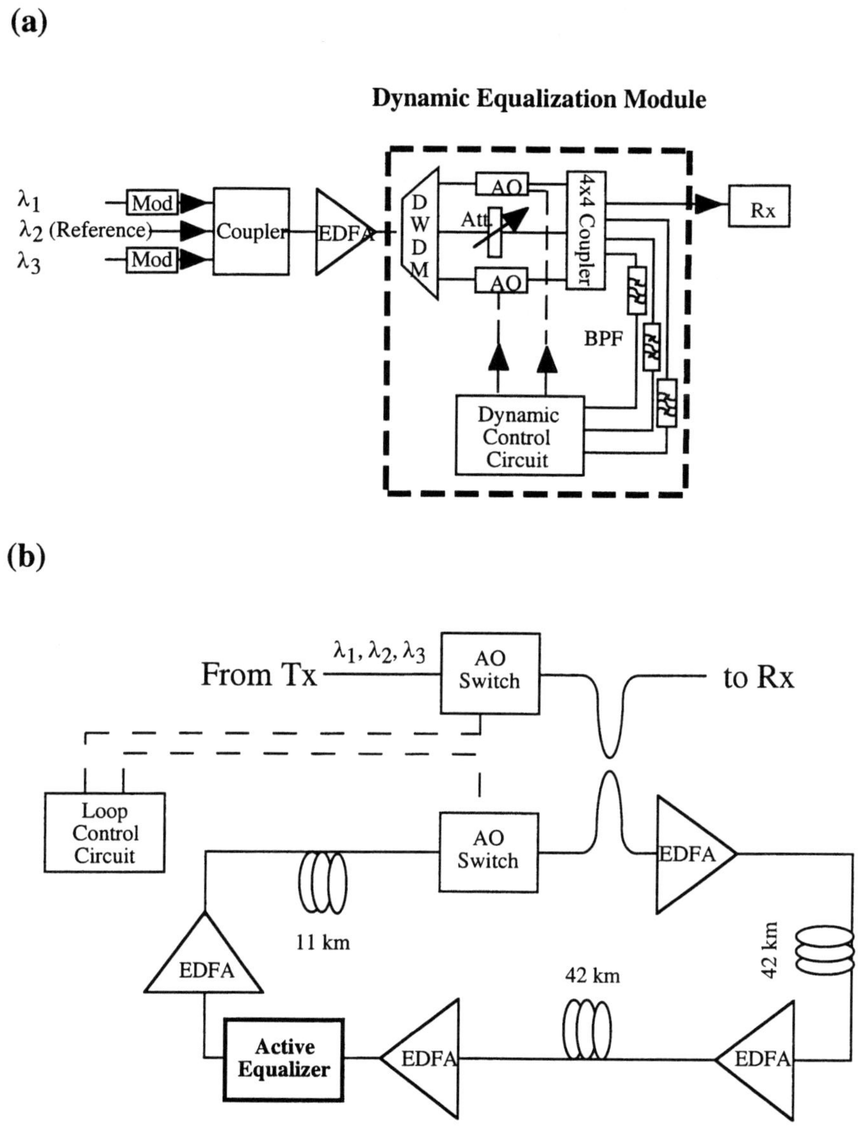

Fig. 6. (a) Experimental setup of the dynamic equalization module. (b) Experimental setup of the recirculating loop using AO modulators for dynamic equalization.

The time response of this dynamic equalization module is shown in Figure 7. Figure 7(a) shows the input signals: the upper one (1560.3 nm) is modulated by a 10 MHz square wave to simulate dynamic power variations, and the lower one (1552.3 nm) is the reference signal and is unmodulated. The optical power of the reference signal is ~5 dB lower than that of the modulated signal. Figure 7(b) shows the output from the equalization module, including the reference signal as well as the modulated signal with and without equalization. Some small modulation residue remains, and the modulated signal and the reference are equalized to within ~ 1 dB. The equalized output signals demonstrate that the module's dynamic response is ~ 0.1 μs.

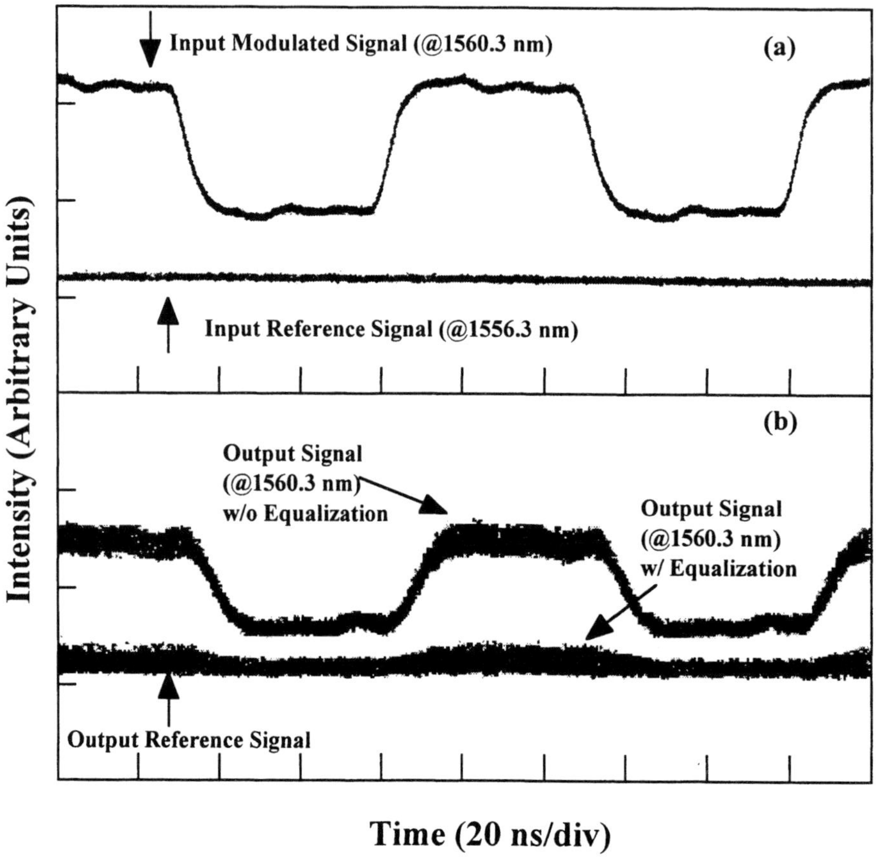

Fig. 7. Time response of the dynamic equalization module. (a) Input of the modulated and reference signals. (b) Output of the equalization module, including the reference signal as well as the modulated signal with and without equalization.

In Figure 8, we further demonstrate the module's performance by showing the input and output optical spectra. Figure 8(a) shows the 3 input WDM channels with ~10 dB power differential. The reference channel is set at 1556.3 nm. Figure 8(b) shows the output spectrum without equalization. The power differential is still ~10 dB, with the sidelobes of the original signals being removed by the filter function of the DWDM. Figure 8(c) shows the output spectrum with equalization, in which the power differential is now equalized to within ~ 1 dB.

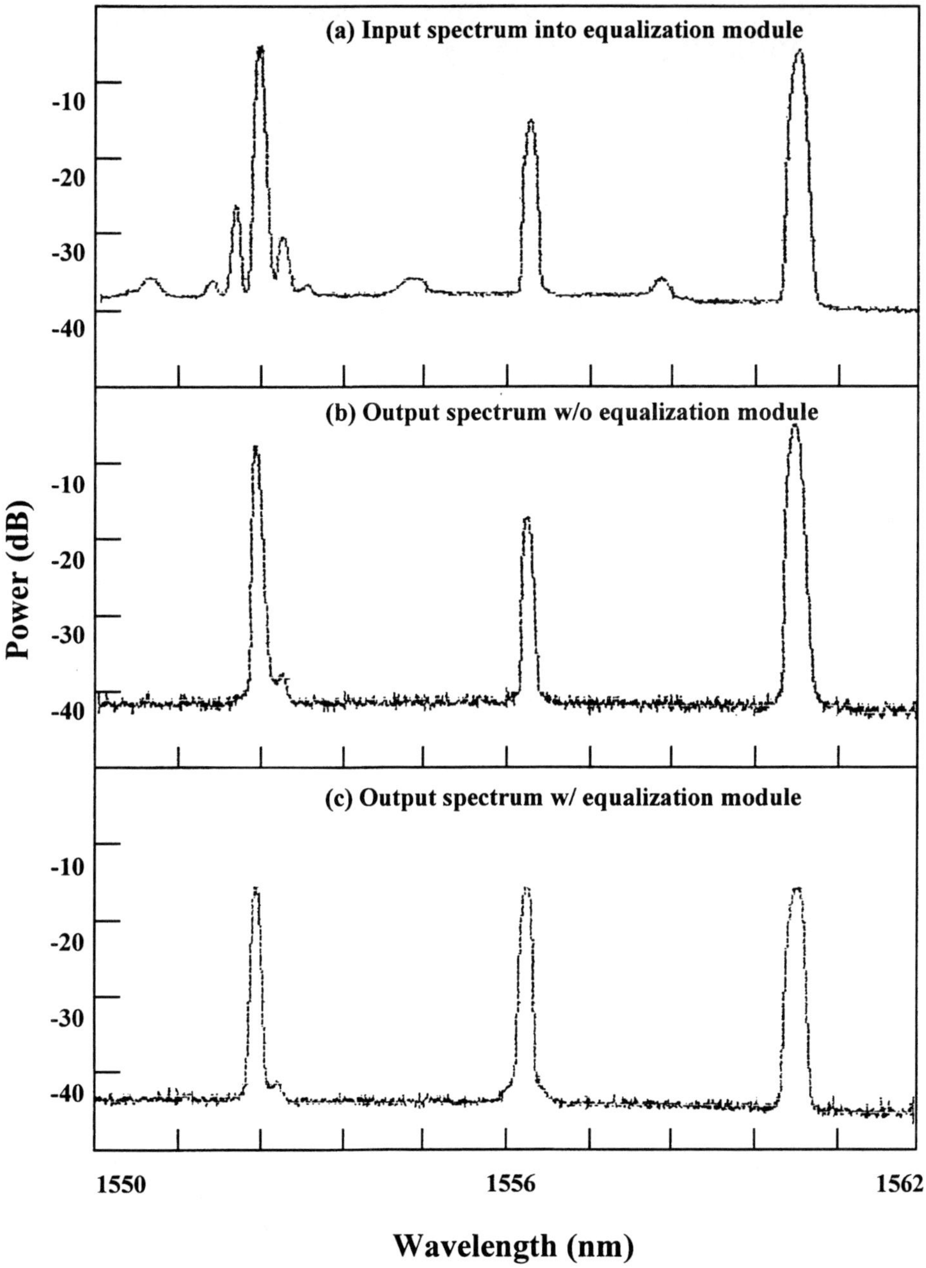

Fig. 8. Optical spectra of the equalization module: (a) input spectrum of the module, (b) output spectrum without equalization, (c) output spectrum with equalization.

4. COMPENSATION OF PMD USING A GRATING WRITTEN INTO A HIGH-BIREFRINGENCE FIBER

By managing fiber dispersion and nonlinearities, the capacity of optical systems has been greatly expanded over the past few years. However, polarization mode dispersion (PMD) has recently emerged as one of the next critical hurdles in achieving high-performance optical transmission systems and networks. This situation has occurred since many fiber segments that were deployed throughout the 1980's have a PMD parameter that is as much as 100-1000 times greater than the more-optimized fiber deployed today. Therefore, it may be highly desirable for ≥10-Gbit/s transmission systems using old embedded fiber to compensate for possible polarization mode dispersion.

In general, PMD is caused by the different transmission speeds of the signal's two states-of-polarization (SOPs) as they propagate along a fiber having a small birefringence. Since the birefringence of a fiber changes randomly along a fiber link, polarization mode dispersion is a statistically random quantity. However, to first order, PMD can be compensated by simply delaying one state-of-polarization with respect to the other (see Figure 9.); this is more accurately known as differential group delay although we will refer to it as PMD for this paper. Because PMD can change both dynamically reconfigurable networks in which the path changes and also for a point-to-point link in which the polarization states change due to temperature fluctuations, it may be quite beneficial to have an adjustable PMD compensator.

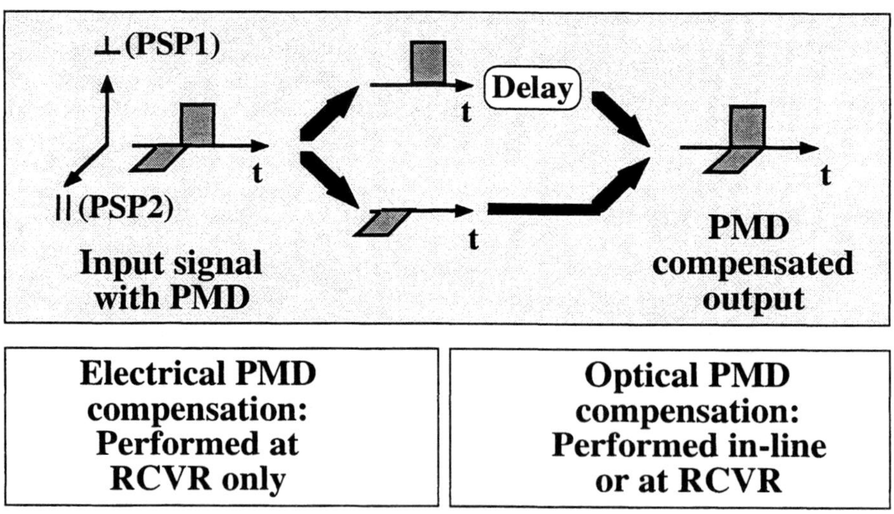

Figure 9. Conceptual diagram of compensation of first order polarization mode dispersion, i.e., differential group delay. Both electrical and optical compensation techniques can be used to provide a differential time delay between the two states of polarization.

Polarization mode dispersion compensation can be accomplished either in the optical or electrical domains. Typically, compensation is achieved by splitting the power of the two polarizations and then delaying one relative to the other. One optical technique is to split the signal into two orthogonal polarization directions and delay one of the polarization components relative to the other using free-space optics, a technique that may be cumbersome. Another optical technique is to use single or multiple pairs of a polarization controllers and a piece of highly-birefringent polarization maintaining (PM) fiber [18, 19]. However, using discrete lengths of PM fiber may have limitations in unability and flexibility. Electrical compensation of delaying the optical power in one polarization is performed after detection at the receiver using phase diversity [20] or electronic filtering [21, 22], but these techniques cannot be performed in-line along a larger optical system.

We use a tunable nonlinearly-chirped fiber Bragg grating (FBG) [23] written into a high-birefringence photosensitive fiber to compensate for varying amounts of polarization mode dispersion. The high-birefringence fiber provides a different time delay for different states-of-polarization, and the nonlinear chirp of the grating provides the selectability of varying amounts of differential polarization time delay. The differential polarization time delay can be continuously tuned by stretching the grating. We use this novel approach to compensate for an induced differential group delay and show good performance at 10 Gb/s signal with 175 ps of adjustability. This all-optical technique has the advantages in that it is continuously adjustable, all-fiber based, and compact.

Figure 10. shows a nonlinearly-chirped fiber Bragg grating that is written into a fiber that has a large refractive index difference between its fast and slow polarization axes, i.e. birefringence. Due to this large refractive index difference, the Bragg reflection from the chirped grating for a given signal wavelength occurs at different locations for different polarizations. In essence, the two signal polarizations "see" two different gratings due to the birefringence. The position difference, ΔL, of the reflection produces a differential time delay, Δt, between the two polarizations. A key feature of our grating is that a nonlinear variation of time delay as a function of wavelength produces a different time delay at different wavelengths. Moreover, since the time delay curves can be shifted to longer wavelengths by stretching the FBG, the time delay can be varied at a *specific* wavelength by using a single mechanical stretcher (Figure 11.). The two optical signals are orthogonally polarized and will therefore combine without interference at the output of the grating.

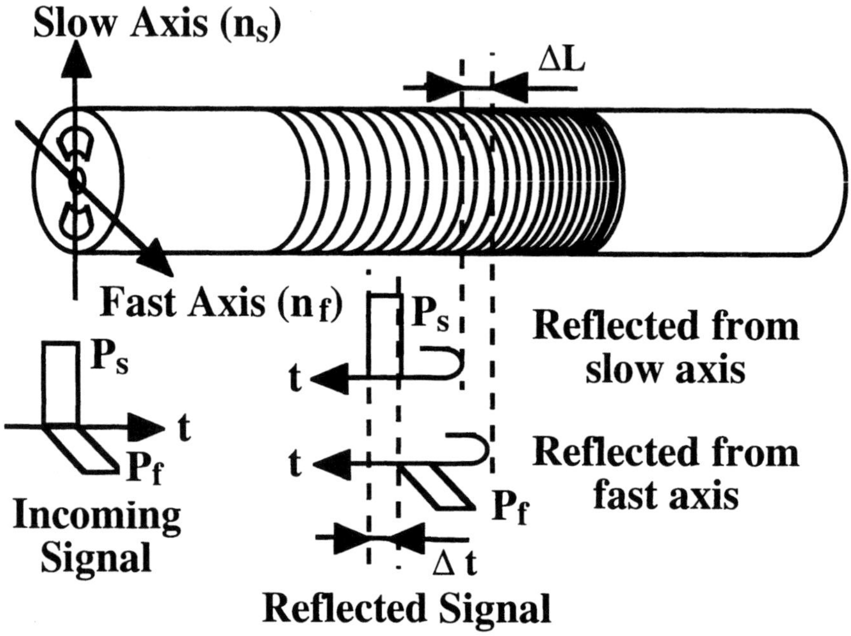

Figure 10. High-birefringence nonlinearly-chirped fiber Bragg grating. The incoming signal has polarization components along both the fiber's fast (P_f) and slow (P_s) axes. The Bragg-reflected signal is seen to have a relative delay Δt between its two polarizations.

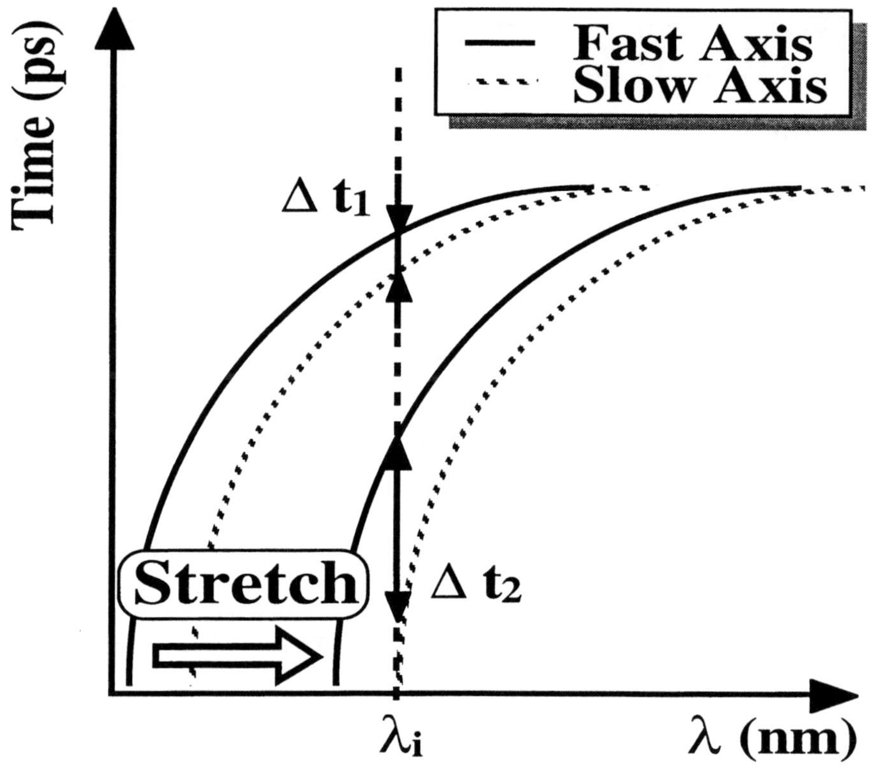

Figure 11. Schematic of the time delay curve of a high-birefringence nonlinearly-chirped fiber Bragg grating. (Δt_1: differential time delay before stretching; Δt_2: differential time delay after stretching, both at wavelength λ_i.)

The nonlinearly-chirped grating is written by exposing a hydrogen-loaded, photosensitive high-birefringence fiber to 300nm (near-UV) light through a nonlinearly-chirped phase mask [4]. The grating measures 15-cm long and is chirped to reflect wavelengths ranging from 1547.2 nm to 1550.5 nm identically for both linear polarization directions.

The fiber's high birefringence produces an almost identical shape for the time delay curves as a function of wavelength for the two axes of polarization, except that the delay curves for the two axes are shifted in wavelength relative to each other [24, 25, 26]. For our photosensitive high-birefringence fiber, the wavelength shift is ~0.6 nm at 1550 nm. Figure 12.(a) shows the time delay curves of the orthogonally-polarized signals reflected from the grating as a function of wavelength. Since the grating is nonlinearly-chirped, the time delay between the two polarizations is a function of wavelength, as shown at Figure 12.(b). The time delay varies from 320 ps to 100 ps when the signal wavelength is varied from 1547 nm to 1550.3 nm.

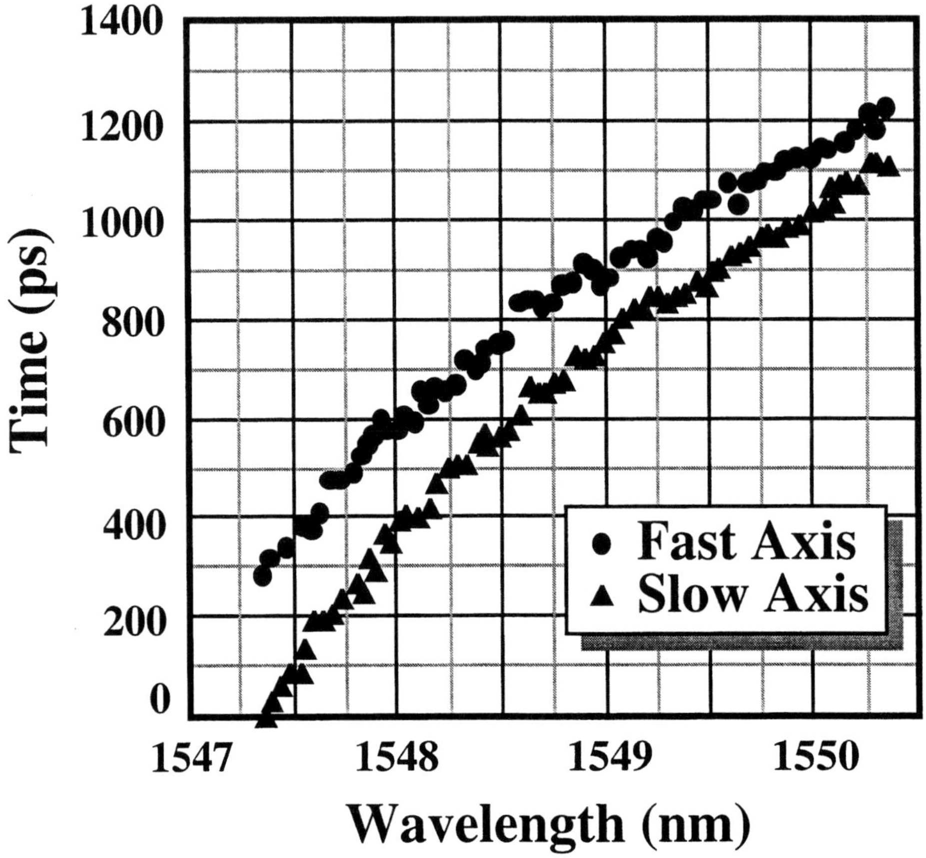
1400
1200
1000
800
600
400
200
0
Time (ps)
1547
1548
1549
1550
Wavelength (nm)
Fast Axis
Slow Axis

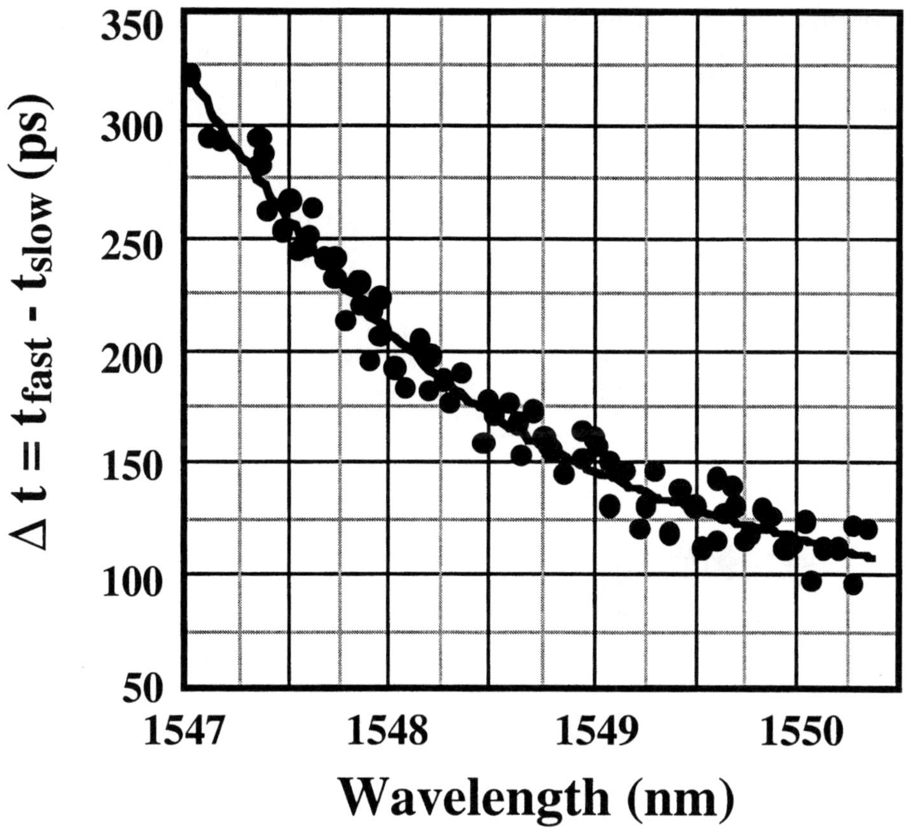

Figure 12. (a) Reflected time delay data for slow axis and fast axis of polarization. (b) Measured difference in time delay between the two orthogonal polarizations. (Solid line is the fit of the differential time delay to our experimental data.

Since stretching the fiber adjusts the differential time delay without altering the polarization, this device has the potential for compensating variable polarization mode dispersion for a long-distance, high-speed optical transmission line. To demonstrate this capability, a 1550.2-nm signal laser is NRZ externally modulated at 10 Gb/s using a 16 GHz electro-optic intensity modulator. Differential group delays of 127 ps and 302 ps are introduced with a PMD emulator. The PMD emulator is composed of two polarizing beam splitters, an optical delay line and a mechanical attenuator. It produces an optical delay for one polarization compared to the other. The ratio of power attenuation in the two parallel paths is adjusted to be identical in order to simulate the worst condition for polarization mode dispersion. A polarization rotator is placed before the high-birefringence nonlinearly-chirped fiber Bragg grating to align the signal's polarization directions to the fast and slow axes of the fiber grating.

Figure 13.(a) shows the baseline eye diagram after the intensity modulator. Figure 13.(b) shows the eye diagram with 127 ps of PMD emulation and after compensation by the unstretched grating; the 127 ps delay value was specifically chosen to match the base compensation provided by our unstretched grating. The eye is completely closed without compensation since 127 ps is larger than one bit period, and the eye is almost completely open after compensation. Figure 13.(c) shows the eye diagram with 302 ps of PMD emulation and after compensation from the grating that was tuned by 0.22% stretching. The eye has completely recovered after compensation.

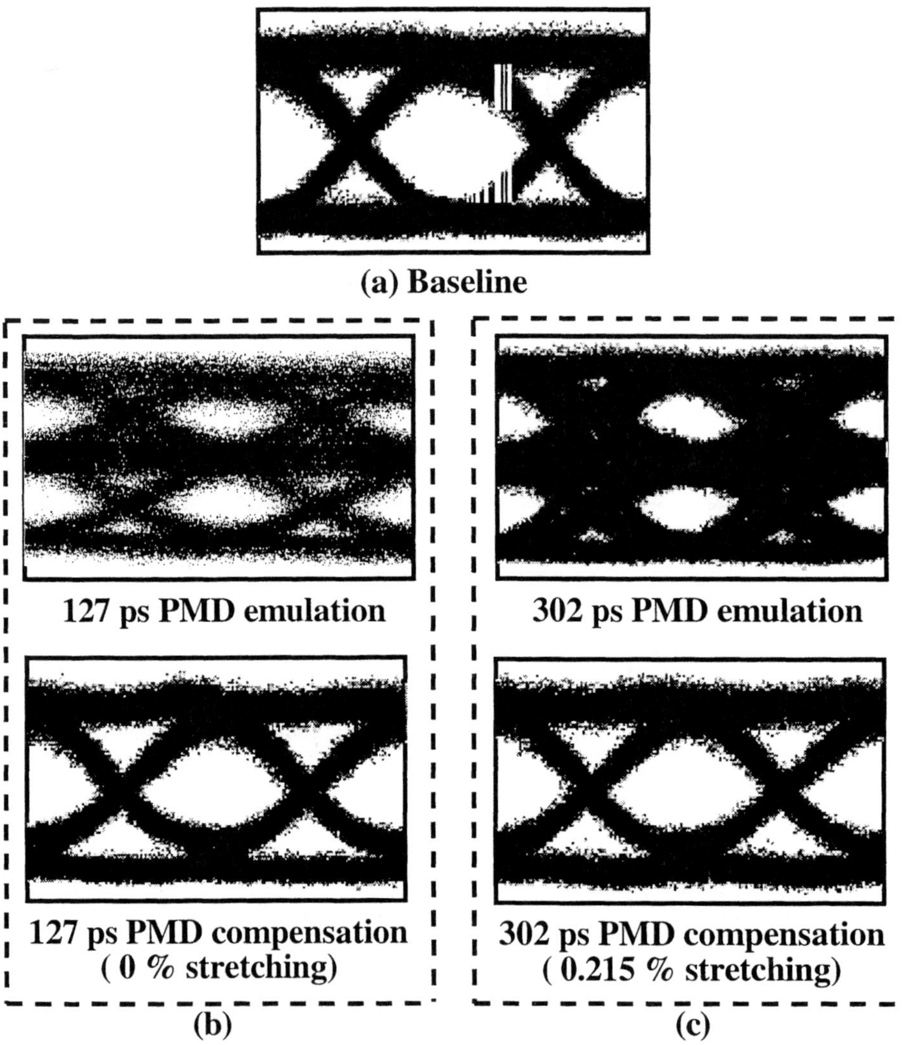

Figure 13. Tunable polarization mode dispersion compensation of a 10 Gb/s data stream using a high-birefringence nonlinearly-chirped grating at 1550.2 nm. (b) 127 ps of polarization mode dispersion is not compensated in the upper trace and is compensated in the lower trace. (c) 302 ps of polarization mode dispersion is not compensated in the upper trace and is compensated in the lower trace.

5. REDUCTION OF COHERENT CROSSTALK IN WDM ADD/DROP MULTIPLEXING NODES BY BIT PATTERN MISALIGNMENT

Add/drop multiplexers are key subsystems in wavelength-division-multiplexed (WDM) networks [27]. ADMs are typically composed of a wavelength demultiplexer followed by a wavelength multiplexer. Ideally, they allow a network node to have access to a single wavelength while allowing all other wavelengths to pass through the node unaffected (Fig. 14a). Unfortunately, wavelength (de)multiplexers [28] are not ideal and allow unwanted crosstalk of any input wavelength onto unintended output ports. For example, a small replica of the signal from λ_1 will appear at all the demultiplexer output ports as shown in Fig. 14a. If a non-ideal demultiplexer and non-ideal multiplexer are placed back-to-back, the small replicas of λ_1 will recombine with the main λ_1 at the multiplexer output. This ADM structure represents a coherent interferometer since it consists of input and output ports connected by optical paths [29,30]. Coherent crosstalk has been shown to be a serious limiting factor in scaleable WDM networks [31].

Here, we propose and analyze a method of reducing the effects of coherent crosstalk in ADMs by purposely introducing fiber delays on the order of bit times between different crosstalk paths. These delays decorrelate the bit pattern of the main signal with the interfering signals and place some of the crosstalk power in the "0" bits of the main signal instead of the "1" bits. For a 16-channel WDM system, this method reduces the device requirement for crosstalk suppression by ~10 dB for RZ and ~5 dB for NRZ transmission.

The modeled system is shown in Fig. 14a. For a specific wavelength, we consider ε_1 to ε_n as the crosstalk level of the combined ADM (demux+mux) coming from the n different replica paths. As the phase and polarization of these signals at the output multiplexer change, the output power of the corresponding channel can vary significantly. Since polarization and relative phases change slowly, the output power will exhibit fading over a period much longer than a bit time [32]. The main degradation occurs on the "1" bits in the main signal. The crosstalk effect on the "0" bits of the main signal is negligible because a "0" bit has a small initial electric field that will not be significantly perturbed.

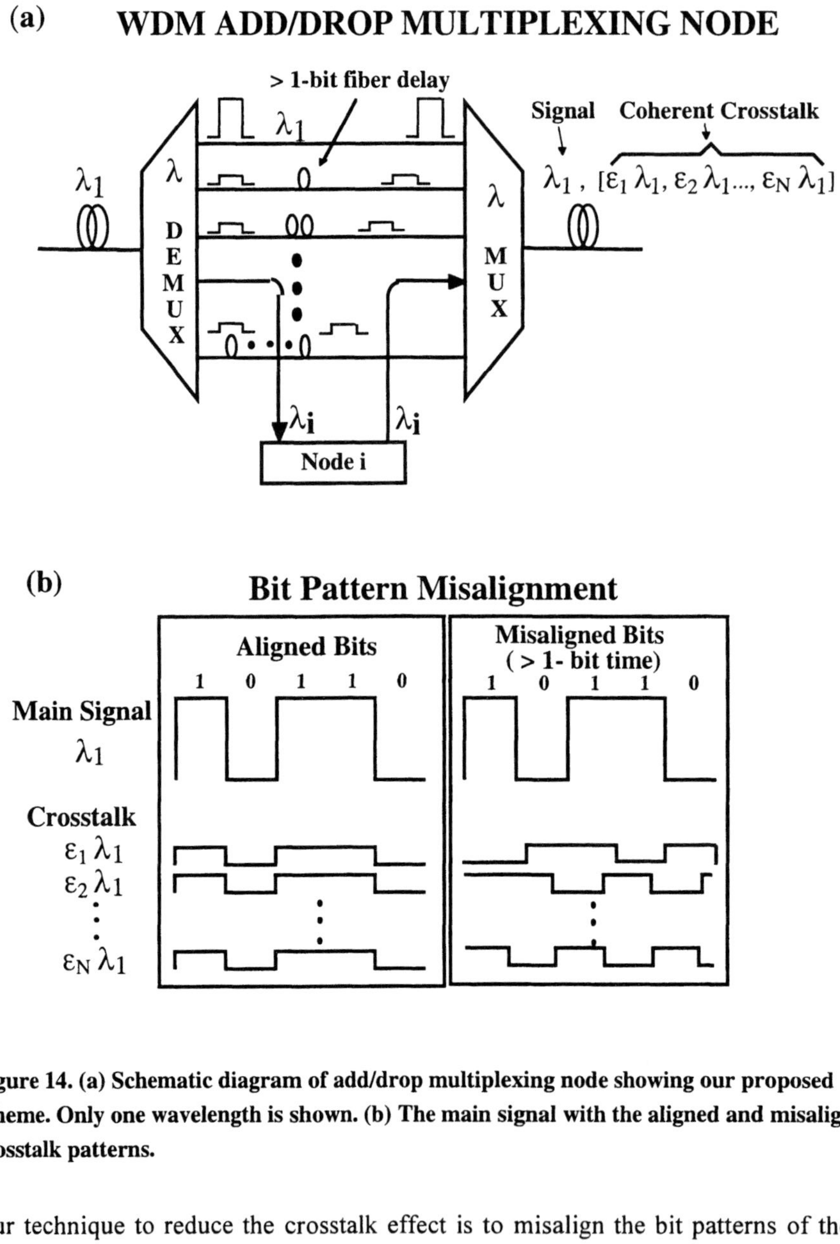

Figure 14. (a) Schematic diagram of add/drop multiplexing node showing our proposed scheme. Only one wavelength is shown. (b) The main signal with the aligned and misaligned crosstalk patterns.

Our technique to reduce the crosstalk effect is to misalign the bit patterns of the main signal and the crosstalk replicas. If we consider a random bit pattern on the original signal, we can misalign all the different replica signals by successively adding longer pieces of fiber to each parallel path. If the length of each path connecting the demultiplexer and multiplexer differs from all other paths by more than a bit length (tens of cm's for Gb/s), then all the patterns from the same input source would be misaligned at the output multiplexer. Figure 14b shows the crosstalk from aligned bit patterns as well as the crosstalk from intentionally misaligned bit patterns. Due to this successive path delay, approximately half of the crosstalk "1" bits are shifted from interfering with the same "1" bit on the main λ_1 signal to interfering with any arbitrary "0" bit on the main signal. Since crosstalk affects "1"s much more than "0"s, the overall effect of crosstalk is reduced.

To evaluate the effectiveness of the proposed technique, we consider equal levels of ε_i for different paths and determine the power penalty for achieving a bit-error-rate (BER) of 10^{-9}. The BER is evaluated as a weighted average over the distribution of the number of interfering crosstalk signals [33]. Figure 15 shows the power penalty as a function of crosstalk for a fixed BER=10^{-9} for aligned and misaligned bits in RZ and NRZ systems. The difference between delay paths is chosen as a bit length plus a fraction of a bit length to take maximum advantage from the "0" levels on the main signal for both RZ and NRZ formats. As can be seen from Fig. 15, RZ format benefits more from our technique than NRZ. This is due to the fact that RZ has a "0" power level for 75% of the total data time whereas NRZ spends 50% in the "0" power state, and it is the "0" power state into which we are "dumping" the crosstalk energy. For a power penalty of 1 dB, a crosstalk device relaxation of ~4 and ~9 dB has been achieved for NRZ and RZ systems, respectively.

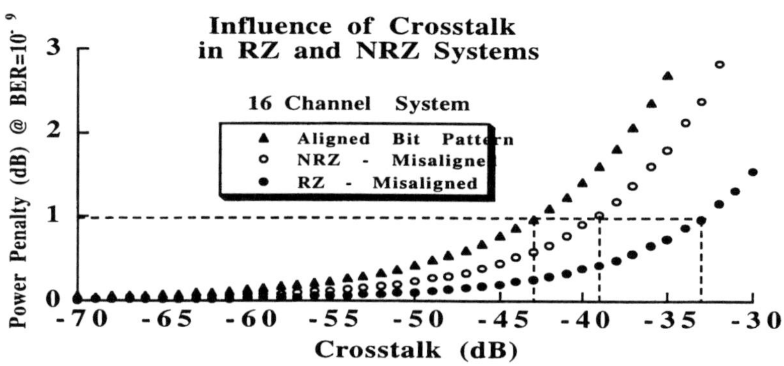

Figure 15. The power penalty versus crosstalk for aligned and misaligned bit patterns in NRZ and RZ systems. Misalignment delays are selected to distribute crosstalk over the bit pattern (slightly more than a bit time difference between each path).

The delay paths need not be exact multiples of a specific length. The only important issue is that the crosstalk power be evenly distributed over the whole bit stream. Thus, a large amount of uncertainty in delays can be tolerated, enabling a robust implementation. Moreover, the bit rate of the incoming signal can be increased arbitrarily since the bits will still remain misaligned as long as the delay is larger than the bit period.

ACKNOWLEDGMENTS

The author wishes to acknowledge the insightful and invaluable collaboration on these projects of the following individuals at USC: Anna Babayan, Liza Babayan, Jinxing Cai, Mustafa Cardakli, Xiaopeng Chen, Jack Feinberg, Kai-Ming Feng, Victor Grubsky, Imran Hayee, Bogdan Hoanca, Xin Jiang, Reza Khosravani, Sanggeon Lee, Dmitry Starodubov, Brandon Whitehead.

REFERENCES

[1] A.E. Willner, "Towards Uniform Channel Performance in Dynamic WDM Systems and Networks," Invited Paper, Conference on Optical Fiber Communications '99, paper ThO5, San Diego, CA, Feb. 1999 (Optical Society of America, Washington D.C., 1999).

[2] K.-M. Feng, J.-X. Cai, V. Grubsky, D.S. Starodubov, M.I. Hayee, S. Lee, X. Jiang, A.E. Willner, and J. Feinberg, "Dynamic Dispersion Compensation in a 10-Gb/s Optical System using Novel Voltage Tuned Nonlinearly-Chirped Fiber Bragg Grating," IEEE Photonics Technology Letters, vol. 11, no. 3, 1999.

[3] J.-X. Cai, K.-M. Feng, A.E. Willner, V. Grubsky, D.S. Starodubov, and J. Feinberg, "Simultaneous Tunable Dispersion Compensation of Many WDM Channels Using a Sampled Nonlinearly-Chirped Fiber-Bragg-Grating," *IEEE Photonics Technology Letters*, accepted for publication: also in *Conference on Optical Fiber Communications* '99, paper FA7, San Diego, CA, Feb. 1999 (Optical Society of America, Washington, D.C., 1999).

[4] D.S. Starodubov, V. Grubsky, J Feinberg, B. Kobrin, and S. Juma, "Bragg grating fabrication in germanosilicate fibers by use of near-UV light: a new pathway for refractive-index changes," *Opt. Lett..*, 22, 1086-1088 (1997).

[5] S. Lee, R. Khosravani, J. Peng, V. Grubsky, D.S. Starodubov, A.E. Willner, and J. Feinberg, "Adjustable Compensation of Polarization Mode Dispersion Using a High-Birefringence, Nonlinearly-Chirped Fiber Bragg Grating," *IEEE Photonics Technology Letters*, vol. 11, no. 10, 1999.

[6] R. Khosravani, M.I. Hayee, B. Hoanca, and A.E. Willner, "Reduction of Coherent Crosstalk in WDM Add/Drop Multiplexing Nodes by Bit Pattern Misalignment," IEEE Photonics Technology Letters, vol. 11, no. 1, 1999.

[7] R.J. Nuyts, Y.K. Park, and P. Gallion, "Performance improvement of 10 Gb/s standard fiber transmission system by using the SPM effect in the dispersion compensating fiber," *IEEE Photonics Technology Letters*, vol. 10, pp. 1406 (1996)

[8] M.J. Cole, H. Geiger, R.I. Laming, S.Y. Set, M.N. Zervas, W.H. Loh, and V. Gusmeroli, "Broadband dispersion-compensating chirped fiber Bragg gratings in 10 Gbit/s NRZ 110 km non-dispersion-shifted fiber link operating at 1.55 μm," *Electron. Lett.* , vol. 33, pp. 70, 1997.

[9] M.M. Ohn, A.T. Alavie, R. Maaskant, M.G. Xu, F. Bilodeau, K.O. Hill, “Tunable fiber grating dispersion using a piezoelectric stack,” *Conference on Optical Fiber Communication*, Paper WJ3, Dallas, TX, Feb. 1997 (Optical Society of America, Washington, DC, 1997)

[10] S.H. Huang, X.Y. Zou, S.-M. Hwang, A.E. Willner, Z. Bao, and D.A. Smith, "Experimental Demonstration of Active Equalization and ASE Suppression of Three 2.5 Gbit/s WDM-Network Channels over 2,500 km Using AOTFs as Transmission Filters," IEEE Photonics Technology Letters, vol. 9, pp. 389-391, 1997.

[11] C. Konishi, T. Yoshida, S. Hamada, K. Asahi, S. Fujita, “Dynamic gain-controlled erbium-doped fiber amplifier repeater for WDM network”, Conference on Optical Fiber Communications ’97, paper TuE1.

[12] *IEEE/OSA J. Lightwave Technol.*, special issue on multiple wavelengths technologies and networks, vol. 14, no. 6, 1996.

[13] A.M. Vengsarkar, J.R. Pedrazzani, J.B. Judkins, P.J. Lemaire, N.S. Bergano and C.R. Davidson, "Long-period fiber-grating-based gain equalizers," *Opt. Lett.,* vol. 21, pp. 336-339, 1996.

[14] A.R. Chraplyvy, J.A. Nagel and R.W. Tkach, "Equalization in amplified WDM lightwave transmission systems," *IEEE Photon. Technol. Lett.* Vol. 4, pp. 920-922, 1992.

[15] J.L. Zyskind, Y. Sun, A. K. Srivastava, J. W. Sulhoff, A.J. Lucero, C. Wolf, and R.W. Tkach, "Fast power transients in optically amplified multiwavelength optical networks," in Optical Fiber Communication Conference, Vol. 2, 1996 OSA Technical Digest Series, Optical Society of America, D.C., San Jose, CA, 1996, Post-deadline paper PD31-1.

[16] C. Dragone, "An NxN optical multiplexer using a planar arrangement of two star couplers," *IEEE Photon. Technol. Lett.*, vol. 3, pp. 812-815, 1991.

[17] W.-H. Lin, H.-F. Li, Y.-J. Chen, M. Dagenais, and D. Stone, "Dual-channel-spacing phased-array wave-guide grating multi/demultiplexers," *IEEE Photon. Technol. Lett.* Vol. 8, pp. 1501-1503, 1996.

[18] T. Takahashi, T. Imai and M. Aiki, “Automatic compensation technique for timewise fluctuating polarisation mode dispersion in in-line amplifier systems,” *IEE Electronics Letters*, vol. 30, pp. 348-349, 1994.

[19] Takeshi Ozeki, Masato Yoshimura, Teruhiko Kudo, Hiroshi Ibe, “Polarization-mode-dispersion equalization experiment using a variable equalizing optical circuit controlled by a pulse-waveform-comparison algorithm,” *Conference on Optical Fiber Communication*, 1994, Technical Digest Series, Paper TuN4 (Optical Society of America, Washington, D.C., 1994).

[20] B. W. Hakki, "Polarization mode dispersion compensation by phase diversity detection," *IEEE/OSA Photonics Tech. Letters*, vol. 9, pp. 121-123, 1997.

[21] H. Bülow, D. Schlump, J. Weber, B. Wedding, R. Heidemann, "Electronic equalization of fiber PMD-induced distortion at 10 Gbit/s," *Optical Fiber Communication Conference*, 1998, Technical Digest Series, Paper WI1 (Optical Society of America, Washington, D.C., 1998).

[22] Jack H. Winters, and Mario A. Santoro, "Experimental equalization of polarization dispersion," *IEEE/OSA Photonics Tech. Letters*, vol. 2, pp. 591-593, 1990.

[23] K.-M. Feng, J.-X. Cai, V. Grubsky, D. S. Starodubov, M. I. Hayee, S. Lee, X. Jiang, A. E. Willner, and J. Feinberg, "Dynamic dispersion compensation in a 10-Gb/s optical system using novel voltage tuned nonlinearly-chirped fiber Bragg grating," *IEEE/OSA Photonics Tech. Letters,* vol. 11, pp. 373-375, 1999.

[24] B. Ortega, L. Dong, W. F. Liu, J. P. de Sandro, L. Reekie, S. I. Tsypina, V. N. Bagratashvili, and R. I. Laming, "High-performance optical fiber polarizers based on long-period grating in birefringent optical fibers," *IEEE/OSA Photonics Tech. Letters*, vol. 9, pp. 1370-1372, 1997.

[25] A. S. Kurkov, M. Douay, O. Duhem, B. Leleu, J. F. Henninot, J. F. Bayon and L. Rivoallan, "Long-period fiber grating as a wavelength selective polarisation element," *IEE Electronics Letters*, vol. 33, pp. 616-617, 1997.

[26] P. Niay, P. Bernage, T. Taunay, M. Douay, E. Delevaque, S. Boj, and B. Poumellec, "Polarization selectivity of gratings written in Hi-Bi fibers by the external method," *IEEE/OSA Photonics Tech. Letters*, vol. 7, pp.391-393, 1995.

[27] R. E. Wagner, R. C. Alferness, A. A. M. Saleh, M. S. Goodman, "MONET: Multiwavelength Optical Networking", J. Lightwave Technology, vol. 14, pp. 1349-1355, 1996.

[28] C. Dragone, C. A. Edwards, R. C. Kistler, "Integrated Optics N X N Multiplexer on Silicon", IEEE Photonics Technology Letters, vol. 3, pp. 896-899, 1991.

[29] A. Arie, M. Tur, E. L. Goldstein, "Probability-density function of noise at the output of a two-beam interferometer," J. Opt. Soc. Am. A, vol. 8, pp. 1936-1941, 1991.

[30] P. J. Legg, D. K. Hunter, I. Andonovic, P.E. Barnsley, "Inter-channel crosstalk phenomena in optical time division multiplexed switching networks", IEEE Photonics Technology Letters, vol. 6, pp. 661-663, 1994.

[31] E. L. Goldstein, L. Eskildsen, "Scaling limitations in transparent optical networks due to low-level crosstalk", IEEE Photonics Technology Letters, vol. 7, pp. 93-94, 1995.

[32] E. L. Goldstein, L. Eskildsen, A. F. Elrefaie, "Performance implications of component crosstalk in transparent lightwave networks", IEEE Photonics Technology Letters, vol. 6, pp. 657-660, 1994.

[33] R. Ramaswami and P. A. Humblet, "Amplifier Induced Crosstalk in Multichannel Optical Networks", IEEE J. Lightwave Technology, vol. 8, pp. 1882-1896, 1990.

Test Issues and Measurements

DWDM components: critical requirement and production test methods

Dilip K. Paul

Gould Fiber Optics Division
1121 Benfield Blvd., Millersville, MD 21108Benfield Blvd., Millersville, MD 21108

SUMMARY

Fiber optic advances made in the 1980s have revolutionized the concept of telecommunications services and applications. With the development of integrated photonics and deployment of optical technology in many diverse applications, inroad of lightwave systems to traditionally captive arena of alternative technologies has been rather impressive. The 90's trend to include technologies such as optical amplifiers, DWDM and gratings combined with a variety of new services such as internet, broadband network, and multimedia, can only make this inroad much deeper in the millennium. An ultimate goal, beyond the point-to-point/-multipoint transmission and distribution capability achieved thus far, is to provide a fault-tolerant, reliable connectivity in multi-user networks where the electrical bottlenecks will be avoided altogether by performing all functions, e.g., switching, mux/demux, etc., transparently in the optical domain. Technology for all-optical networking is still evolving with much development and engineering work remaining. These advanced networks based on WDM technology can only be realized to its fullest capability provided appropriate components are developed and tested, and the integrated systems meet specified performance goals. Therefore, the requirements for the production test facility and performance specifications of advanced DWDM networks have become increasingly complex and stringent. This paper reviews the critical requirements and production test issues relevant to DWDM components and networks. Impact of high-tech systems specifications on the quality of components and testability is also addressed.

1. INTRODUCTION

Early Wavelength Division Multiplex (WDM) systems employed two wavelengths (1300 and1550 nm). In early eighties, there were a few experimental demonstration of WDM systems operating @ 830 and 1310 nm wavelength bands. These systems typically used discrete components designed for the specific wavelength region of interest. The only critical component in these early systems was the demultiplexer (DEMUX) used to select/discriminate the wavelength of interest at the output port. Typically, dichroic filter based devices were used as the optical filter/discriminator element for the WDM systems.

Since those early years, many technology advancements have been reported in the overall discipline of photonics and optical fiber communications networks. Particularly in the last decade, the rapid strides made in the WDM technology and relevant data/communications networking concepts have led to the development/demonstration/commercial deployment of several WDM networks and services. Now, the trend in communications networking is to move on to dense wavelength division multiplex (DWDM) technologies and systems which could space the channels even closer (to 50-GHz). To make these systems field-deployable by assuring a low-maintenance and long lifetime performance, reliable fiber optics are being developed at a considerable effort. It is true that extensive use of fiber optics in both terrestrial and submarine communications has made even the skeptics believe in the reliability of lightwave systems. However, the concern for reliability still remains in yet unexplored areas such as in space and other harsh environments and for new products and applications of new technologies such as DWDM. Furthermore, developing high reliability components and fail-safe systems at a competitive cost remains a daunting challenge.

The DWDM technology advances with eventual applications in multi-media telecommunications systems make the performance specifications and test requirements of each product considerably tighter as well as complex. Consequently, the search continues for reliable and high accuracy test methods and standards for critical DWDM components. In this overview paper, several critical WDM test issues and their relevance to high-density WDM products, systems, and networks are addressed.

The technology and market trends of DWDM components, networks, and services are briefly reviewed in section 2. Several critical DWDM components and test issues are introduced in the following section. The subject of polarization dependent loss (PDL) and polarization mode dispersion (PMD) in both active and passive components is discussed in detail in section 4. Also included in this section are the topics on thermal dependence of PMD, high accuracy measurement of small PMDs as well as the effect of PMD in high-speed networking. Starting with a comparison of standard polarimetric (PM) and interferometic (IM) methods for PMD testing used today, a newly developed technique that makes use photon pairs is introduced in this section. Production test requirements and facilities that are considered useful to test the performance of advanced DWDM components and systems are presented in section 5. Importance of test sequencing and following a well-documented test plan and procedure is stressed to highlight the test yield issues in production testing environment. Sections 6, 7, and 8 present WDM networking where critical technology items (section 6), complex system requirements (section 7) and competing optical network architecture (section 8) are introduced. Testing of WDM networks to its specification is quite involved and it requires a different set of considerations that are geared more for performance testing at systems level. Here, parameters such as signal-to-noise ratio, bit error rate (BER), noise figure, inter-channel interference, cross-talk, isolation, PMD, phase noise, jitter, non-linearity, etc. are more relevant. Relationship of these systems level parameters with component specifications is discussed in these sections.

2. DWDM TECHNOLOGY & MARKET

Several in-depth market research studies in opto-electronics and fiber optics sectors have indicated the worldwide demand for optical fiber and photonics products growing at a healthy pace in 1998, and the trend is expected to continue into the millennium. A 45% of this demand is in North America with a distribution of 25% in CATV, 35% in CLECs, 15% in long haul, 5% in premises, and 20% in the remaining market segments. A combination of greater demand for bandwidth, increased Internet usage, the expanding number of telecommunications carriers and system developers and the impact of deregulation, both national and international, will continue to drive the deployment of fiber for enhanced and/or new services. Specifically, the market growth will be driven by factors such as:

- Migration of dense wavelength division multiplexing (DWDM) technology into metropolitan networks
- Adoption of DWDM networking technology in Europe and other developing countries
- Continued transition of traditional electrical functions to be accomplished optically, and
- Decreasing cost of bandwidth with advances in fiber optic technology (a factor of ten cost reduction from $155 per Mbps in 1996 @ OC-3 level to $15 per Mbps @ OC-192 level in 2000).

In this context, the world DWDM system market has grown from 310 M$ in 1996 to 1,560 M$ in 1998. It is projected at 2.22 B$ in 1999 reaching 3.21 B$ in 2000. The US market is around 70% of the world market.

With regard components/product opportunities, mention must be made of Er-doped fiber amplifier, dispersion compensation module, special fibers, optical add/drop modules (OADMs), variable optical attenuators, ultra high reliability couplers and splitters, narrow linewidth high-stability single mode lasers, demultiplexer and many other advanced photonic products used by DWDM systems developers.

Current trend: Metro DWDMs and OADMs

A few major suppliers such as Lucent, Nortel, Cienna, and Alcatel, and a few small ones such as NEC, Pirelli, and Fujitsu currently share the US DWDM market. The trend is to integrate advanced DWDM technology with bandwidth management software for bandwidth-efficient, on-demand, cost-effective transmission of data, voice, video, and Internet protocol and SONET services via a single fiber. A large number of vendors have already demonstrated several prototypical DWDM systems in recent Supercomm '99 held in Atlanta, GA. A brief summary of the products and relevant technology is provided below.

ALCATEL: Model 1690 for metro DWDM (access devices) and Model 16986 for large, regional DWDM rings -- add/drop units with 4 to 8 wavelength channels on the metro side. Model 1686 is a MUX unit that allows carriers to create regional rings (large size, 40-150 km long rather than small ones @ 40-km length). ALCATEL Model 1690 OADM has 1-16 wavelength channels @ OC-48 connectivity. Model 1686 OADM has 1-16 channels @ OC-48 or OC-192 speeds (to be upgraded to 32 channels of OC-192 by end of 1999). Model 1690 cost @$25K per channel. The competitors include LUCENT (Wavestar OLS 400G), NORTEL Networks, OSICOM Technologies--each having their own system.

OSICOM's system, a third generation DWDM evolution, includes enhanced management capability, optical service channel, optical probe, and network management platforms. CORVIS products include *Gateway* (provides access to DWDMs @ OC-12, OC-48, OC-192 and with SONET Mux and ATM switches); *optical add/drop mux* (OADMs: 1-160 channels @ OC-48 and interface with SONET); *router* (optical, no elec. power required for signal routing/switching); and *optical amplifier* (handle up to 160 channels @ OC-48 speed).

3. DWDM COMPONENTS & TESTS ISSUES

WDM has enabled explosive growth in fiber optic communications systems and networks. 2-channnel WDM systems (1310 and 1550 nm) developed in the early years now systematically provide 96 channels or so in a 35 nm window of the 1550-nm wavelength region available in Er-doped fiber amplifiers. For the ITU-grid of optical communications channels at 1550-nm wavelength band that are spaced 50-GHz apart, this demands a very narrow (0.4 nm) separation between adjacent optical channels. At 100-Ghz spacing, the adjacent channels separated by 0.8 nm, is still a very small number. On top of this finesse, the systemic performance requirements such as isolation between adjacent channels, cross-talk, return loss, polarization-induced effects, SNR, BER, etc., demand a highly exacting performance from the devices that could be used in practical DWDM systems and networks. Any infraction of these components and subsystems from the specified performance severely impacts both overall systems capacity and design margin, and thus, cost-competitiveness. The need to develop accurate measurement techniques and related test facilities to qualify DWDM components can never be overstated. Recently, DWDM technology has found its way into large communications networks where signal processing functions such as, transmission, routing, switching are all increasingly performed in the optical domain. Multimedia transmission technology using IP, high-speed, high capacity data, voice, and video communications in a single optical link has become a hot topic this year.

WDM tests and measurements requirements for the advanced DWDM systems that are being proposed are briefly reviewed. Impact of high-tech systems specifications on the quality of components and relevant test issues is discussed.

A general configuration of a DWDM systems architecture (see Figure 1) has identified all passive and active components that are deemed critical for the systems performance. The components include light sources, transmission medium; concentrator/multiplexer of several light sources; high sensitivity, low noise photo-receivers; and connectivity/configuration of network nodes and management and control of the resources. In a DWDM network where multiple carriers are separated by only a few nanometer spacing, optical amplifiers (which tend to have optical gain performance in relatively narrow ranges around 1550 nm) are typically used. The optical amplifier increases the carrier output and can improve the SNR of the system.

Most typically, a WDM network designer would like to accommodate as many channels as possible; implement the channel upgrades in a modular fashion with an economical maintenance of the network resources and scheduled services, thereby evenly distributing the cost of technology development and field deployment in successive phases. Any discussion of test issues would therefore include WDM testing at component, subsystem, system, and network levels. Management of the networking functions and a dynamic control/reconfiguration of the network architecture become critical, particularly in the context of an extended communications network.

The specific DWDM components considered are Isolators, Tap/Splitter couplers; WDMs (MUX/DEMUX); Optical Amplifiers; other active components such as lasers, ELEDs, photo-detectors. optical modulators, etc. Specifications of DWDM components/technology, various test methods used for production testing and their limitations evolve with the state-of-the art technology advances. As one will note, the ultimate performance limitations of a DWDM system is controlled by either of the following parameters:

- Laser linewidth, stability of wavelength, degree of polarization, and optical power output; and ease of control to maintain the laser line (peak wavelength) within a specified channel width.
- Low-noise, high sensitivity photo-detector/receiver with a high dynamic range and/or wavelength discrimination capability (tunability) of use as DEMUX.
- Spectral characteristics of passive components (such as cut-off edges, pass-band features, attenuation, in-band ripple, phase, polarization, uniformity, reproducibility, etc.)
- Noise in OE components, spectral emissions, impact on systems level performance
- Polarization induced effects (PDL, PDL in components and impact on systems)
- Isolation, back-reflection, directivity, etc. of optical components
- Effect of the component specifications on systems performance (noise, jitter, pulse spreading in high speed networks, reliability and redundancy issues)

In optical testing of DWDMs, the critical issues include sensitivity, dynamic range, noise, resolution, repeatability and reproducibility. Often, the measurements are based on the insertion loss of the DUT. In this regard, the critical test parameters include the center wavelength and the pass band, slope accuracy, isolation, return loss, directivity, PDL and thermal effects over the useful bandwidth of the DUT. Also, the definition of these parameters and the ports where they are measured need to be standardized. However, two features of importance are the absolute value of the parameter being measured and its relative changes with varying test conditions (e.g., wavelength, temperature, time, etc.). Therefore, both high dynamic range and high resolution and sensitivity are at premium. The importance of selecting appropriate components/instruments for the test facility with systematic characterization of the complete facility for measurement repeatability and reproducibility of the test data can never be over-emphasized.

Major points to be made in reference to production testing of optical components include the following:

- Actual tests (test set-up, data collection and analysis) must take very little time
- Automated test facility requiring minimum operator skill is recommended
- Must ensure repeatability of the test data on the same product measured at different workstations.
- The adopted test procedures must be simple enough.
- Documentation of the test data as well as frequent maintenance and calibration schedules of the test facility must be strictly observed.

4. MEASUREMENT OF PDL AND PMD OF PDL AND PMD

Polarimetric techniques such as JME and Poincare' sphere methods, permit the PMD to be fully characterized by determining the optical frequency dependence of the polarization dispersion vector.[1] The magnitude of this vector is the DGD and its orientation yields the PSP of the DUT. The PMD is defined as the mean or rms value of the DGD averaged over all optical frequencies. In the case of PS method, the output state of polarization (SOP) describes a curve in the Poincare' sphere, whose local angular velocity determines the DGD. Both the JME and PS techniques require the measurement of the three Stokes parameters describing the SOP as a function of the optical frequency, a time consuming process to carry out using a tunable laser and a polarimeter. Other commonly used approaches to PMD measurement include the interferometric[2] and wavelength scanning fixed analyzer[3] (WSFA) methods, where the projection of the output SOP along one axis characteristic of the measurement instrument, is used for rapid determination of the PMD. However, the information lost by discarding the other two components severely hampers testing of small (< 50 fs) PMD values, particularly for the medium exhibiting intermediate or strong mode coupling.

It has been observed that the two main methods of measuring PMD in birefringent fibers may give different results when applied to elements with both PMD and PDL. This difference in behavior can be explained in terms of the birefringence properties of optical fiber systems, which is becoming increasingly important in modern telecommunications networks. The two principal states of polarization (PSP) in fibers, which are defined as the polarization states for which the output polarization is independent of the frequency to first order is central to the study of the concept of PSP and PMD. For fibers with no PDL, the two PSPs are orthogonal. A light pulse polarized along a PSP propagates without dispersion (assuming there is negligible chromatic dispersion or non-linear effects). The two orthogonal PSPs represent the *slowest* and the *fastest* propagating pulses. Any other pulse will broaden during the propagation. The maximum broadening is given by the differential group delay (DGD) which is the differential delay between the two PSPs. For real fiber systems, however, the PSPs are unstable. Therefore, the parameter which is characteristic of the fiber link quality is the mean or average or rms value of the DGD. It is well known that this time ensemble is equivalent to a frequency ensemble of the DGD. This frequency (wavelength) average is known as the polarization mode dispersion (PMD) of the fiber. It is the basic parameter used to estimate the birefringence-induced distortions in fiber optic systems. An alternative method to measure PMD is based on low coherence optical interferometry, which provides direct data on the pulse spreading and therefore, the PMD.

Although the interferometric (IM) method is not related to polarimetric (PM) method (i.e. the PSP concept), the two methods yield equivalent results for cases without PDL. The mere fact that one particular test set-up creates this discrepancy between the two measurement methods calls for a careful assessment of PMD characterization when PDL is present.

PMD in Optical Devices with PDL

The basic concepts of PSP and DGD still hold for the case of concatenated optical fiber links with both PMD and PDL. However, the interaction of PMD and PDL produces an overall DGD that is larger than the sum of the DGDs of all the trunks. Also, the two PSPs are not necessarily orthogonal nor do they characterize the slowest and fastest pulse. Furthermore, like the PMD, PDL is also a statistical quantity, varying with the wavelength. For practical systems, where the polarization couplings vary with time, PDL also becomes a time-dependent quantity. The relevant characterization parameter is then its average, either in time or in frequency (wavelength). It has been shown that a "zero" DGD does not necessarily imply "zero" PMD, particularly for systems with PDL.[4] In this case, the presence of the non-zero PMD can be confirmed by IM test method.

Comparison of Interferometric and Polarimetric PMD test Methods

Comparison between IM and PM methods to measure PMD has represented an important issue since the early development of the two techniques.[5,6] However, the time and wavelength dependent random fluctuations of the PMD and the difference between the theoretical approaches of the two measurement methods made this task difficult.

In particular, the fiber impulse response (IR) due to PMD is a fundamental element in the IM theory that models the fiber as a concatenation of linearly birefringent wave plates with random orientations. On the other hand, the IR has never been calculated on the basis of PSP theory, which is the basis of PM techniques. It is demonstrated that the IR has a two-lobe shape (per PSP theory) and, in general, is different from the Gaussian hypothesis used otherwise in IM analysis. The calculation of the IR starting from the PM measurement allows a direct comparison with IM measurements, not only in terms of DGD of the PSPs, but also in terms of full interferometric pattern.

A new comparison technique between IM and PM PMD measurement methods based on the calculation of the impulse response of a fiber affected by PMD according to the theory of PSPs has been developed.[7] The method allows a direct comparison of the interferometric pattern with the auto-correlation function of the impulse response calculated from PM measurements. Experimental results indicate that, despite time fluctuations of the interferometric patterns, a quite satisfactory agreement between the two methods can be found. A further improvement of the technique could be achieved by measuring the wavelength dependence of the Stokes vectors representing the output PSPs, that is at present, not allowed by commercial polarimeters. Finally, a revision of the data analysis algorithm of interferometric patterns seems to be advisable to fit the behavior of the experimental data that are, at the same time, non-Gaussian like and time fluctuating.

PMD in Optical Amplifier

The usual way to measure the PMD on a fiber consists in averaging the DGD over some dl and is based on the assumption that the PMD in a fiber is statistically distributed. For measuring PMD in active elements, such as optical amplifiers, which contain several elements (fibers, isolators, wavelength selective couplers, etc.), the DGD tests are restricted to short $\Delta\lambda$ intervals. Moreover, the gain spectra of the optical amplifiers further limits this region to the typical operational bandwidth (1530-1570 nm) for EDFAs and EDFFAs; to 1270-1330 nm for PDFFAs. Furthermore, for 3-level fiber amplifiers, *PMD must be measured under pumping conditions selected to assure a significant degree of polarization states during the test.* Therefore, the PMD for optical amplifiers must be defined under specific operating condition. A typical tunable laser, polarization controller, DUT, and polarimeter test set-up is shown in Figure 2. To assure accurate measurements, systematic errors as well as the PMDs contributed by the test elements are estimated first. In an optical amplifier, the DGD is usually very small without the pump power. With pumping, however, the ASE may be quite large to perturb the test data. The probe power is therefore increased such that the output signal is larger than the ASE. Measuring the degree of polarization, DOP, (maintained between 60 and 100%) easily checks this condition by increasing the input power. Jones Matrix Eigen (JME) Analysis, especially with the pump power, warrants caution. In JME method, the measurement includes analysis of successively three linear output states of polarization. When the state of polarization (SOP) of the probe is orthogonal to one of the three polarizers, only the unpolarized ASE is analyzed instead of the expected amplified signal.[8,9] Three essential conditions for correct measurement of PMD using standard polarimetric method are:

- Keep pump power on to avoid depolarization effect
- Set large enough probe intensity (signal larger that ASE) for DOP > 60%
- Use circular input polarization to maintain same probe intensity after any of the three polarizers

Testing of Grating based DWDMs/Couplers

Since their invention some ten years ago, fiber Bragg gratings (FBGs) have found rapid applications in practical systems which include optical communication, sensing, and other optical signal processing systems. With the advent of DWDMs in recent years, the role of FBGs has become crucial with commensurate demand on grating fabrication, their performance and testing. The specific test issues are of course dependent on the variety of applications contemplated using these FBGs. Major applications of FBGs include filters, MUX/DEMUX, dispersion compensator, and narrow linewidth tunable laser sources among others. To meet the specifications for these applications, the optical response of the FBGs are measured and the data analyzed via appropriate inverse scattering calculations. Specific tests include:

Reflection and Transmission Spectra: Typical set up uses a broadband source such as ELEDs or doped fiber lasers together with optical spectrum analyzer (OSA). It is very fast, but the resolution is limited to ~0.05 nm. Measurements at high resolution (~0.001 nm) are possible using Fabry-Perot etalon and step-tunable external cavity laser and optical power meter. These units are expensive and test time is rather long for high resolution testing. An additional limitation with scanning source method is the relative lack of absolute wavelength calibration. Use of optical wavemeter can greatly improve the calibration, but will add to the test facility cost.

Measurement of narrow spectral features requires availability of high-stability, narrow linewidth tunable lasers and high-sensitivity, low-noise optical detector/power meters. Several injection-locking techniques with optical PLL have been reported to produce stable, narrow linewidth lasers. Also, commercial units are now available from several vendors.

Measurement of grating side bands will often require a test set up with high dynamic range and low noise. Use of OSA will limit such capability to ~30 dB. The range can be improved to ~60 dB by using narrow bandwidth, sharp cut-off filters. All these additional components must be well characterized to estimate the residual error test data.

Group Velocity Dispersion: Uses differential phase measurement of the reflected and straight through light from the FBG device measured in a vector voltmeter or a network analyzer.[10] It is also possible to use optical phase shifter to compensate the phase dispersion in an interferometric arrangement. A RF modulated tunable optical source will enable the dispersion to be measured as a function of wavelength. The network analyzer measurements can be accomplished either in the optical or electrical domain. Commercial units, albeit expensive, are available in both flavors.

Measurements on Grating Structure, Spatial Profile: For direct measurement of grating structures, two simple yet interesting methods include testing for the index modulation depth[11] and side scatter[12] from the grating resonant structures. The first technique is used to determine index profile, while the second demonstrates the grating features such as chirp, phase discontinuity and apodization.

Measurement of Polarization Properties: The photo-induced index change in the fiber grating is generally anisotropic which significantly impacts the polarimetric characteristics of the FBGs. Both transmission and reflection ellipsometry can be used to determine the polarization behavior of the grating and also to asses the geometrical asymmetry of the writing process and FBG manufacturing.[13] Typical test setup will include polarizing beam splitters and quarter wave plates in conjunction with an appropriate optical source and detector. Several commercial ellipsometers are available for such measurements.

PMD in Chirped FBGs

Chirped fiber gratings written into optical fibers are finding many diverse, yet unique applications: dispersion compensators, variable time-delay units for use in phased-array antenna, etc. Their use in short lengths of fibers (as typical in phased –array antenna feed networks) does not affect the PMD of the overall system. However, in long-haul communication networks spewing data at very high speed, use of such devices are only beneficial if they accurately compensate chromatic dispersion without other impairments considered detrimental to systems performance. PMD is a critical measure of the systems performance in high-speed networks. Chirped FBGs typically show a high level of polarization mode dispersion, which can be a significant factor to the total PMD in a long-haul fiber communications network. Therefore, an accurate estimation of the PMD in chirped FBGs is required. The physical mechanism causing the PMD in chirped FBGs is attributed to the DGD resulting from the difference in propagation times for the parallel and perpendicular (to the birefringent axis of the fiber) modes of lightwaves reflected by the chirped FBG. For a linearly chirped FBG, successive wavelengths are reflected from progressively further positions in the grating thereby producing a time delay, which varies linearly with the incident wavelengths. This is synonymous to frequency or wavelength independent (that is, constant chromatic dispersion) operation. For a linearly chirped BG written on a fiber with wavelength independent birefringence, the DGD coefficient is $\Delta=\beta/c$, where β is the birefringence of the fiber. The DGD for the straight through and the reflected signals in the FBG is related by $\Delta\tau = D\lambda\Delta c/n$, where D is the chromatic dispersion. *It is evident that the DGD for the reflected signal is a large multiple of the DGD for the straight through signal.* [14]

A convenient test set-up using standard polarimetric method is used to measure the PMD. Prior to making measurements on the FBG, the polarization state controllers are adjusted for each wavelength to launch the required polarization states and the stokes vector on the surface of the Poincare' sphere analyzed to determine the DGD and the principle states of polarization for each wavelength step. Also, PDL and second-order PMD can be estimated from this measurement.

The measurement accuracy will depend on the PMD of each individual optical element used in the test set-up. Certain precautions such as covering the wavelength band of interest in a few short steps and selecting as many polarization states as deemed necessary to enhance the DGD sensitivity

Temperature dependence of PMD measurements

PMD can degrade the performance of high capacity optical communication systems drastically. In an NRZ system, the PMD penalty is mainly due to inter-symbol interference (ISI) caused by the DGD between the two PSPs associated with a single-mode fiber link. The DGD of a fiber has a random value, which varies with both wavelength and time. The DGD distribution is usually *assumed* to be *stationary at fixed temperature with a Maxwellian probability density function*. The time dependence of the DGD at a particular wavelength stems from environmental conditions, which in turn vary with time. In particular, uncontrolled temperature variation can contribute to changes in DGD. It is therefore important to determine how thermal effects impact DGD. This change can be measured in terms of a PMD correlation temperature[15],

$$C_{T,\lambda 0}(\Delta T) = \Sigma\{(\Delta\tau(\lambda_0, T) - \underline{\Delta\tau})(\Delta\tau(\lambda_0,T+\Delta T) - \underline{\Delta\tau})dT\} / \Sigma\{(\Delta\tau(\lambda_0, T) - \Delta\tau)^2 dT\}$$

The above summation is evaluated for infinite temperature range at each wavelength. However, the fact that the DGD process is stationary enables the DGD temperature correlation to be averaged over an entire range of wavelengths to compensate for the loss of accuracy due to the limited temperature range over which the test is conducted. The correlation temperature, T_c, separating the independent instances of PMD is defined to be the full width at half power maximum (FWHM) of the temperature correlation function described above. The test method can be based on polarimetric technique and the analysis done by Jones-Matrix Eigenanalysis (JME) to obtain the DGD values. Based on the measurements of a variety of fiber optic devices and several spans of SMFs, the correlation temperature was found to follow an inverse square-law dependence[15] on the fiber length, l.: [$T_c = 334/l^{0.5}$]

Measuring small PMD via the Poincare' sphere method using wavelength scanning data6

A new method[16] has been proposed to use a variant of the PS technique where only two polarimetric parameters deduced from WS measurements with different linear axes analysis. The frequency-domain polarimetric analysis method is reported to be much faster than the standard polarimetric method and can measure subpicosecond PMDs from a limited number of WS tests (only two acquisitions for DUTs exhibiting weak mode coupling).

Accurate PDL measurement

Non-mechanical approach[17,18] (see Figure) uses deterministic fixed static method at specified polarization states. The input polarization states can be set by the retarders (either manually or electrically controlled using voltage-controlled liquid crystal retarder devices) To enhance measurement sensitivity, synchronous detection scheme would allow 0.001 dB or better resolution

The critical components in the set up that would affect the test uncertainties include the light source, the resolution of the polarimeter, and overall system noise. The sources of uncertainties are:

- Polarization state -- of the calibrating polarimeter, typically ~0.1 to 0.8% of the signal
- Thermal dependence of polarimeter -- depends on the retarders and other optical components in the polarimeter, variation in signal power, polarization states, etc. The total effect could be 0.2% of the signal
- Systems internal PDL -- Residual PDL of the system, although small (0.1 to 0.2% of the signal, can be eliminated by ratiometric measurement
- Test system uncertainty -- could be as high as 1-2% of the PDL
- Measurement repeatability -- typical $1\sigma \sim 0.0005$ dB
- Overall repeatability -- root sum square of various uncertainty factors

- Miscellaneous factors -- attributed to connectorized pigtails, alignment errors, etc.
- For PDLs <0.1 dB, the test uncertainties are in the range of +3% to +5% of the PDL

PMD Testing using Modulation Phase Shift Method

A common feature of the current PMD test methods is the use of a broad optical bandwidth facility. With increasing use of WDMs and narrow channelization, the need to make PMD tests within a *very narrow optical band pass* region is becoming the order of the day. One technique, which is well suited to narrow band PMD testing, is the Modulation Phase-shift technique (MPS).[19] The basic principle is to directly measure the DGD of light traveling through each of the principal states of polarization of the device under test (DUT). As shown in the test set-up (Figure 3), light from a tunable laser diode with a RF broadened linewidth (50-500 MHz) undergoes an external intensity modulation at GHz rate by an electro-optic Mach-Zehender (M-Z) modulator. The modulated light passes through a polarization controller and the DUT before being detected by a high speed InGaAs photodetector and filtered with a narrowband filter centered at the modulating frequency. The phase difference between the RF electrical modulation signal and the optical signal output from the DUT is a measure of the group delay (DGD) through the optical system: [$\Delta\tau = (\phi_+ - \phi_-)/360*f$, where the delay is in seconds, ϕ in degrees, and modulation frequency f in Hz].

The uncertainty in the MPS measurement comes primarily from two sources; resolution limit of the vector voltmeter and the degree to which the incident light waves can be launched exactly on the principal axes. RF broadening the source can significantly reduce F-P effects in the test system, and the residual random phase jitters are attributed to the noise of the vector voltmeter. The uncertainty due to launching non-principal states is a source of systematic error. The measured phase difference between two orthogonal launch states is a function of the alignment of the launch states with respect to principal axes. The fractional DGD error encountered for a given alignment error (which is the angle on the Poincare' sphere corresponding to the launch polarization state and the slow principal axis of the DUT) is indeed small for small angular errors (e.g., a 5-degree misalignment angle will yield a measured DGD which is only 0.4% below the true value). Differential phase detection technique can reduce the measurement noise of the system and automated test facility could eliminate the tedium in manual testing and the likely sources of human error due to fatigue or skill levels.

Two-photon techniques

Characterization of single-mode fiber optic devices requires a basis for time, wavelength and polarization. Parametric down conversion in non-linear crystals naturally provides pairs of photons, which are extremely highly correlated in time, energy (thus wavelength/frequency) and polarization. Now that photon pair sources can be made cheap and compact enough to offer practical alternatives for many other traditional measurement schemes, a proposal has been made to measure the PMD using twin-photon techniques.[20]

The intrinsic time-energy-polarization correlation of the down-converted photons make them a natural candidates for dispersion measurements (PMDs). The photon pairs are produced by type II parametric down-conversion which permits a strong time-energy correlation and also entangled polarization. This method is essentially similar to the to the white light interferometric technique, but the wide spectrum of the down converted photons (~ 300 nm in the current demonstration) allowed a resolution of 0.01 fs in contrast with the 50-100 fs resolution of the standard interferometric technique based on LEDs.

Effect of PMD in High-speed Network

In the last few years, PMD has attracted serious attention of system designers, particularly due to its potential limitation on the performance of optically transparent high-speed, long haul transmission networks. PMDs can cause systems penalties or even signal distortions, and thus, unacceptable BERs when operating the next generation systems such as OC-192 systems. The PMD induced pulse distortion at each signal wavelength can be manipulated between a maximum and a minimum value by changing the SOP of the light launched into the fiber. Thus an almost undistorted pulse transmission can be realized even when the DGD exceeds the pulse width. The maximum achievable pulse distortion is correlated to the DGD at the signal wavelength. This wavelength dependence may yield different signal degradation for the channels of a WDM system. In a recent field trial @ 10Gb/s with more than 40-ps PMD, the BER varied from 10^{-6} to 10^{-12} without polarization control.[21,22] The *system could be stabilized*, however, at the *minimum BER just by controlling the input SOP in sequences of minutes*. Use of polarization controllers are therefore recommended in high-speed networks.

5. PRODUCTION TEST REQUIREMENTS & FACILITY

Production Test Capacity and Facility for DWDM Components and Systems

Production testing of components must consider parameters such as availability of adequate test equipment, test plan/procedure, and skilled operators that are configured to produce high-yield testing with a minimum effort and least cost to the manufacturer. For a typical device such as a WDM/coupler, the insertion loss, polarization dependent loss, and temperature specify its optical performance and wavelength dependent effects on insertion loss, isolation, return loss and directivity. Mechanical strength/integrity of the device needs to be assured by pull testing. For certain applications, some parameters may be more critical than others. Consequently, the tests are customer-specific and product-sensitive. However, facilities must be provided to test each of these parameters.

Before components can be shipped to customers, all devices such as couplers, splitters, WDMS in their final form, including products that are repackaged or built into an assembly and connectorized by assembly services must undergo 100% performance testing. To keep pace with manufacturing capacity increases for both standard and customized products (bare fiber and connectorized ports), additional automated test stations for both bare fiber devices and repackaged and connectorized devices must be planned.

Automated Optical Test Station

The optical performance of the couplers is tested at a computerized test station specifically designed to measure the optical characteristics of the devices in a production environment. A typical test station (Figure 4) includes various light sources, optical switches, attenuators, polarization dependent loss (PDL) analyzer, the device under test, and photodetectors interfaced with a fast computer using GPIB bus. Many discrete wavelengths may be routed to the test station via a 1xN optical switch located upstream of the test station. Light from each discrete laser source passes through a polarization controller (PC) prior to reaching the test station. The PC, under computer control, varies the states of polarization of the output optical signal. Capabilities to test a variety of devices such as couplers and WDMs at both short and long wavelengths have been developed. For the PDL analyzers, both home-built as well as commercial units optimized for specified wavelength bands are used.

At Gould, currently a 16-bit A/D board is used to measure the detector outputs, providing a measurement resolution of one part in 65,535. This translates to a 0.0004% minimum detectable change in the coupling ratio for a 1% tap coupler. The test computer automatically grades the optical performance of the device against the applicable specifications. This grading operation is performed using a customized file created per customer specifications shortly after a work order is released for the manufacture of the couplers. The test data is stored in a centralized database. Paper copies of the test reports and customer data sheet are also generated at the computerized test stations after the devices have been graded. The proprietary software used in the data analysis has been developed at Gould Fiber Optics.

Connectorized Product Test Station

Facilities required to test connectorized products include high stability multi-wavelength optical sources, high sensitivity, high dynamic range optical heads and power meters, low loss optical switches (NxM ports), and patch cords with assorted fiber types and connectors matching the connectorized DUTs. Recent products tested include 1x4, 1x8, 1x16, 1x32, 4x4, 16x16 hybrid devices with LC connectors (optical ports) and SMA (electrical ports). Commercial test platforms are also available for testing of these products.

Temperature Cycle Station & Thermal Sensitivity

Thermal sensitivity, as defined by the peak-to-peak variation in coupling ratio, is measured over a specified temperature range. A suitable arrangement of heating/cooling to the applicable temperature range and a reasonable means of inserting the DUT in the rapid thermal cycling station with an automated data acquisition of the coupling ratio and analysis are required. The test capability has been upgraded from 2-port device stations to 3- and 4-port device stations. The test fixture can accommodate several couplers at a time, thereby improving the turn around time on multiport components. The station is programmed to accept a variety of temperature profile (ramp rate) and temperature range. An average rate of 15 degree/min and temperature cycling between –20 to 125 C are typical. During the rapid thermal cycling test, the computer monitors device temperature, coupling ratio, and test time. Typically, 3000 data points are collected per minute of test time. At the conclusion of the test, the computer calculates the difference between the maximum and minimum values of the coupling ratio recorded during the test. Currently, the temperature range is extended to –55 C.

Thermal Soak/Cycle Burn-in Oven

In production environment, the manufactured components are first burnt-in before any meaningful performance tests are conducted. Depending on the production volume and customer requirements, enormous capability for the burn-in tests of couplers, splitters, and WDMs may be needed. These thermal chambers must be environmentally controllable and programmable for different soak/cycling profile and handle large volume burn-in/screening tests.

WDM Component Test Facility

To accommodate the development and commercialization efforts for advanced DWDM components and applications, which include products such as wavelength division multiplexer/demultiplexer (WDM) and Mach-Zehender Interferometric (MZI) devices using fused fiber and glassolder technologies for narrow band (~ 5-nm channel spacing) applications and other DWDM products using FBT and/or planar I/O devices and thin film filters for add/drop applications within ITU grid, must be tested to their specifications. To meet the stringent specifications of WDM products, which include components at 980, 980/1550, 1475/1550, 1550 nm and longer wavelength bands, special test equipment are required. To expedite production testing of large volume WDM/DWDMs, the test facility must cover appropriate wavelength bands using carefully selected, top of the line components and test equipment. Gould DWDM test facility (Figure 5) includes the following equipment:

- Fiber Bragg Grating (FBG) Stabilized Discrete Wavelength Single-mode Lasers
- Optical Switch Matrix
- Polarization Controller
- Optical Loss Analyzer
- Variable & Fixed Optical Attenuators
- Back Reflection Meter @ 980, 1480, and 1550 nm
- High Dynamic Range Optical Detectors
- Optical Spectrum Analyzer (0.05 nm to 0.01 nm resolution)

Skilled Personnel

The ability to produce components to exacting specifications depends on a variety of factors, the most important one is the skill level of the personnel. The variety of fiber types, splitting ratios, lead lengths, operating wavelengths, and packaging that customers specify require the development and documentation of customized test plans. Scheduling both the workflow at each test station and test personnel in ever-dynamic production environment is a challenging task requiring full-time attention. Consequently, supervisory positions with increased responsibility to establish and document quality and test procedures and to ensure that test operators have the continuous training needed to accomplish the productivity goals must be planned. Also, staffing will be needed to handle the increasingly complex job of test plan development of advanced products, data analysis, productivity, quality, and capacity improvement, and maintenance of test equipment to specified measurement accuracy. To ensure a reasonable production test yield, the above elements must all be in place.

6. CRITICAL TECHNOLOGY DEVELOPMENT AND FUTURE PROSPECTS

The above discussion has identified several WDM components and their production test methods. Selection of the most favorable option among the three types of couplers of interest (i.e., WDM coupler with interference filters, WDM coupler with diffraction gratings, and active WDM coupler) for a specified application will depend on the physical characteristics of the coupler and on certain economic aspects. Compact configurations requiring a minimum number of elements that can be mass-produced and minimum adjustment efforts in field-use are expected to have favorable acceptance in future applications. Another important criterion is the reliability of the device under both static and dynamic operating conditions while subjected to benign/harsh environments specific to the application. The test requirements and facilities will have to address these challenging issues as and when needed.

In consideration of the linewidth narrowing of single-frequency, single-mode laser diodes, two other areas of single-mode fiber optic multiplexing technology need careful assessment: thermal stabilization and state-of-polarization maintenance. For example, the temperature dependence of tunable WDM devices will require a more critical evaluation. Presently, thermo-optic reflectors and switches have been demonstrated which actively utilize the refractive index deviation at the surface of the device induced by thermal changes. For multilayer WDM devices, an evaluation of temperature-independence must take into consideration both material combination and device/structural design.

A polarization controller, which is essential to maintain the state-of-polarization in coherent optical communications, is also needed in optical Mux/Demux devices and single-mode fiber optic links, particularly for those applications employing polarization multiplexing. The development of endless-polarization control systems with fewer control elements should enhance system performance.

In order to be able to mass-produce and test reliable WDM components at a cost-competitive price, various device processing and packaging technologies must be addressed. For high channel density, single-mode applications, the extent to which advances in integrated optics can be exploited for WDM is seriously investigated. Development of arrayed waveguide technology is an example of this effort. For the ultimate utilization of single-mode fiber bandwidth capability, provision of coherent optical detection for each wavelength channel might be contemplated for future generations of high channel density WDM systems.

7. WDM SYSTEMS REQUIREMENTS AND ISSUES

The first step in designing a multiplexed network system is the compilation of a detailed set of requirements, constraints, and performance parameters for existing multiplex equipment and system configurations. In addition, a number of systems issues can affect the performance and implementation of multiplex equipment. The following issues are considered important in the design of multiplexed architecture:

- Asynchronous vs. synchronous combining of lower-level multiplex streams
- Routing of control (signaling) and data/com signals
- Framing structure
- Network timing
- Range resolution (between the lowest and highest information rates and minimum capacity increment)
- Reconfiguration flexibility (at port and higher-level multiplex)
- Definition of the group multiplex equipment family in terms of functionality and input/output signal characteristics

The second step involves the development of candidate system architecture incorporating optical multiplex equipment. As a minimum, the new optical architecture and multiplex equipment used should provide the functional performance equivalent to that of existing RF/coaxial systems. Several hierarchical architectures can be conceived for multiplexing at various levels, i.e., peer-level, direct, and hybrid multiplexing. These configurations provide the possibility of multiplexing and demultiplexing different individual signals, groups of multiplexed signals, and their combinations at various levels. In another configuration, the incoming multiplexed signal may first be demultiplexed before remultiplexing with other individual or multiplexed signals. Such a network architecture has the potential of add/drop/multiplex (ADM) capability at the node. It should be emphasized that optical WDM allows both analog and digital signals to be carried over the same fiber. This feature is considered in the development of appropriate hierarchical architecture. A qualitative comparison of these architectures is provided in Table.

8. OPTICAL DISTRIBUTION NETWORKS: CONCEPTS, ARCHITECTURES, AND TRADE ISSUES

Efficient distribution (routing/switching) of signals among various nodes in a complex system is very much dependent on the imbedded network architecture. Simple hardwiring concept such as physically disconnecting, repositioning, and reconnecting coaxial cables is neither an elegant nor a workable solution, particularly for handling flexible, mixed signals in a high-capacity dynamically reconfigurable information system. In advanced applications, traffic reconfiguration may be accomplished by a switching architecture to reroute the traffic and a distribution network to interconnect all the nodes in the system.

The input signals to the distribution network may be in digital and/or analog form and divided into the nodal interfaces, each contributing a certain bandwidth at a specified information rate. It is assumed that both transmit and receive functions are provided at each interface, and as typical in certain onboard systems, the input and output interfaces may not be too far apart (couple of hundred meters). A selected configuration should have the following features:

- Modular design via the network interfaces, each operating at a fixed information rate, permits expansion of the system to accommodate future traffic growth
- A mixture of analog and digital interfaces can be served
- Overall network cost can potentially be reduced

The use of a fiber optic distribution network could be attractive for this scenario for several reasons. First, the system requiring a rather high total information bandwidth (several GHz) growing to higher values in future years can be easily accommodated. Second, optical fiber cables are small in cross section and lightweight. They could be easily fitted into existing ducts of the crowded platforms. Finally, unlike coaxial cables, optical fibers do exhibit a very high immunity to EMI/EMP, crosstalk, and do not have ground loop problems. Several useful network architectures can be conceived for interconnecting the nodes: *Time-Division Multiplex Bus*, *Star Distribution Network*, *Space-Division Multiplex Bus*, *Wavelength-Division Multiplex (WDM) Bus*, *Ring Bus*, *Multilevel Fiber Optic Transmission Network*, etc. The choice of architecture determines the complexity of the electronic equipment and electro-optic hardware to build a working system. Some of these distribution systems are relatively straightforward to implement, while others may require advances in the state-of-the-art. Table 4 provides a qualitative comparison of these distribution networks. These schemes need to be evaluated with respect to their optical power requirements, synchronization requirements, electrical and optical component costs, development costs, implementation complexity, and functional maintenance/reliability.

Further details of the components/devices used, need for specialty components and signal regeneration, and design modularity of each architectures are summarized in the Table. In this table, the cost and development efforts are also qualitatively compared on the basis of certain assumptions such as:

- the factors that determine the cost of a given architecture are the combined costs of electrical and optical components and the development costs incurred to implement any architecture, which does not use off-the-shelf components. These include high-speed electronics, ASICs, Hi-REL optical and electronic components, optical/electrical regeneration, etc.
- the development effort necessary for implementation of a given architecture depends on the signal bandwidth of constituent optical links and on the multiplexing technique.

These characteristic parameters will be used to compare the architectures with respect to technical feasibility, cost, and the development effort necessary for a timely and reliable insertion. It should be noted, however, that for a successful implementation with a specified reliability/redundancy requirement, the constraints of the payload size and prime power may override the cost reasonability criterion.

Table 5. Characteristics of Various Optical Distribution Architectures

Item	TDM Bus	Star Bus	SDM Bus	WDM Bus	Ring Bus	Multilevel Network
Data Rate*	N.R	R	R	R	N.R	4R
No. of Opt. Tx.	N	2N	N	N	2N	1
No. of Opt. Rx	N.R	2N	R	R	N.R	4R
Fiber Type	SMF	MMF	MMF	MMF	SMF	MMF
Light Source	Diode Laser	LED	Diode Laser	Diode Laser	Diode Laser	Diode Laser
Photo-Detector	APD	PIN	PIN	APD	APD	APD
Active Repeater	Yes	No	No	Yes	No	No
Special Equipmt./ Component	Highest Quality	No	No	Opt. Filter banks, Grating demux	Highest quality	High-quality
Full Modular Design	No	Yes	Yes	No	No	No
Relative Cost	High	Low	Mode-rate	High	High	Mode-rate
Develop. Effort/ Implement. Complexity	Highest	Low	Low	Moderate-to-high	Moderate-to-high	Mode-rate

This section has provided an overview of the DWDM network design, including selection of components, network architecture for specifies performance, and cost. Several trade-off features are also identified. Once such system is configured and hardwired to designed specifications, the performance testing must be conducted to verify if the design has met its goal. The test parameters at systems level usually include optical power budget, rise time and dispersion, PDL, PMD, jitter, BER/SNR, cross talk, isolation, etc. To accomplish these tests in a production environment, a systemic approach is required. Often times, commercial test facilities can be acquired for automated test and data acquisition and analysis of the test data for derived performance parameters. As the total DWDM system comprises many individual components with their own specifications, the impact of deviations from the specifications of individual components will be many times magnified for the complete system. Therefore, selection of components for a desired system performance must take into consideration all these factors. Also, development of test plans and procedures for systems level testing must start early on, at the time of selecting/procuring components for the network.

9. CONCLUSION

The critical requirement and production test issues relevant to DWDM components and networks are reviewed in this paper. Impact of the performance requirements of advanced systems on the quality of individual components and their testability is also addressed.

10. REFERENCES

1 B. Huttner, IEEE Photon. Tech. Lett, Vol. 4, pp. 1066 (1992).
2. N. Geiser, et. al. IEEE J. Lightwave Tech., Vol. 9, p. 821 (1991).
3. C. D. Poole, et. al., IEEE J. Lightwave Tech. Vol. 12, p. 917 (1994).
4. B. Huttner, et. al., Polarization-induced pulse spreading in birefringent optical fibers with zero differential group delay, NIST Special Publication 930, Tech. Digest Symp. On Optical Fiber Measurements, 1998, Boulder CO, pp.49-52. See also, B. Huttner, et. al. In Optics Lett., Vol. 22, p. 504 (1997).
5. N. Grisin, et. al., Experimental comparison between two different methods for measuring PMD in single-mode fobers, Electron. Lett., Vol. 27, No. 24, pp. 2292-93 (1991); See also, ibid, Vol. 32, No. 7, pp. 680-81 (19996).
6. C. D. Poole, et. al., Phenomenological approach to polarization dispersion in single-mode fiber, Electron. Lett., Vol. 22, No. 19, pp. 1029-30 (1986).
7. M. Schiano, Comparison between interferometric and polarimetric PMD measurement techniques based on fiber impulse response, NIST Special Publication 930, Tech. Digest Symp. On Optical Fiber Measurements, 1998, Boulder CO, pp.41-44.
8. C. Geiser, et. al., Measurement of polarization dispersion in fiber amplifiers, NIST Special Publication 930, Tech. Digest Symp. On Optical Fiber Measurements, 1998, Boulder CO, pp.19-22.
9. R. Leners, et. al., Analytical model of polarization dependent gain in Er-doped fiber amplifier, Conf. on Optical Amplifiers and their Applications, 1994, Yokohama, Japan, Tech. Digest, Post-deadline paper, pp. PD-109.
10. S. Ryu, et. al., Novel chromatic dispersion measurement method over continuous GHZ tuning range, IEEE J. Lightwave Tech., Vol. JLT 7, pp. 1177-80 (1989).

11. P. A. Krug, et. al., Measurement of index modulation by an optical fiber Bragg grating, Optics Lett., Vol. 20, pp. 1767-69 (1995).
12. J. Canning, et. al. Direct measurement of grating chirp using resonant side scatter spectra, Electron. Lett., Vol. 32, pp. 1608-10 (1996).
13. D. Stepanov, et. al., Ellipsometry of fiber Bragg gratings, Proc. Australian Conf. on Optical Fiber Technology, pp. 164-67 (1995).
14. D. Ives, Polarization mode dispersion in chirped fiber Bragg gratings, NIST Special Publication 930, Tech. Digest Symp. On Optical Fiber Measurements, 1998, Boulder CO, pp.15-18.
15. S. Corbeil, et. al., Temperature dependence of PMD Measurements, NIST Special Publication 930, Tech. Digest Symp. On Optical Fiber Measurements, 1998, Boulder CO, pp. 53-56.
16. N. Cyr, Determination of small PMD values via the Poincare' sphere method using wavelength scanning data, NIST Special Publication 930, Tech. Digest Symp. On Optical Fiber Measurements, 1998, Boulder CO, pp.45-48
17. R. Craig, et. al., Accurate PDL measurement and calibration standard development, NIST Special Publication 930, Tech. Digest Symp. On Optical Fiber Measurements, 1998, Boulder CO, pp.5-8.
18. B. Nyman, et. al., Automated system for measuring PDL, Conf. Ofiber Communication, OFC'94, San Jose, CA, Tech. Digest, pp. 230-31.
19. P. Williams, et. al., Narrowband measurement of PMD using the modulation phase shift method, NIST Special Publication 930, Tech. Digest Symp. On Optical Fiber Measurements, 1998, Boulder CO, pp.23-26.
20. N. Gisin, et. al., Twin-photon technique for fiber measurement, NIST Special Publication 930, Tech. Digest Symp. On Optical Fiber Measurements, 1998, Boulder CO, pp. 35-40.
21. R. Leppla, et. al., Impact of signal wavelength and input state of polarization on PMD-induced pulse distortion in > 10 Gb/s transmission system, NIST Special Publication 930, Tech. Digest Symp. On Optical Fiber Measurements, 1998, Boulder CO, pp.27-30.
22. N. Robinson, et. al., 4xSONET OC-192 Field-installed dispersion managed soliton system over450 km single-mode fiber in th e1550-nm Er-band, Conf. Optical Fiber Communication, OFC'98, Post Deadline Paper, San Jose, 1998.

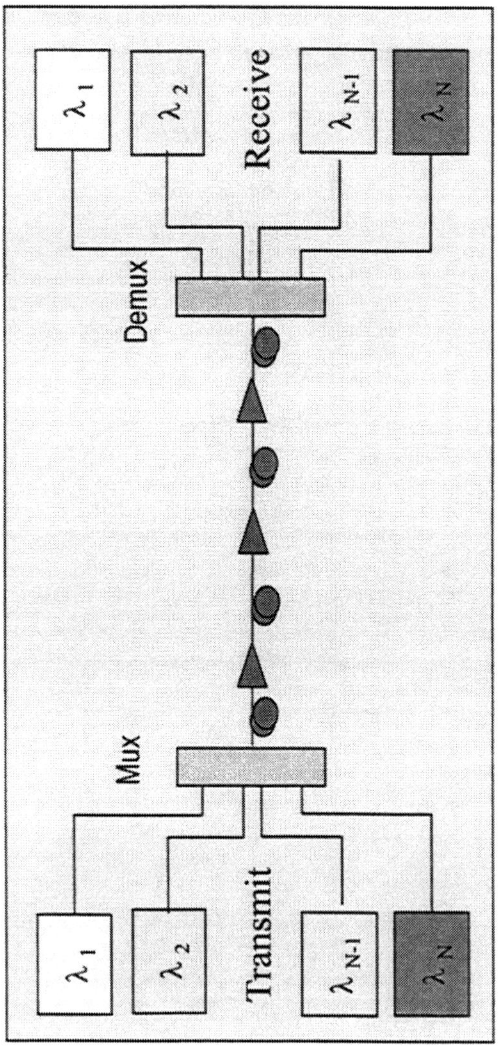

Fig. 1 DWDM System: General Configuration

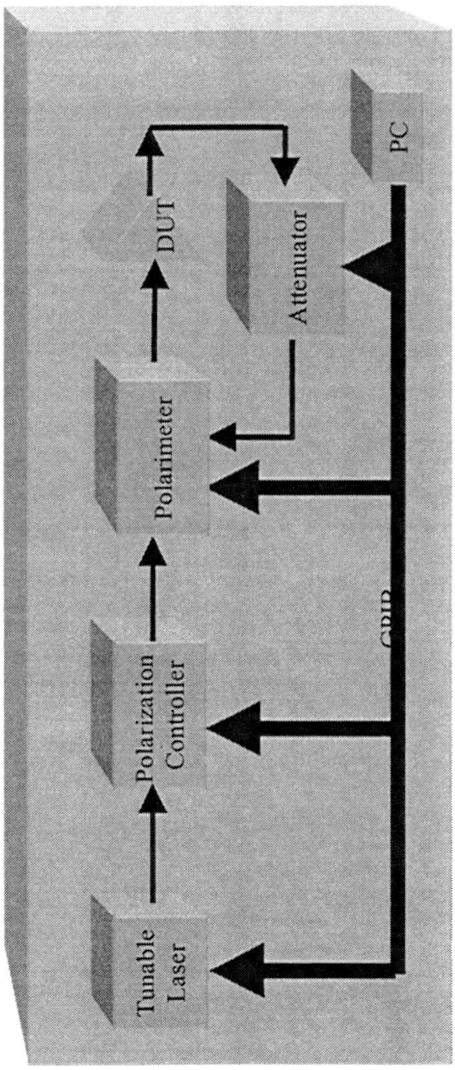

Fig. 2 Test set-up for PMD measurement in Optical Amplifier

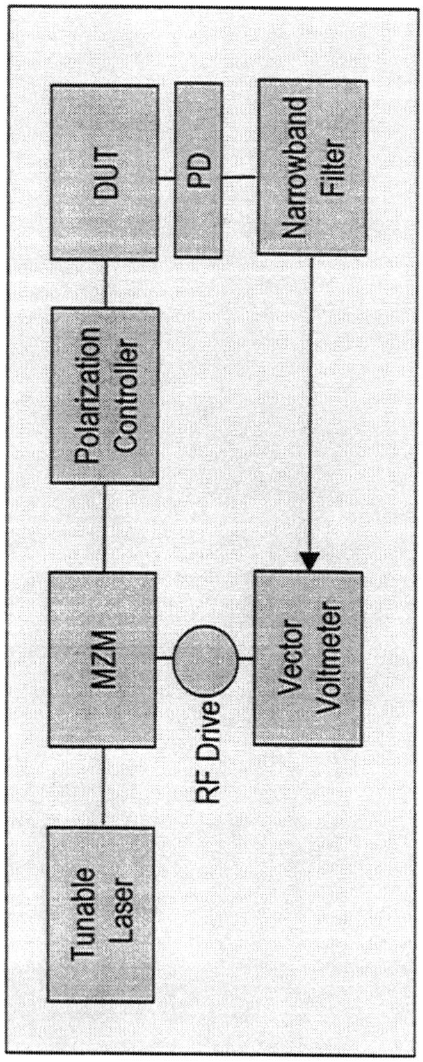

Fig. 3 PMD Testingusing Modulation Phase Shift Method

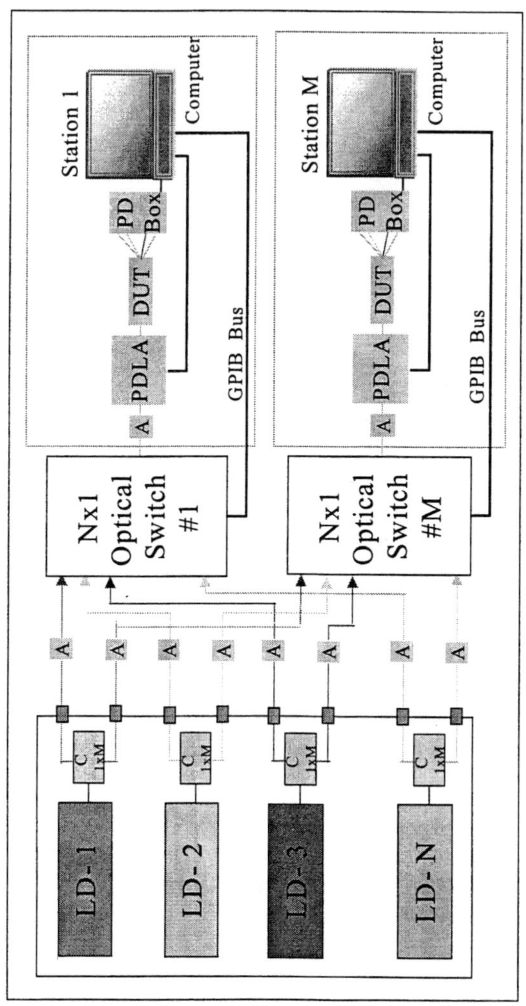

Fig. 4 Production Test Automated Optical Station Configuration

Figure 5. WDM Devices Test Facility

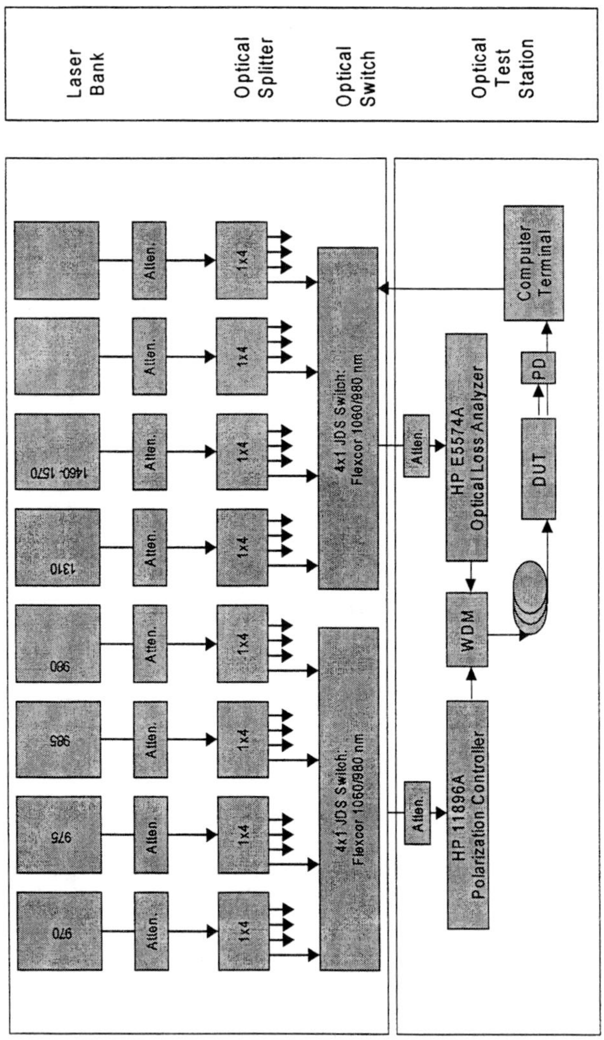